Birkhäuser

Frontiers in Mathematics

Yuming Qin • Xin Liu • Taige Wang

Global Existence and Uniqueness of Nonlinear Evolutionary Fluid Equations

 Birkhäuser

Yuming Qin
Department of Applied Mathematics
Donghua University
Shanghai, China

Xin Liu
Business Information Management School
Shanghai Institute of Foreign Trade
Shanghai, China

Taige Wang
Department of Mathematics
Virginia Tech
Blacksburg, VA, USA

ISSN 1660-8046
Frontiers in Mathematics
ISBN 978-3-0348-0593-3
DOI 10.1007/978-3-0348-0594-0

ISSN 1660-8054   (electronic)

ISBN 978-3-0348-0594-0   (eBook)

Library of Congress Control Number: 2015932229

Mathematics Subject Classification (2010): 76-XX, 76A05, 76D05, 76Wxx

Springer Basel Heidelberg New York Dordrecht London
© Springer Basel 2015

Printed on acid-free paper

Springer Basel AG is part of Springer Science+Business Media (www.birkhauser-science.com)

To our Parents

Contents

Preface

This book is aimed at presenting some recent results on global wellposedness of nonlinear evolutionary fluid equations.

Most of the material of this book is based on the research carried out by the authors and their collaborators in recent years. Some of the material has been previously published only in original papers, while some of it has never been published until now.

There are 8 chapters in this book. Chapter 1 concerns the global existence and asymptotic behavior of solutions to a 1D magnetohydrodynamics (MHD) fluid system. Chapter 2 concerns the global existence and exponential stability of solutions to a 1D compressible and radiative MHD flow model. In Chapter 3 we study the global existence and exponential stability of solutions to a 1D thermally radiative MHD with self-gravitation. Chapter 4 investigates the global existence of solutions for a 1D self-gravitating viscous radiative and reactive gas model. Chapter 5 deals with the global existence and exponential stability of solutions to a compressible viscous micropolar fluid model. Chapter 6 will deal with the global existence and exponential stability of solutions to a compressible viscous micropolar fluid model. In Chapter 7 we establish the global existence and exponential stability of solutions to a full non-Newtonian fluid model ($p > 2$), which is very different from those Newtonian fluid models ($p = 2$) discussed in Chapters 3–5 in Qin and Huang [102]. Moreover, Chapter 5 in Qin and Huang [102] mainly deals with a model of Newtonian radiative fluids where the radiative effect is accounted for with different constitutive relations from those in Chapter 7 in this book, where only the non-Newtonian fluid without radiative effect is considered. To deal with such a non-Newtonian model, we need to design more delicate and more complicated estimates than those for the Newtonian model in Chapters 3–5 in Qin and Huang [102]. Chapter 8 is a continuation of Chapter 1 in Qin and Huang [102], in which the global existence of solutions in H^i ($i = 1, 2, 4$) has been obtained. In this chapter, we further establish the exponential stability of spherically symmetric solutions for nonlinear non-autonomous compressible Navier-Stokes equations based on the uniform estimates derived in Chapter 1 of Qin and Huang [102].

We sincerely hope that the reader will become familiar with the main ideas and essence of the basic theories and methods for establishing the global wellposedness and the asymptotic behavior of solutions for the models considered in this book. We also hope that the reader can be stimulated by some ideas from this book and undertake further study and research after having read the related references and bibliographic comments in this book.

We wish to express our gratitude to Dr. Thomas Hempfling and Dr. Barbara Hellriegel from Springer Basel AG for their great efforts to publish this book.

We also want to take this opportunity to thank all the people who concern about us including our teachers, colleagues and collaborators, etc. Yuming Qin sincerely thanks his mathematical teacher, Tonghe Xie, in the primary and middle school, who led him into the magic mathematical field. Qin also thanks his former Ph. D. advisor, Professor Songmu Zheng, for his constant encouragement, great help and support. Qin would like to thank Professors Daqian Li (Ta-tsien Li), Boling Guo, Jiaxing Hong, Weixi Shen, Ling Hsiao, Shuxing Chen, Zhenting Hou, Long-an Ying, Guangjun Yang, Guowang Chen, Jinhua Wang, Junning Zhao, Tiehu Qin, Yongji Tan, Sining Zheng, Zhouping Xin, Tong Yang, Hua Chen, Chaojiang Xu, Jingxue Yin, Liqun Zhang, Weike Wang, Mingxin Wang, Fahuai Yi, Song Jiang, Chengkui Zhong, Xuguang Lu, Yinbin Deng, Daoming Cao, Xiaoping Yang, Cheng He, Yi Zhou, Xiangyu Zhou, Quansui Wu, Daoyuan Fang, Ping Zhang, Changjiang Zhu, Changxing Miao, Kunyu Guo, Feimin Huang, Huijiang Zhao, Zheng-an Yao, Changzheng Qu, Yaping Wu, Zhaoli Liu, Huicheng Yin, Xiaozhou Yang, Shu Wang, Yaguang Wang, Zhong Tan, Xingbin Pan, Feng Zhou, Baojun Bian, Shengliang Pan, Wen-an Yong, Boxiang Wang, Lixin Tian, Shangbin Cui, Shijin Ding, Xi-nan Ma, Huaiyu Jian, Yachun Li, Benjin Xuan, Ting Wei, Quansen Jiu, Hailiang Li, Jiabao Su, Kaijun Zhang, Peidong Lei, Yongqian Zhang, Zhaoyang Yin, Wenyi Chen, Zhigui Lin, Xiaochun Liu, Yeping Li, Hao Wu, Ting Zhang, Zhenhua Guo, Chunpeng Wang, Zhifei Zhang, Hongjun Yu, Caidi Zhao and Yawei Wei for their constant help. Also Qin would like to thank Herbert Amann, Michel Chipot from Switzerland, J.A. Burns, Taiping Liu, Guiqiang Chen, D. Gilliams, Irena Lasiecka, Joel Spruck, M. Slemrod, Yisong Yang, Zhuangyi Liu, T. H. Otway, Shouhong Wang, Yuxi Zheng, Chun Liu, Changfeng Gui, Shouchuan Hu, Jianguo Liu, Hailiang Liu, Tao Luo from the USA, Roger Temam, Alain Miranville, D. Hilhorst, Vilmos Komornik, Mokhtar Kirane, Patrick Martinez, Fatiha Alabau-Boussouira, Yuejun Peng and Bopeng Rao from France, Hugo Beirao da Veiga, Maurizio Grasselli, Cecilia Cavaterra from Italy, Bert-Wolfgang Schulze, Ingo Witt, Reinhard Racke, Michael Reissig, Jürgen Sprekels, H.-D. Alber from Germany, Enrique Zuazua, Peicheng Zhu form Spain, Jaime E. Muñoz Rivera, Tofu Ma, Alexandre L. Madureira, Jinyun Yuan, D. Andrade, M. M. Cavalcanti, Frédéric G. Christian Valentin from Brazil, and Tzon Tzer Lü, Jyh-Hao Lee, Sui Sun Cheng from Chinese Taiwan for their constant help.

Yuming Qin also acknowledges the NNSF of China for its support. Currently, this book project is being supported by the NNSF of China with contracts no. 11031003, no. 11271066 and no. 10871040, by a grant of Shanghai Education Commission with contract no. 13ZZ048, and by the Sino-German cooperation grant "Analysis of partial differential equations and applications" with contracts no. 446 CHV 113/267/0-1, no.11111130182 and no. 11211130031.

Last but not least, Yuming Qin wants to express his deepest thanks to his parents (Zhenrong Qin and Xilan Xia), sisters (Yujuan Qin and Yuzhou Qin), brother (Yuxing Qin), wife (Yu Yin) and son (Jia Qin) for their great help, constant concern and advice in his career, and Xin Liu and Taige Wang take this opportunity to express thanks to their parents for their great support in their career.

Professor Yuming Qin
Department of Applied Mathematics
College of Science
Donghua University
Shanghai 201620, China
e-mail: yuming_qin@hotmail.com

Xin Liu
Business Information Management School
Shanghai Institute of Foreign Trade
Shanghai 201620, P. R. China
e-mail: xinliu120@hotmail.com

Taige Wang
460 McBryde Hall
Department of Mathematics
Virginia Tech
Blacksburg, VA 24061-0123, USA
e-mail: tigerwtg@math.vt.edu

Chapter 1

Global Existence and Asymptotic Behavior of Solutions to the Cauchy Problem for the 1D Compressible Magnetohydrodynamic Fluid System

1.1 Main Results

In this chapter, we shall study the global existence and large-time behavior of H^i-global solutions ($i = 1, 2, 4$) to the 1D MHD compressible system. The MHD system describes the interaction between intense magnetic fields and fluid conductors of electricity (see, e.g., [70]). The appearance of the electrically conducting fields grants this system with physically theoretic background of astrophysics, plasma physics, etc. In Lagrangian coordinates, the system can be written

$$\eta_t = v_x, \tag{1.1.1}$$

$$v_t = \left(\frac{\lambda v_x}{\eta} - P - \frac{1}{2} |\mathbf{b}|^2 \right)_x, \tag{1.1.2}$$

$$\mathbf{w}_t = \left(\frac{\mu \mathbf{w}_x}{\eta} \right)_x + \mathbf{b}_x, \tag{1.1.3}$$

$$(\eta \mathbf{b})_t = \left(\frac{\nu \mathbf{b}}{\eta} \right)_x + \mathbf{w}_x, \tag{1.1.4}$$

$$e_t = \left(\frac{\hat{\kappa} \theta_x}{\eta} \right)_x - P v_x + \frac{\lambda v_x^2 + \mu |\mathbf{w}_x|^2 + \nu |\mathbf{b}_x|^2}{\eta}, \tag{1.1.5}$$

where for $(x, t) \in \mathbb{R} \times \mathbb{R}_+ = (-\infty, +\infty) \times [0, +\infty)$ is the Lagrangian mass coordinate. The unknown quantities η, v, $\mathbf{w} \in \mathbb{R}^2$, $\mathbf{b} \in \mathbb{R}^2$, and e are the specific volume, the longitudinal velocity, the transverse velocity, the transverse magnetic field and the internal energy, respectively. Moreover, the absolute temperature θ appears as a variable in the pressure function $P = P(\eta, \theta)$ and $e = e(\eta, \theta)$.

We consider (1.1.1)–(1.1.5) subject to the initial condition

$$(\eta(x,0), v(x,0), \mathbf{w}(x,0), \mathbf{b}(x,0), \theta(x,0))$$
$$= (\eta_0(x), v_0(x), \mathbf{w_0}(x), \mathbf{b_0}(x), \theta_0(x)), \; \forall\, x \in \mathbb{R}. \tag{1.1.6}$$

Also, $\hat{\kappa}(x,t)$ is the heat conductivity and λ, μ, ν, etc., are also physical constants, representing the various viscosity coefficients. In this chapter, we also arrange $H^i = W^{i,2}$ ($i = 1, 2, 4$), $\|\cdot\|$ and $C^{k,\alpha}(\mathbb{R})$ to denote the norm in $L^2(\mathbb{R})$, and the space of functions whose derivatives are Hölder continuous with exponent α and order of differentials from 0 to k, respectively.

Generally the Stefan-Bolzmann law (the radiative gas model) holds as follows:

$$P(\eta, \theta) = R\theta/\eta + a\theta^4/3, \quad e(\eta, \theta) = C_V \theta + a\eta\theta^4, \tag{1.1.7}$$

where R, a and C_V are physical constants.

We also assume the physical constant $C_V = 1$, that $\hat{\kappa}(x,t)$ is a positive constant $\hat{\kappa}$ for simplicity and that for $i = 1, 2, 3, 4$ the positive constants C_i are dependent respectively on generic constants and the initial data's H^i norms, but independent of time $t > 0$.

We now state our main results in this chapter.

Theorem 1.1.1. *Assume that* $\eta_0 - \overline{\eta}$, v_0, $\mathbf{w_0}$, $\mathbf{b_0} - \overline{\mathbf{b}}$, $\theta_0 - \overline{\theta} \in H^1(\mathbb{R})$ *with* $\eta_0(x)$, $\theta_0(x) > 0$ *on* $\mathbb{R}$. *Define*

$$e_0^2 := \|\eta_0 - \overline{\eta}\|_{L^\infty}^2 + \int_{\mathbb{R}} (1 + x^2)^\alpha \Big[(\eta_0(x) - \overline{\eta})^2 + v_0^2(x) + |\mathbf{w_0}(x)|^2$$

$$+ |\mathbf{b_0}(x) - \overline{\mathbf{b}}|^2 + |\mathbf{b_0}(x) - \overline{\mathbf{b}}|^4 + (\theta_0(x) - \overline{\theta})^2 + v_0^4(x) \Big] dx$$

where $\alpha > \frac{1}{2}$ *is an arbitrary, but fixed parameter. Then if* $e_0 \le \epsilon_0$, *where* $\epsilon_0 \in (0, 1]$, *the problem* (1.1.1)–(1.1.7) *has a unique* H^1-*global solution* $(\eta(t), v(t), \mathbf{w}(t), \mathbf{b}(t),$ $\theta(t)) \in L^\infty(\mathbb{R}_+, H^1(\mathbb{R}))$ *and the following estimates hold:*

$$0 < C_1^{-1} \le \eta(t,x) \le C_1^{-1} \quad \text{on } \mathbb{R} \times \mathbb{R}_+, \tag{1.1.8}$$

$$0 < C_1^{-1} \le \theta(t,x) \le C_1^{-1} \quad \text{on } \mathbb{R} \times \mathbb{R}_+, \tag{1.1.9}$$

$$\|\eta(t) - \overline{\eta}\|_{H^1}^2 + \|v(t)\|_{H^1}^2 + \|\theta(t) - \overline{\theta}\|_{H^1}^2 + \|\mathbf{w}(t)\|_{H^1}^2 + \|\mathbf{b}(t) - \overline{\mathbf{b}}\|_{H^1}^2$$

$$+ \int_0^t \Big[\|\eta_x\|^2 + \|v_x\|^2 + \|\theta_x\|^2 + \|\mathbf{w}_x\|^2 + \|\mathbf{b}_x\|^2 + \|v_{xx}\|^2 + \|\theta_{xx}\|^2$$

$$+ \|\mathbf{w}_{xx}\|^2 + \|\mathbf{b}_{xx}\|^2 + \|v_t\|^2 + \|\theta_t\|^2 + \|\mathbf{w}_t\|^2 + \|\mathbf{b}_t\|^2 \Big] (s)ds \le C_1. \tag{1.1.10}$$

Moreover, as $t \to +\infty$,

$$\|(\eta - \overline{\eta}, v, \mathbf{w}, \mathbf{b} - \overline{\mathbf{b}}, \theta - \theta)(t)\|_{L^\infty} + \|(\eta_x, v_x, \mathbf{w}_x, \mathbf{b}_x, \theta_x)(t)\| \to 0. \tag{1.1.11}$$

Theorem 1.1.2. *Assume that* $\eta_0 - \overline{\eta}, v_0, \mathbf{w}_0, \mathbf{b}_0 - \overline{\mathbf{b}}, \theta_0 - \overline{\theta} \in H^2(\mathbb{R})$ *and* $\eta_0(x) > 0, \theta_0(x) > 0$ *on* $\mathbb{R}$ *and other assumptions, same as those of Theorem 1.1.1, hold. Then for any* $t > 0$, *the Cauchy problem* (1.1.1)–(1.1.7) *has a unique* H^2-*global solution* $(\eta(t), v(t), \mathbf{w}(t), \mathbf{b}(t), \theta(t)) \in L^\infty(\mathbb{R}_+, H^2(\mathbb{R}))$ *and the following estimate holds:*

$$\|\eta(t) - \overline{\eta}\|_{H^2}^2 + \|\eta(t) - \overline{\eta}\|_{W^{1,\infty}}^2 + \|\eta_t(t)\|_{H^1}^2 + \|v(t)\|_{H^2}^2 + \|v(t)\|_{W^{1,\infty}}^2$$
$$+ \|v_t(t)\|^2 + \|\mathbf{w}(t)\|_{H^2}^2 + \|\mathbf{w}(t)\|_{W^{1,\infty}}^2 + \|\mathbf{w}_t(t)\|^2 + \|\mathbf{b}(t) - \overline{\mathbf{b}}\|_{W^{1,\infty}}^2$$
$$+ \|\mathbf{b}(t) - \overline{\mathbf{b}}\|_{H^2}^2 + \|\mathbf{b}_t(t)\| + \|\theta(t) - \overline{\theta}\|_{H^2}^2 + \|\theta(t) - \overline{\theta}\|_{W^{1,\infty}}^2 + \|\theta_t(t)\|^2$$
$$+ \int_0^t \Big[\|\eta_x\|_{H^1}^2 + \|\eta_x\|_{L^\infty}^2 + \|\eta_t\|_{H^2}^2 + \|v_x\|_{H^2}^2 + \|v_x\|_{W^{1,\infty}}^2 + \|v_t\|_{H^1}^2 + \|\mathbf{w}_x\|_{H^2}^2$$
$$+ \|\mathbf{w}_x\|_{W^{1,\infty}}^2 + \|\mathbf{w}_t\|_{H^1}^2 + \|\mathbf{b}_x\|_{H^2}^2 + \|\mathbf{b}_x\|_{W^{1,\infty}}^2 + \|\mathbf{b}_t\|_{H^1}^2 + \|\theta_x\|_{H^2}^2$$
$$+ \|\theta_x\|_{W^{1,\infty}}^2 + \|\theta_t\|_{H^1}^2 \Big](s)ds \le C_2. \tag{1.1.12}$$

Moreover, as $t \to +\infty$,

$$\|\eta_t(t)\|_{H^1} + \|\eta_t(t)\|_{L^\infty} + \|v_t(t)\| + \|\mathbf{w}_t(t)\| + \|\mathbf{b}_t(t)\| + \|\theta_t(t)\| \to 0, \tag{1.1.13}$$

$$\|(\eta - \overline{\eta}, v, \mathbf{w}, \mathbf{b} - \overline{\mathbf{b}}, \theta - \overline{\theta})(t)\|_{W^{1,\infty}} + \|(\eta_x, v_x, \mathbf{w}_x, \mathbf{b}_x, \theta_x)(t)\|_{H^1} \to 0. \tag{1.1.14}$$

Theorem 1.1.3. *Assume that* $\eta_0 - \overline{\eta}_0, v_0, \mathbf{w}_0, \mathbf{b}_0 - \overline{\mathbf{b}}, \theta_0 - \overline{\theta} \in H^4(\mathbb{R})$ *and* $\eta_0(x) > 0, \theta_0(x) > 0$ *on* $\mathbb{R}$ *and that the other assumptions of Theorem 1.1.2 hold. Then for any* $t > 0$, *the Cauchy problem* (1.1.1)–(1.1.7) *admits a unique* H^4-*global solution* $(\eta(t), v(t), \mathbf{w}(t), \mathbf{b}(t), \theta(t)) \in L^\infty(\mathbb{R}_+, H^4(\mathbb{R}))$ *and the following estimates hold:*

$$\|\eta(t) - \overline{\eta}\|_{H^4}^2 + \|\eta(t) - \overline{\eta}\|_{W^{3,\infty}}^2 + \|\eta_t(t)\|_{H^3}^2 + \|\eta_{tt}(t)\|_{H^1}^2 + \|v(t)\|_{H^4}^2$$
$$+ \|v(t)\|_{W^{3,\infty}}^2 + \|v_t(t)\|_{H^2}^2 + \|v_{tt}(t)\|^2 + \|\mathbf{w}(t)\|_{H^4}^2 + \|\mathbf{w}(t)\|_{W^{3,\infty}}^2 + \|\mathbf{w}_t(t)\|_{H^2}^2$$
$$+ \|\mathbf{w}_{tt}(t)\|^2 + \|\mathbf{b}(t) - \overline{\mathbf{b}}\|_{H^4}^2 + \|\mathbf{b}(t) - \overline{\mathbf{b}}\|_{W^{3,\infty}}^2 + \|\mathbf{b}_t(t)\|_{H^2}^2 + \|\mathbf{b}_{tt}(t)\|^2$$
$$+ \|\theta(t) - \overline{\theta}\|_{H^4}^2 + \|\theta(t) - \overline{\theta}\|_{W^{3,\infty}}^2 + \|\theta_t(t)\|_{H^2}^2 + \|\theta_{tt}(t)\|^2 \le C_4, \tag{1.1.15}$$

$$\int_0^t \Big[\|\eta_x\|_{H^3}^2 + \|\eta_t\|_{H^4}^2 + \|\eta_{tt}\|_{H^2}^2 + \|\eta_{ttt}\|^2 + \|\eta_x\|_{W^{2,\infty}}^2 + \|v_x\|_{H^4}^2 + \|v_t\|_{H^3}^2$$
$$+ \|v_{tt}\|_{H^1}^2 + \|v_x\|_{W^{3,\infty}}^2 + \|\mathbf{w}_x\|_{H^4}^2 + \|\mathbf{w}_t\|_{H^3}^2 + \|\mathbf{w}_{tt}\|_{H^1}^2 + \|\mathbf{w}_x\|_{W^{3,\infty}}^2$$
$$+ \|\mathbf{b}_x\|_{H^4}^2 + \|\mathbf{b}_t\|_{H^3}^2 + \|\mathbf{b}_{tt}\|_{H^1}^2 + \|\mathbf{b}_x\|_{W^{3,\infty}}^2 + \|\theta_x\|_{H^4}^2 + \|\theta_t\|_{H^3}^2$$
$$+ \|\theta_{tt}\|_{H^1}^2 + \|\theta_x\|_{W^{3,\infty}}^2 \Big](s)ds \le C_4. \tag{1.1.16}$$

Moreover, as $t \to +\infty$,

$$\|(\eta_x, v_x, \mathbf{w}_x, \mathbf{b}_x, \theta_x)(t)\|_{H^3} + \|\eta_t(t)\|_{H^3} + \|\eta_t(t)\|_{W^{2,\infty}}$$
$$+ \|v_t(t)\|_{H^2} + \|v_t(t)\|_{W^{1,\infty}} + \|\mathbf{w}_t(t)\|_{H^2} + \|\mathbf{w}_t(t)\|_{W^{1,\infty}}$$
$$+ \|\mathbf{b}_t(t)\|_{H^2} + \|\mathbf{b}_t(t)\|_{W^{1,\infty}} + \|\theta_t(t)\|_{H^2} + \|\theta_t(t)\|_{W^{1,\infty}} \to 0, \tag{1.1.17}$$

$$\|\eta_{tt}\|_{H^1} + \|v_{tt}(t)\| + \|\mathbf{w}_{tt}(t)\| + \|\mathbf{b}_{tt}(t)\| + \|\theta_{tt}(t)\|$$
$$+ \|(\eta_x, v_x, \mathbf{w}_x, \mathbf{b}_x, \theta_x)(t)\|_{W^{2,\infty}} \to 0. \tag{1.1.18}$$

Corollary 1.1.1 *The H^4-global solution $(\eta(t), v(t), \mathbf{w}(t), \mathbf{b}(t), \theta(t))$ obtained in Theorem 1.1.3 is actually a classical solution. Precisely, $(\eta(t), v(t), \mathbf{w}(t), \mathbf{b}(t), \theta(t)) \in C^{3,\frac{1}{2}}(\mathbb{R})$ and, as $t \to +\infty$,*

$$\|(\eta_x(t), v_x(t), \mathbf{w}_x(t), \mathbf{b}_x(t), \theta_x(t))\|_{C^{2,\frac{1}{2}}} + \|\eta_t(t)\|_{C^{2,\frac{1}{2}}}$$
$$+ \|(v_t(t), \mathbf{w}_t(t), \mathbf{b}_t(t), \theta_t(t))\|_{C^{1,\frac{1}{2}}} + \|\eta_{tt}(t)\|_{C^{\frac{1}{2}}} \to 0. \tag{1.1.19}$$

1.2 Global Existence and Asymptotic Behavior in $H^1(\mathbb{R})$

In this section, we shall establish global H^1 estimates for solutions $(\eta, v, \mathbf{w}, \mathbf{b}, \theta)$ to the system. Here C and $\widetilde{C}$ will stand for some generic constants (≥ 1) which might depend on systematic constants such as R, etc., for the most.

At first, we suppose that

$$|\eta(x,t) - \overline{\eta}| + |\mathbf{b}(x,t) - \overline{\mathbf{b}}| + \phi(t)|\theta(x,t) - \overline{\theta}| \leq \min\{\overline{\eta}, |\overline{\mathbf{b}}|, \overline{\theta}\}/2, \tag{1.2.1}$$

for all $(x,t) \in \mathbb{R} \times \mathbb{R}_+$, where $\phi(t) = \min\{t, 1\}$. It is obvious that $|\mathbf{b}(x,t) - \overline{\mathbf{b}}| \leq C$.

Lemma 1.2.1. *For all $t > 0$,*

$$\frac{1}{2}\int_{\mathbb{R}} v^2 dx + \frac{1}{2}\int_{\mathbb{R}} |\mathbf{w}|^2 dx + \frac{1}{2}\int_{\mathbb{R}} \eta|\mathbf{b}|^2 dx + \overline{\theta}\int_{\mathbb{R}}\left(\frac{\theta}{\overline{\theta}} - \log\frac{\theta}{\overline{\theta}} - 1\right) dx \tag{1.2.2}$$

$$+ R\overline{\eta}\int_{\mathbb{R}}\left(\frac{\eta}{\overline{\eta}} - \log\frac{\eta}{\overline{\eta}} - 1\right) dx + \int_0^t\int_{\mathbb{R}}\left(\frac{\lambda v_x^2 + \mu|\mathbf{w}|_x^2 + \nu|\mathbf{b}_x|^2}{\theta\eta} + \hat{\kappa}\frac{\theta_x^2}{\eta\theta^2}\right) dxds \leq C.$$

Proof. Multiplying equations (1.1.1) to (1.1.5) by $R(1 - \overline{\eta}/\eta)$, v, $\mathbf{w}$, $\mathbf{b}$, $1 - \overline{\theta}/\theta$ respectively, integrating the results on $\mathbb{R} \times \mathbb{R}_+$ and summing them together, we get (1.2.2) directly. $\qquad\square$

We then estimate the L^2 norms of v, $\mathbf{w}$, $\eta - \overline{\eta}$, $\mathbf{b} - \overline{\mathbf{b}}$, $\theta - \overline{\theta}$, using a weighted L^2-norm as $\|\cdot\|_w = (\int_{\mathbb{R}}(1 + x^2)^\alpha|\cdot|^2 dx)^{1/2}$, which is basic for H^1-global estimates. Let us introduce

$$\psi(x) = (1 + x^2)^\alpha \quad \left(\alpha > \frac{1}{2}\right)$$

as a weight function.

Lemma 1.2.2. *Under the hypotheses of Theorem 1.1.1 except, for the moment, for the condition e_0, we have the following estimate for all $t > 0$:*

$$\|v(t)\|^2 + \|\eta(t) - \overline{\eta}\|^2 + \|\mathbf{w}(t)\|^2 + \|\mathbf{b}(t) - \overline{\mathbf{b}}\|^2 + \|\theta(t) - \overline{\theta}\|^2$$

$$+ \int_0^t\left[\|v_x\|^2 + \|\mathbf{w}_x\|^2 + \|\mathbf{b}_x\|^2 + \|\theta_x\|^2\right](s)ds \leq Ce_0^2, \tag{1.2.3}$$

where $e_0 \leq 1/(2C)$.

Proof. Multiplying (1.1.2) by $\psi(x)v$ and integrating over $\mathbb{R}$ we can see, using integration by parts that

$$\frac{1}{2}\frac{\partial}{\partial t}\int_{\mathbb{R}}\psi v^2 dx = -\int_{\mathbb{R}}\frac{v_x}{\eta}(\psi v)_x dx - \int_{\mathbb{R}}\left(\frac{R\theta}{\eta}+\frac{|\mathbf{b}|^2}{2}\right)_x(\psi v)dx$$

$$= -\int_{\mathbb{R}}\frac{v_x}{\eta}(\psi_x v + \psi v_x)dx + \int_{\mathbb{R}}\left(\frac{R\theta}{\eta}-\frac{R\bar{\theta}}{\bar{\eta}}\right)(\psi_x v + \psi v_x)dx$$

$$+\frac{1}{2}\int_{\mathbb{R}}|\mathbf{b}-\bar{\mathbf{b}}|^2(\psi_x v + \psi v_x)dx + \bar{\mathbf{b}}\cdot\int_{\mathbb{R}}(\mathbf{b}-\bar{\mathbf{b}})(\psi_x v + \psi v_x)dx.$$

Next, using the inequality $|\psi_x| \le C|\psi|$ with (1.2.1) and the mean value theorem for the function $f(\eta,\theta) = R\theta/\eta - R\bar{\theta}/\bar{\eta}$, and then integrating this equation over $(0,t)$, $t \in [0,1]$ and employing Young's inequality, we get

$$\int_{\mathbb{R}}\psi v^2(x,t)dx + \int_0^t\int_{\mathbb{R}}\psi v_x^2\,dxds \tag{1.2.4}$$

$$\le Ce_0^2 + C\int_0^t\int_{\mathbb{R}}\left[\psi((\eta-\bar{\eta})^2 + v^2 + |\mathbf{b}-\bar{\mathbf{b}}|^4 + |\mathbf{b}-\bar{\mathbf{b}}|^2 + (\theta-\bar{\theta})^2)\right]dxds,$$

for all $t \in [0,1]$. Multiplying (1.1.1) by $\psi(x)(\eta-\bar{\eta})$ and integrating in the same way, we can see that

$$\int_{\mathbb{R}}\psi(\eta-\bar{\eta})^2(x,t)dx$$

$$\le Ce_0^2 + C\int_0^t\int_{\mathbb{R}}\psi\left((\eta-\bar{\eta})^2 + v^2 + (\theta-\bar{\theta})^2\right)dxds, \quad \forall t \in [0,1]. \tag{1.2.5}$$

Adding (1.2.4) to (1.2.5) gives

$$\int_{\mathbb{R}}\psi\left((\eta-\bar{\eta})^2 + v^2\right)(x,t)dx + \int_0^t\int_{\mathbb{R}}\psi v_x^2\,dxds \tag{1.2.6}$$

$$\le Ce_0^2 + C\int_0^t\int_{\mathbb{R}}\psi\left((\eta-\bar{\eta})^2 + v^2 + |\mathbf{b}-\bar{\mathbf{b}}|^4 + |\mathbf{b}-\bar{\mathbf{b}}|^2 + (\theta-\bar{\theta})^2\right)dxds,$$

for all $t \in [0,1]$. In the same way, we dot-multiply (1.1.3) and (1.1.4) by the vectors $\psi(x)\mathbf{w}$ and $\psi(x)(\mathbf{b}-\bar{\mathbf{b}})$, respectively, and use (1.2.6) to obtain

$$\int_{\mathbb{R}}\psi|\mathbf{w}|^2(x,t)dx + \int_0^t\int_{\mathbb{R}}\psi|\mathbf{w}_x|^2dxds$$

$$\le Ce_0^2 + C\int_0^t\int_{\mathbb{R}}\psi(|\mathbf{w}|^2 + |\mathbf{b}-\bar{\mathbf{b}}|^2)dxds, \tag{1.2.7}$$

$$\int_{\mathbb{R}} \psi \eta |\mathbf{b} - \overline{\mathbf{b}}|^2(x,t)dx + \int_0^t \int_{\mathbb{R}} \psi |\mathbf{b}_x|^2 dxds$$

$$\leq Ce_0^2 + C \int_0^t \int_{\mathbb{R}} \psi \Big[(\eta - \overline{\eta})^2 + v^2 + |\mathbf{w}|^2 + |\mathbf{b} - \overline{\mathbf{b}}|^4$$

$$+ |\mathbf{b} - \overline{\mathbf{b}}|^2 + (\theta - \overline{\theta})^2 \Big] dxds, \quad \forall t \in [0,1]. \tag{1.2.8}$$

Using (1.2.1), we also derive from (1.1.2), (1.1.4) and (1.1.5) for $t \in [0,1]$ and sufficiently small $\delta > 0$ that

$$\int_{\mathbb{R}} \psi v^4(x,t)dx + \int_0^t \int_{\mathbb{R}} \psi v^2 v_x^2 dxds \tag{1.2.9}$$

$$\leq Ce_0^2 + \left(C + \max_{s \in [0,t]} \int_{\mathbb{R}} \psi v^2 dx \right) \int_0^t \int_{\mathbb{R}} \psi \Big[v^2 + (\eta - \overline{\eta})^2$$

$$+ |\mathbf{b} - \overline{\mathbf{b}}|^2 + |\mathbf{b} - \overline{\mathbf{b}}|^4 + (\theta - \overline{\theta})^2 \Big] dxds,$$

$$\int_{\mathbb{R}} \psi \eta |\mathbf{b} - \overline{\mathbf{b}}|^4(x,t)dx + \int_0^t \int_{\mathbb{R}} \psi |\mathbf{b}_x|^2 |\mathbf{b} - \overline{\mathbf{b}}|^2 dxds$$

$$\leq Ce_0^2 + \int_0^t \int_{\mathbb{R}} \psi(|v_x||\mathbf{b} - \overline{\mathbf{b}}|^4 + |v_x||\mathbf{b} - \overline{\mathbf{b}}|^3 + |\mathbf{w}_x||\mathbf{b} - \overline{\mathbf{b}}|^3) dxds$$

$$\leq Ce_0^2 + \delta \int_0^t \int_{\mathbb{R}} v_x^2 dxds + \delta \int_0^t \int_{\mathbb{R}} \mathbf{w}_x^2 dxds \tag{1.2.10}$$

$$+ C \int_0^t \int_{\mathbb{R}} \psi \Big[v^2 + (\eta - \overline{\eta})^2 + |\mathbf{w}|^2 + |\mathbf{b} - \overline{\mathbf{b}}|^2 + |\mathbf{b} - \overline{\mathbf{b}}|^4 + (\theta - \overline{\theta})^2 \Big] dxds,$$

and

$$\int_{\mathbb{R}} \psi((\theta - \overline{\theta})^2 + v^4)(x,t)dx + \int_0^t \int_{\mathbb{R}} \psi(\theta_x^2 + v^2 v_x^2) dxds$$

$$\leq Ce_0^2 + \left(C + \max_{s \in [0,t]} \int_{\mathbb{R}} \psi v^2 dx \right) \int_0^t \int_{\mathbb{R}} \psi \Big[v^2 + (\eta - \overline{\eta})^2 + |\mathbf{w}|^2 \tag{1.2.11}$$

$$+ |\mathbf{b} - \overline{\mathbf{b}}|^2 + |\mathbf{b} - \overline{\mathbf{b}}|^4 + (\theta - \overline{\theta})^2 \Big] dxds + C \int_{\mathbb{R}} \psi(\theta - \overline{\theta})^2 dx \int_0^t \max_{\mathbb{R}} (\theta - \overline{\theta})^2 ds.$$

Set

$$h(t) = \sup_{s \in [0,t]} \int_{\mathbb{R}} \psi(x) \Big[v^2 + (\theta - \overline{\theta})^2 \Big](x,s)dx. \tag{1.2.12}$$

Now using the interpolation inequality $\max_{\mathbb{R}} (\theta - \overline{\theta})^2 \leq C\|\theta - \overline{\theta}\|\|\theta_x\|$ in (1.2.11), summing (1.2.6)–(1.2.11) together and applying the generalized Gronwall inequality we have that, for all $t \in [0,1]$,

$$\int_{\mathbb{R}} \psi \Big((\eta - \overline{\eta})^2 + (\theta - \overline{\theta})^2 + v^2 + v^4 + |\mathbf{w}|^2 + |\mathbf{b} - \overline{\mathbf{b}}|^2 + |\mathbf{b} - \overline{\mathbf{b}}|^4 \Big)(x,t)dx$$

$$+ \int_0^t \int_{\mathbb{R}} \psi\left(\theta_x^2 + v_x^2 + v^2 v_x^2 + |\mathbf{w}_x|^2 + |\mathbf{b}_x|^2 + |\mathbf{b}_x|^2 |\mathbf{b} - \overline{\mathbf{b}}|^2\right) dx ds$$
$$\leq C(e_0^2 + h^3(t)) \exp(h(t)). \tag{1.2.13}$$

By the definition of $h(t)$, $h(t) > 0$ and $\exp(h(t)) > 1$. Assuming

$$h(t) \leq \min\{\log(4/3), \, 1/(2\sqrt{C})\},$$

we derive $1/\exp(h(t)) - Ch^2(t) \geq 1/2$, we have $h(t) \leq 2Ce_0^2$, and so $h(t) \leq e_0$ if $e_0 \leq 1/(2C)$. As a result, (1.2.13) can be improved under $e_0 \leq 1/(2C)$ to

$$h(t) \leq \min\left\{\log\frac{4}{3}, \, 1/2\sqrt{C}, \, Ce_0^2\right\} \leq Ce_0^2, \text{ for } t \in [0,1]. \tag{1.2.14}$$

Repeating what we have done in (1.2.13), we arrive at

$$\int_{\mathbb{R}} \psi\left((\eta - \overline{\eta})^2 + (\theta - \overline{\theta})^2 + v^2 + v^4 + |\mathbf{w}|^2 + |\mathbf{b} - \overline{\mathbf{b}}|^2 + |\mathbf{b} - \overline{\mathbf{b}}|^4\right)(x,t)dx$$
$$+ \int_0^t \int_{\mathbb{R}} \psi\left(\theta_x^2 + v_x^2 + v^2 v_x^2 + |\mathbf{w}_x|^2 + |\mathbf{b}_x|^2 + |\mathbf{b}_x|^2 |\mathbf{b} - \overline{\mathbf{b}}|^2\right) dx ds$$
$$\leq Ce_0^2, \quad t \in [0,+\infty), \tag{1.2.15}$$

provided that $e_0 \leq 1/(2C)$, i.e., (1.2.3). This completes the proof. $\square$

We shall now obtain the global H^1 estimates for $(\eta, v, \mathbf{w}, \mathbf{b}, \theta)$. We define

$$H(t) := \sup_{0 \leq s \leq t} \left\{\|\eta - \overline{\eta}\|_{L^\infty}^2 + \phi^2\left[\|v_x\|^2 + \|\mathbf{w}_x\|^2\right.\right.$$
$$\left.+ \|\mathbf{b}_x\|^2 + \|(\mathbf{b} - \overline{\mathbf{b}})\mathbf{b}_x\|^2\right] + \phi^4\|\theta_x\|^2\right\}(s) + \int_0^t \left[\phi^2\|v_t\|^2 + \phi^4\|\theta_t\|^2\right.$$
$$\left.+ \|v_x\|^2 + \phi^2\|\mathbf{w}_t\|^2 + \phi^2\|\mathbf{b}_t\|^2 + \phi^2\|\mathbf{b}_t \cdot (\mathbf{b} - \overline{\mathbf{b}})\|^2\right](s)ds. \tag{1.2.16}$$

Lemma 1.2.3. *Under the hypotheses of Lemma 1.2.2, we have the estimate*

$$H(t) \leq e_0, \tag{1.2.17}$$

where $e_0 < \min\{1/(2C), \, 1/(2\tilde{c})\}$*, and* $\tilde{c} \geq 1$ *depends on some physical constants.*

Proof. Multiplying (1.1.2) by $\phi^2 v_t$ and integrating the result over $\mathbb{R} \times \mathbb{R}_+$, we arrive at

$$\int_0^t \int_{\mathbb{R}} \phi^2 v_t^2 \, dx ds + \int_0^t \int_{\mathbb{R}} \left(\frac{\phi^2 v_x^2}{\eta}\right)_t dx ds = -\int_0^t \int_{\mathbb{R}} \left(\frac{R\theta}{\eta}\right)_x \phi^2 v_t \, dx ds$$
$$- \int_0^t \int_{\mathbb{R}} \phi^2 \mathbf{b} \cdot \mathbf{b}_x v_t \, dx ds + \int_0^t \int_{\mathbb{R}} \frac{2\phi\phi_t \cdot v_x^2}{\eta} dx ds - \int_0^t \int_{\mathbb{R}} \frac{\phi^2 v_x^3}{\eta} dx ds.$$

From the definition of $\phi(t)$, we derive that $|\phi| \leq 1$, $|\phi_t| \leq 1$. If we now use (1.2.1) and (1.2.15) and apply Young's inequality, we obtain

$$\phi^2 \int_{\mathbb{R}} v_x^2 \, dx + \int_0^t \int_{\mathbb{R}} \phi^2 v_t^2 \, dx ds \leq Ce_0^2 + C \int_0^t \int_{\mathbb{R}} \phi^4 v_x^4 \, dx ds + CH^2(t), \quad t \geq 0.$$
(1.2.18)

Thanks to (1.2.1), we see

$$\int_0^t \int_{\mathbb{R}} \phi^4 v_x^4 \, dx ds \leq C \int_0^t \phi^4 \left[\max_{\mathbb{R}} v_x^2 \int_{\mathbb{R}} v_x^2 \, dx \right] ds$$

$$\leq Ce_0^2 + C \int_0^t \left[\phi^4 \max_{\mathbb{R}} \left(\frac{v_x}{\eta} - R\frac{\theta}{\eta} + R\frac{\overline{\theta}}{\overline{\eta}} - \frac{|\mathbf{b}|^2}{2} \right)^2 \int_{\mathbb{R}} v_x^2 \, dx \right] ds$$

$$\leq Ce_0^2 + C \int_0^t \phi^4 (\|v_x\|^2 + \|\eta - \overline{\eta}\|^2 + \|\theta - \overline{\theta}\|^2 + \|\mathbf{b} - \overline{\mathbf{b}}\|^2 + \|v_t\|^2) \|v_x\|^2 ds$$

$$+ C \int_0^t \phi^4 \|v_x\|^2 ds$$

$$\leq C(e_0^2 + H^2(t)),$$
(1.2.19)

whence

$$\phi^2(t) \int_{\mathbb{R}} v_x^2 \, dx + \int_0^t \int_{\mathbb{R}} \phi^2 v_t^2 \, dx ds \leq C(e_0^2 + H^2(t)), \quad \forall t \geq 0.$$
(1.2.20)

Similarly, we can get from (1.1.3) by multiplying (1.2.15) with $\phi^2 \mathbf{w}_t$, that

$$\phi^2(t)\|\mathbf{w}_x(t)\|^2 + \int_0^t \phi^2(s)\|\mathbf{w}(s)\|^2 ds \leq Ce_0^2 + CH^2(t), \quad \forall t \geq 0.$$
(1.2.21)

In an analogous manner we infer from (1.1.4), that

$$\int_{\mathbb{R}} \phi^2 \eta |\mathbf{b}_x|^2 dx + \int_0^t \int_{\mathbb{R}} \phi^2 |\mathbf{b}_t|^2 dx ds$$

$$\leq Ce_0^2 + \int_0^t \int_{\mathbb{R}} \phi^2 |\mathbf{b}_t|^2 |\mathbf{b} - \overline{\mathbf{b}}|^2 dx ds + \int_0^t \phi^2 \left(\|\mathbf{b}_t\|^2 + \|v_x\|^2 + \|\mathbf{w}_x\|^2 \right) \|\mathbf{b}_x\|^2 ds$$

$$\leq Ce_0^2 + CH^2(t), \quad \forall t \geq 0,$$
(1.2.22)

and

$$\int_{\mathbb{R}} \phi^2 |\mathbf{b}_x|^2 |\mathbf{b} - \overline{\mathbf{b}}|^2 dx + \int_0^t \int_{\mathbb{R}} \phi^2 |\mathbf{b}_t|^2 |\mathbf{b} - \overline{\mathbf{b}}|^2 dx ds \leq Ce_0^2 + CH^2(t), \quad \forall t \geq 0.$$
(1.2.23)

Further, it follows from (1.1.5) that

$$\phi^4(t)\|\theta_x(t)\|^2 + \int_0^t \|\theta_t(s)\|^2 \phi^4(s) ds$$

$$\leq Ce_0^2 + C \int_0^t \int_{\mathbb{R}} \phi^4 \left(v_x^4 + |\mathbf{w}_x|^4 + |\mathbf{b}_x|^4 + |v_x|\theta_x^2 \right) dx ds + \frac{1}{2} \int_0^t \int_{\mathbb{R}} \phi^4 \theta_t^2 \, dx ds$$

$$\leq Ce_0^2 + CH^2(t) + C \int_0^t \left[\phi^8 \max_{\mathbb{R}} v_x^2 \int_{\mathbb{R}} \theta_x^2 \, dx ds \right] + \frac{1}{2} \int_0^t \int_{\mathbb{R}} \phi^4 \theta_t^2 \, dx ds$$

$$\leq Ce_0^2 + CH^2(t) + \frac{1}{2} \int_0^t \int_{\mathbb{R}} \phi^4 \theta_t^2 \, dx ds, \quad \forall t \geq 0,$$

i.e.,

$$\phi^4(t)\|\theta_x(t)\|^2 + \int_0^t \|\theta_t(s)\|^2 \phi^4(s) ds \leq Ce_0^2 + CH^2(t), \tag{1.2.24}$$

for all $t \geq 0$.

We now estimate the bounds for $\eta - \overline{\eta}$. Using (1.1.1), we rewrite (1.1.2) as

$$(\log \eta)_{xt} = v_t + R \left(\frac{\theta}{\eta} - \frac{\overline{\theta}}{\overline{\eta}} \right)_x + \left[\frac{|\mathbf{b} - \overline{\mathbf{b}}|^2}{2} + \overline{\mathbf{b}} \cdot (\mathbf{b} - \overline{\mathbf{b}}) \right]_x. \tag{1.2.25}$$

Integrating (1.2.25) over $(-\infty, x) \times (0, t)$ ($t \in [0, 1]$) and then taking the absolute value and using (1.2.1), (1.2.15) and the weight $\alpha > 1/2$, we obtain

$$|\eta - \overline{\eta}| \leq C|\log \eta/\overline{\eta}|$$

$$\leq C|\eta_0 - \overline{\eta}| + C \int_{-\infty}^x (|v| + |v_0|) dy + C \int_0^t (|\eta - \overline{\eta}| + |\theta - \overline{\theta}|) ds$$

$$+ C \int_0^t |\mathbf{b} - \overline{\mathbf{b}}|^2 ds + C \int_0^t |\mathbf{b} - \overline{\mathbf{b}}| ds$$

$$\leq Ce_0 + C\|\psi^{\frac{1}{2}} v\| \, \|\psi^{-\frac{1}{2}}\| + C \int_0^t |\eta - \overline{\eta}| ds + C \int_0^t \|\theta - \overline{\theta}\|_{H^1} ds$$

$$+ C \int_0^t \|\mathbf{b} - \overline{\mathbf{b}}\|^2 ds + C \left(\int_0^t \|\mathbf{b} - \overline{\mathbf{b}}\|_{H^1}^2 ds \right)^{\frac{1}{2}}$$

$$\leq Ce_0 + C \int_0^t |\eta - \overline{\eta}| ds + C \left(\int_0^t \|\theta - \overline{\theta}\|_{H^1}^2 ds \right)^{\frac{1}{2}}$$

$$+ C \int_0^t \|\mathbf{b} - \overline{\mathbf{b}}\|^2 ds + C \left(\int_0^t \|\mathbf{b} - \overline{\mathbf{b}}\|_{H^1}^2 ds \right)^{\frac{1}{2}}$$

$$\leq Ce_0 + C \int_0^t |\eta - \overline{\eta}| ds, \quad \forall t \in [0, 1]. \tag{1.2.26}$$

Applying the Gronwall inequality to (1.2.26), we get

$$|\eta(x, t) - \overline{\eta}| \leq Ce_0, \quad x \in \mathbb{R}, \forall t \in [0, 1]. \tag{1.2.27}$$

When $t \geq 1$, we denote $F := (v_x/\eta) - R(\eta/\theta) + R(\overline{\eta}/\overline{\theta})$. Obviously $[1/\eta - 1/\overline{\eta}]_t + R\theta/\eta \cdot [1/\eta - 1/\overline{\eta}] = -F/\eta - R(\theta - \overline{\theta})/\eta$. Multiplying this equality by

$1/\eta - 1/\bar{\eta}$, using (1.2.1), (1.1.2), (1.2.15), and the interpolation inequality, we obtain

$$\left[\frac{1}{\eta} - \frac{1}{\bar{\eta}}\right]_t^2 + C^{-1}\left[\frac{1}{\eta} - \frac{1}{\bar{\eta}}\right]^2 \leq C\left(\|F\|_{L^\infty}^2 + \|\theta - \bar{\theta}\|_{L^\infty}^2\right)$$
$$\leq C\left(\|F\|_{H^1}^2 + \|\theta - \bar{\theta}\|_{H^1}^2\right)$$
$$\leq Ce_0^2 + C\|(v_x, v_t, \mathbf{b}_x, \theta_x)\|^2, \quad \forall t \geq 1,$$

which, combined with (1.2.15) and (1.2.18), yields for $x \in \mathbb{R}$, $t \geq 1$,

$$|\eta(x,t) - \bar{\eta}|^2 \tag{1.2.28}$$
$$\leq Ce_0^2 + C|\eta(x,1) - \bar{\eta}|^2 + C\int_1^t \|(v_x, v_t, \mathbf{b}_x, \theta_x)\|^2(s)ds \leq C(e_0^2 + H^2(t)).$$

Combining the definition of $H(t)$, (1.2.20)–(1.2.24), (1.2.27) and (1.2.28), we obtain $H(t) \leq \tilde{c}[e_0^2 + H^2(t)]$ for $t \in [0, +\infty)$, where $\tilde{c} \geq 1$ depends on C and other physical constants. Similarly to the estimate for $h(t)$, we still assume $H(t)$ so small that $1 - \tilde{c}H(t) \geq 1/2$, and then $H(t) \leq 2\tilde{c}e_0^2 \leq e_0$ provided that $e_0 < \min\{1/(2C), 1/(2\tilde{c})\}$. The proof is complete. □

Proof of Theorem 1.1.1. From Lemmas 1.2.2–1.2.3, we derive immediately that for $x \in \mathbb{R}$, $t \geq 0$,

$$|\eta(x,t) - \bar{\eta}| + |\mathbf{b}(t) - \bar{\mathbf{b}}| + \phi(t)|\theta(x,t) - \bar{\theta}|$$
$$\leq H^{1/2}(t) + C\|\mathbf{b} - \bar{\mathbf{b}}\|^{1/2}\|\mathbf{b}_x\|^{1/2} + C\|\theta - \bar{\theta}\|^{1/2}\|\theta_x\|^{1/2}$$
$$\leq H^{1/2}(t) + Ce_0H^{1/4}(t) \leq \sqrt{2\tilde{c}}\,e_0(1 + \tilde{C}\sqrt{e_0}) < \frac{1}{3}\min\{\bar{\eta}, |\bar{\mathbf{b}}|, \bar{\theta}\}. \tag{1.2.29}$$

Actually, we assume that e_0 is small enough to ensure that $1 + \tilde{C}\sqrt{e_0} < 4/3$, and then $e_0 < 1/(4\sqrt{2})$. So we set $e_0 < \min\{1/(6\sqrt{\tilde{c}}), 1/(3\tilde{C})^2, 1/(2\tilde{c}), 1/(2\tilde{C})\} := \epsilon_0$, under the hypotheses discussed in Lemmas 1.2.2–1.2.3.

We can see that the upper bound of $|\eta - \bar{\eta}| + |\mathbf{b} - \bar{\mathbf{b}}| + \phi(t)|\theta - \bar{\theta}|$ in the verification is strictly smaller than that in condition (1.2.1). We conclude that

$$\|\eta - \bar{\eta}\|_{L^\infty}^2 + \|v_x\|^2 + \|\mathbf{w}_x\|^2 + \|\mathbf{b}_x\|^2 + \|\theta_x\|^2$$
$$+ \int_0^t \left(\|v_t\|^2 + \|\theta_t\|^2 + \|\mathbf{w}_t\|^2 + \|\mathbf{b}_t\|^2\right)(s)ds \leq \epsilon_0^2 \leq C_1. \tag{1.2.30}$$

Further, from (1.2.29) with arguments similar to those used in [112], we can get (1.1.12) and (1.1.13). From $(\eta - \bar{\eta}, v, \mathbf{w}, \mathbf{b} - \bar{\mathbf{b}}, \theta - \bar{\theta}) \in H^1(\mathbb{R})$, we rewrite (1.2.25) again as

$$\left(\frac{\eta_x}{\eta}\right)_t = v_t + \left(\frac{R\theta}{\eta}\right)_x + \left(\frac{|\mathbf{b} - \bar{\mathbf{b}}|^2}{2} + \bar{\mathbf{b}}\cdot\mathbf{b}\right)_x, \tag{1.2.31}$$

multiply it by η_x/η, integrate the result over $\mathbb{R} \times [0,t]$ and use (1.1.13) to get

$$\|\eta_x(t)\|^2 + \int_0^t \int_{\mathbb{R}} \theta \eta_x^2 \, dx ds \tag{1.2.32}$$

$$\leq C_1 + \frac{1}{2} \int_0^t \int_{\mathbb{R}} \theta \eta_x^2 \, dx ds + C \int_0^t \left(\|v_t\|^2 + \|\theta_x\|^2 + \|\mathbf{b} - \overline{\mathbf{b}}\|^2 \right) ds \leq C_1.$$

Using again from (1.1.13), we get

$$\|\eta_x(t)\|^2 + \int_0^t \|\eta_x(s)\|^2 ds \leq C_1. \tag{1.2.33}$$

We can rewrite (1.1.2) as

$$v_t = \frac{v_{xx}}{\eta} - \frac{v_x \eta_x}{\eta^2} - \frac{R\theta_x}{\eta} + \frac{R\theta \eta_x}{\eta^2} - \mathbf{b} \cdot \mathbf{b}_x,$$

then we infer that

$$\|v_{xx}(t)\|^2 \leq C_1 \left(\|v_t(t)\|^2 + \|v_x(t)\|_{L^\infty}^2 + \|\theta_x(t)\|^2 + \|\eta_x(t)\|^2 + \|\mathbf{b}_x(t)\|^2 \right),$$

and conclude immediately from (1.1.13), Lemmas 1.2.2–1.2.3 and the interpolation inequalities that

$$\int_0^t \|v_{xx}(s)\|^2 ds \leq C_1, \quad \|v_{xx}(t)\| \leq C_1 \|v_t(t)\| + C_1.$$

Similarly, one can also verify that

$$\int_0^t \left(\|\mathbf{w}_{xx}\|^2 + \|\mathbf{b}_{xx}\|^2 + \|\theta_{xx}\|^2 \right)(s) ds \leq C_1,$$

$$\|\mathbf{w}_{xx}(t)\| \leq C_1 \|\mathbf{w}_t(t)\| + C_1,$$
$$\|\mathbf{b}_{xx}(t)\| \leq C_1 \|\mathbf{b}_t(t)\| + C_1,$$
$$\|\theta_{xx}(t)\| \leq C_1 \|\theta_t(t)\| + C_1.$$

Combining with (1.2.30), (1.1.12) and (1.1.13), we complete the proof of (1.1.14) of Theorem 1.1.1.

Integrating by parts, we have

$$\frac{1}{2} \frac{d}{dt} \int_{\mathbb{R}} v_x^2 \, dx = \int_{\mathbb{R}} v_x v_{xt} \, dx, \tag{1.2.34}$$

$$\frac{1}{2} \frac{d}{dt} \int_{\mathbb{R}} |\mathbf{w}_x|^2 dx = \int_{\mathbb{R}} \mathbf{w}_x \cdot \mathbf{w}_{xt} \, dx, \tag{1.2.35}$$

$$\frac{1}{2} \frac{d}{dt} \int_{\mathbb{R}} |\mathbf{b}_x|^2 dx = \int_{\mathbb{R}} \mathbf{b}_x \cdot \mathbf{b}_{xt} \, dx, \tag{1.2.36}$$

$$\frac{1}{2}\frac{d}{dt}\int_{\mathbb{R}}\theta_x^2\,dx = \int_{\mathbb{R}}\theta_x\theta_{xt}\,dx, \tag{1.2.37}$$

and then (1.1.1) and (1.1.15) yield

$$\int_0^{+\infty}\left|\frac{d}{dt}\left(\|(\eta_x,\,v_x,\,\mathbf{w}_x,\,\mathbf{b}_x,\,\theta_x)\|^2\right)(t)\right|dt \leq C_1. \tag{1.2.38}$$

Combining this with (1.2.34)–(1.2.38), we immediately conclude that

$$\|(\eta_x,\,v_x,\,\mathbf{w}_x,\,\mathbf{b}_x,\,\theta_x)(t)\| \to 0, \quad \text{as } t \to +\infty. \tag{1.2.39}$$

Finally, employing the interpolation inequality, we also arrive at

$$\|(\eta - \bar{\eta},\, v,\, \mathbf{w},\, \mathbf{b} - \bar{\mathbf{b}},\, \theta - \bar{\theta})(t)\|_{L^\infty} \to 0, \quad \text{as } t \to +\infty. \tag{1.2.40}$$

This completes the proof of Theorem 1.1.1. $\square$

1.3 Global Existence and Asymptotic Behavior in $H^2(\mathbb{R})$

In this section, we shall complete the proof of Theorem 1.1.2. We begin with the following lemma to summarize the estimates in $H^1(\mathbb{R})$.

Lemma 1.3.1. *If the assumptions of Theorem 1.1.1 are valid, then the H^1-generalized global solution $(\eta(t), v(t), \mathbf{w}(t), \mathbf{b}(t), \theta(t))$ to the Cauchy problem (1.1.1)–(1.1.7) verifies (1.1.12)–(1.1.15) for any $t > 0$,*

$$\|\eta(t) - \bar{\eta}\|_{H^1}^2 + \|v(t)\|_{H^1}^2 + \|\mathbf{w}(t)\|_{H^1}^2 + \|\mathbf{b}(t)\|_{H^1}^2 + \|\theta(t) - \bar{\theta}\|_{H^1}^2 + \|\eta_t(t)\|^2$$

$$+ \int_0^t \left(\|v_x\|_{H^1}^2 + \|\mathbf{w}_x\|_{H^1}^2 + \|\mathbf{b}_x\|_{H^1}^2 + \|\theta_x\|_{H^1}^2 + \|\eta_x\|^2 + \|v_t\|^2\right.$$

$$\left. + \|\mathbf{w}_t\|^2 + \|\mathbf{b}_t\|^2 + \|\theta_t\|^2\right)(s)ds \leq C_1, \tag{1.3.1}$$

$$\|\eta(t) - \bar{\eta}\|_{L^\infty}^2 + \|v(t)\|_{L^\infty}^2 + \|\mathbf{w}(t)\|_{L^\infty}^2 + \|\mathbf{b}(t)\|_{L^\infty}^2 + \|\theta(t) - \bar{\theta}\|_{L^\infty}^2$$

$$+ \int_0^t \left(\|\eta_t\|_{H^1}^2 + \|v_x\|_{L^\infty}^2 + \|\mathbf{w}_x\|_{L^\infty}^2 + \|\mathbf{b}_x\|_{L^\infty}^2 + \|\theta_x\|_{L^\infty}^2\right)(s)ds \leq C_1. \tag{1.3.2}$$

Proof. Estimate (1.3.1) is just (1.1.10). By interpolation theory, we get

$$\|v(t)\|_{L^\infty} \leq C\|v(t)\|_{H^1}, \qquad \|\mathbf{w}(t)\|_{L^\infty} \leq C\|\mathbf{w}(t)\|_{H^1}, \tag{1.3.3}$$

$$\|\mathbf{b}(t)\|_{L^\infty} \leq C\|\mathbf{b}(t)\|_{H^1}, \qquad \|\theta(t) - \bar{\theta}\|_{L^\infty} \leq C\|\theta(t) - \bar{\theta}\|_{H^1}, \tag{1.3.4}$$

$$\|v_x(t)\|_{L^\infty} \leq C\|v_x(t)\|_{H^1}, \qquad \|\mathbf{w}_x(t)\|_{L^\infty} \leq C\|\mathbf{w}_x(t)\|_{H^1}, \tag{1.3.5}$$

$$\|\mathbf{b}_x(t)\|_{L^\infty} \leq C\|\mathbf{b}_x(t)\|_{H^1}, \qquad \|\theta_x(t)\|_{L^\infty} \leq C\|\theta_x(t)\|_{H^1}. \tag{1.3.6}$$

Also yields (1.1.1),

$$\|\eta_t(t)\|_{H^1} = \|v_x(t)\|_{H^1}. \tag{1.3.7}$$

Thus estimate (1.3.2) follows from Theorem 1.1.1, (1.3.1) and (1.3.3)–(1.3.7). The proof is complete. $\square$

Lemma 1.3.2. *Under the assumptions of Theorem 1.1.2, the following estimates hold for any $t > 0$:*

$$\|\theta_t(t)\|^2 + \|v_t(t)\|^2 + \|\mathbf{w}_t(t)\|^2 + \|\mathbf{b}_t(t)\|^2$$

$$+ \int_0^t (\|v_{xt}\|^2 + \|\mathbf{w}_{xt}\|^2 + \|\mathbf{b}_{xt}\|^2 + \|\theta_{xt}\|^2)(s)ds \leq C_2, \tag{1.3.8}$$

$$\|v_x(t)\|_{L^\infty}^2 + \|v_{xx}(t)\|^2 + \|\mathbf{w}_x(t)\|_{L^\infty}^2 + \|\mathbf{w}_{xx}(t)\|^2 + \|\mathbf{b}_x(t)\|_{L^\infty}^2$$

$$+ \|\mathbf{b}_{xx}(t)\|^2 + \|\theta_x(t)\|_{L^\infty}^2 + \|\theta_{xx}(t)\|^2 \leq C_2, \tag{1.3.9}$$

$$\|v(t)\|_{H^2}^2 + \|\mathbf{w}(t)\|_{H^2}^2 + \|\mathbf{b}(t)\|_{H^2}^2 + \|\theta(t) - \bar{\theta}\|_{H^2}^2 + \|\eta_t(t)\|_{H^1}^2 \leq C_2. \tag{1.3.10}$$

Proof. Differentiating (1.1.2) with respect to t, then multiplying the resulting equation by v_t in $L^2(\mathbb{R})$ and using Lemma 1.3.1, we get

$$\frac{d}{dt}\|v_t(t)\|^2 + C_1^{-1}\|v_{xt}(t)\|^2$$

$$\leq \frac{1}{2C_1}\|v_{xt}(t)\|^2 + C_2(\|v_x(t)\|^3\|v_{xx}(t)\| + \|\theta_t(t)\|^2 + \|\mathbf{b}_t(t)\|^2 + \|v_x(t)\|^2)$$

$$\leq \frac{1}{2C_1}\|v_{xt}(t)\|^2 + C_2(\|v_x(t)\|^2 + \|\theta_t(t)\|^2 + \|\mathbf{b}_t(t)\|^2 + \|v_{xx}(t)\|^2) \tag{1.3.11}$$

which in turn yields

$$\|v_t(t)\|^2 + \int_0^t \|v_{xt}(s)\|^2 ds \leq C_2 + C_1 \int_0^t (\|v_x\|^2 + \|\theta_t\|^2 + \|\mathbf{b}_t\|^2 + \|v_{xx}\|^2)(s)ds$$

$$\leq C_2. \tag{1.3.12}$$

Hence, by (1.1.2), Lemma 1.3.1, the interpolation inequalities and Young's inequality, we have

$$\|v_{xx}(t)\| \leq C_1 \left(\|v_t(t)\| + \|v_x(t)\| + \|\eta_x(t)\| + \|\mathbf{b}_x(t)\| + \|v_x(t)\|^{1/2}\|v_{xx}(t)\|\right)$$

$$\leq \frac{1}{2}\|v_{xx}(t)\| + C_1 \left(\|v_t(t)\| + \|v_x(t)\| + \|\mathbf{b}_x(t)\| + \|\eta_x(t)\|\right)$$

which combined with (1.3.12), (1.3.1) and (1.3.2) leads to

$$\|v_{xx}(t)\| \leq C_1 (\|v_t(t)\| + \|v_x(t)\| + \|\mathbf{b}_x(t)\| + \|\eta_x(t)\|) \leq C_2, \quad \forall t > 0, \tag{1.3.13}$$

$$\|v_x(t)\|_{L^\infty}^2 \leq C_1\|v_x(t)\| \|v_{xx}(t)\| \leq C_2, \quad \forall t > 0. \tag{1.3.14}$$

Similarly, we derive from (1.1.3) the bound

$$\frac{d}{dt}\|\mathbf{w}_t(t)\|^2 + C_1^{-1}\|\mathbf{w}_{xt}(t)\|^2 \leq \frac{1}{2C_1}\|\mathbf{w}_{xt}(t)\|^2 + C_2 \left(\|v_x(t)\|^2 + \|\mathbf{b}_t(t)\|^2\right) \tag{1.3.15}$$

which, combined with Lemma 1.3.1, gives

$$\|\mathbf{w}_t(t)\|^2 + \int_0^t \|\mathbf{w}_{xt}(s)\|^2 ds \leq C_2, \quad \forall t > 0. \tag{1.3.16}$$

Further, from (1.1.3), the interpolation inequalities and Young's inequality yield

$$\|\mathbf{w}_{xx}(t)\| \le C_1 \left(\|\mathbf{w}_t(t)\| + \|\mathbf{w}_x(t)\| \|\eta_x(t)\| + \|\mathbf{b}_x(t)\| \right),$$

whence

$$\|\mathbf{w}_{xx}(t)\| \le C_2, \quad \forall t > 0, \tag{1.3.17}$$

$$\|\mathbf{w}_x(t)\|_{L^\infty} \le C_1 \|\mathbf{w}_x(t)\| \|\mathbf{w}_{xx}(t)\| \le C_2, \quad \forall t > 0. \tag{1.3.18}$$

Similarly, (1.1.4) yields

$$\frac{d}{dt}\|\mathbf{b}_t(t)\|^2 + C_1^{-1}\|\mathbf{b}_{xt}(t)\|^2$$

$$\le \frac{1}{2C_1}\|\mathbf{b}_{xt}(t)\|^2 + C_2 \left(\|v_x(t)\|^2 + \|\mathbf{b}_t(t)\|^2 + \|\mathbf{w}_t(t)\|^2 \right) \tag{1.3.19}$$

whence

$$\|\mathbf{b}_t(t)\|^2 + \int_0^t \|\mathbf{b}_{xt}(s)\|^2 ds \le C_2, \quad \forall t > 0, \tag{1.3.20}$$

and also

$$\|\mathbf{b}_{xx}(t)\| + \|\mathbf{b}_x(t)\|_{L^\infty} \le C_2, \quad \forall t > 0. \tag{1.3.21}$$

Finally from (1.1.5) it follows that

$$\frac{d}{dt}\|\theta_t(t)\|^2 + C_1^{-1}\|\theta_{xt}(t)\|^2 \le \frac{1}{2C_1}\|\theta_{xt}(t)\|^2 + C_2 \left(\|\theta_x(t)\|^2 + \|v_x(t)\|^2 + \|\theta_t(t)\|^2 \right.$$

$$\left. + \|v_{tx}(t)\|^2 + \|\mathbf{w}_{xt}(t)\|^2 + \|\mathbf{b}_{xt}(t)\|^2 \right), \tag{1.3.22}$$

which, combined with Lemma 1.3.1, gives

$$\|\theta_t(t)\|^2 + \int_0^t \|\theta_{xt}(s)\|^2 ds \le C_2, \quad \forall t > 0, \tag{1.3.23}$$

and

$$\|\theta_{xx}(t)\| + \|\theta_x(t)\|_{L^\infty} \le C_2. \tag{1.3.24}$$

Thus estimates (1.3.8)–(1.3.10) follow from (1.1.1), (1.3.12)–(1.3.14), (1.3.16)–(1.3.18), (1.3.20)–(1.3.21), (1.3.23)–(1.3.24) and Lemma 1.3.1. The proof is complete. □

Lemma 1.3.3. *Under the assumptions of Theorem 1.1.2, the following estimates hold for any $t > 0$*

$$\|\eta_{xx}(t)\|^2 + \|\eta_x(t)\|_{L^\infty}^2 + \int_0^t \left(\|\eta_{xx}\|^2 + \|\eta_x\|_{L^\infty}^2 \right)(s)ds \le C_2, \tag{1.3.25}$$

$$\int_0^t \left(\|v_{xxx}\|^2 + \|\mathbf{w}_{xxx}\|^2 + \|\mathbf{b}_{xxx}\|^2 + \|\theta_{xxx}\|^2 \right)(s)ds \le C_2. \tag{1.3.26}$$

Proof. Differentiating (1.1.2) with respect to x and using equation (1.1.1), we get

$$\frac{\partial}{\partial t}\left(\frac{\eta_{xx}}{\eta}\right) + \frac{R\theta\eta_{xx}}{\eta^2} = v_{tx} + \frac{R\theta_{xx}}{\eta} + \frac{2v_{xx}\eta_x - 2R\theta_x\eta_x}{\eta^2}$$
$$+ \frac{(2R\theta - 2\mu v_x)\eta_x^2}{\eta^3} + |\mathbf{b}_x|^2 + \mathbf{b}\cdot\mathbf{b}_{xx}. \qquad (1.3.27)$$

Multiplying (1.3.27) by η_{xx}/η in $L^2(\mathbb{R})$, and using Lemmas 1.3.1–1.3.2, we deduce that

$$\frac{d}{dt}\left\|\frac{\eta_{xx}}{\eta}(t)\right\|^2 + C_1^{-1}\|\eta_{xx}(t)\|^2 \leq \frac{1}{2C_1}\|\eta_{xx}(t)\|^2 + C_2(\|\theta_x(t)\|^2 + \|\eta_x(t)\|^2$$
$$+ \|v_{xx}(t)\|^2 + \|\theta_{xx}(t)\|^2 + \|v_{tx}(t)\|^2$$
$$+ \|\mathbf{b}_x(t)\|_{L^\infty}^2\|\mathbf{b}_x(t)\|^2 + \|\mathbf{b}_{xx}(t)\|^2) \qquad (1.3.28)$$

which, together with Lemma 1.3.2, implies that, for any $t > 0$,

$$\|\eta_{xx}(t)\|^2 + \int_0^t \|\eta_{xx}(s)\|^2 ds \leq C_2, \qquad (1.3.29)$$

$$\|\eta_x(t)\|_{L^\infty}^2 \leq C\|\eta_x(t)\|\|\eta_{xx}(t)\| \leq C_2, \qquad (1.3.30)$$

$$\int_0^t \|\eta_x(s)\|_{L^\infty}^2(s)ds \leq C\int_0^t \left(\|\eta_x(s)\|^2 + \|\eta_{xx}(s)\|^2\right)(s)ds \leq C_2. \qquad (1.3.31)$$

Differentiating (1.1.2), (1.1.3), (1.1.4) and (1.1.5) with respect to x and using Lemmas 1.3.1–1.3.2, we deduce that, for any $t > 0$,

$$\|v_{xxx}(t)\| \leq C_2(\|v_t(t)\| + \|v_{tx}(t)\| + \|v_{xx}(t)\| + \|\eta_{xx}(t)\| + \|v_x(t)\|$$
$$+ \|\theta_{xx}(t)\| + \|\theta_x(t)\| + \|\eta_x(t)\| + \|\mathbf{b}_x(t)\| + \|\mathbf{b}_{xx}(t)\|), \qquad (1.3.32)$$

$$\|\mathbf{w}_{xxx}(t)\| \leq C_2(\|\mathbf{w}_{tx}(t)\| + \|\eta_{xx}(t)\| + \|\mathbf{b}_{xx}(t)\|), \qquad (1.3.33)$$

$$\|\mathbf{b}_{xxx}(t)\| \leq C_2(\|\mathbf{b}_{tx}(t)\| + \|\mathbf{b}_t(t)\| + \|\eta_{xx}(t)\| + \|\mathbf{w}_{xx}(t)\|), \qquad (1.3.34)$$

$$\|\theta_{xxx}(t)\| \leq C_2(\|\theta_t(t)\| + \|\theta_{tx}(t)\| + \|\theta_{xx}(t)\| + \|\eta_{xx}(t)\| + \|v_{xx}(t)\|$$
$$+ \|\theta_x(t)\| + \|\mathbf{w}_{xx}(t)\| + \|\mathbf{b}_{xx}(t)\|). \qquad (1.3.35)$$

Thus estimates (1.3.25)–(1.3.26) follow from (1.3.29)–(1.3.35) and Lemmas 1.3.1–1.3.2. □

In order to obtain the asymptotic behavior of global solutions, we will need the following lemma:

Lemma 1.3.4. (The Shen-Zheng Inequality) *Suppose y and h are nonnegative functions on $[0, +\infty)$, y' is locally integrable, and y, h satisfy*

$$\forall t > 0: \quad y'(t) \leq A_1 y^2(t) + A_2 + h(t),$$

$$\forall T > 0: \quad \int_0^T y(s)ds \le A_3, \int_0^T h(s)ds \le A_4,$$

with A_1, A_2, A_3, A_4 being positive constants independent of t and T. Then for any $r > 0$

$$\forall t \ge 0: \quad y(t+r) \le \left(\frac{A_3}{r} + A_2 r + A_4\right) e^{A_1 A_2}.$$

Moreover,

$$\lim_{t \to +\infty} y(t) = 0.$$

Proof. See, e.g., [120] and also [101], p. 21, Theorem 1.2.4. □

Lemma 1.3.5. *Under the assumptions of Theorem 1.1.2, the H^2-generalized global solution $(\eta(t), v(t), \mathbf{w}(t), \mathbf{b}(t), \theta(t))$ obtained in Lemmas 1.3.1–1.3.3 to the Cauchy problem (1.1.1)–(1.1.7) satisfies (1.1.18) and (1.1.19).*

Proof. We start arguing as in Lemma 1.3.1, by differentiating the equations (1.1.2)–(1.1.5) with respect to t, then multiplying the results with v_t, $\mathbf{w}_t$, $\mathbf{b}_t$, θ_t, respectively, resulting in

$$\frac{d}{dt}\|v_t(t)\|^2 + (2C_1)^{-1}\|v_{tx}(t)\|^2 \le C_2 \left(\|v_x(t)\|^2 + \|v_{xx}(t)\|^2 + \|\theta_t(t)\|^2 + \|\mathbf{b}_t(t)\|^2\right),$$
$$(1.3.36)$$

$$\frac{d}{dt}\|\mathbf{w}_t(t)\|^2 + (2C_1)^{-1}\|\mathbf{w}_{tx}(t)\|^2 \le C_2 \left(\|\eta_x(t)\|^2 + \|\mathbf{b}_t(t)\|^2\right), \qquad (1.3.37)$$

$$\frac{d}{dt}\|\mathbf{b}_t(t)\|^2 + (2C_1)^{-1}\|\mathbf{b}_{tx}(t)\|^2 \le C_2 \left(\|\mathbf{b}_x(t)\|^2 + \|\mathbf{b}_t(t)\|^2 + \|\mathbf{w}_t(t)\|^2\right), \quad (1.3.38)$$

$$\frac{d}{dt}\|\theta_t(t)\|^2 + (2C_1)^{-1}\|\theta_{tx}(t)\|^2 \le C_2 \big(\|v_x(t)\|^2 + \|\theta_x(t)\|^2 + \|\theta_t(t)\|^2 + \|v_{tx}(t)\|^2$$
$$+ \|\mathbf{w}_{tx}(t)\|^2 + \|\mathbf{b}_{tx}(t)\|^2\big). \qquad (1.3.39)$$

From (1.3.28), we also derive

$$\frac{d}{dt}\left\|\frac{\eta_{xx}}{\eta}(t)\right\|^2 + (2C_1)^{-1}\|\eta_{xx}(t)\|^2 \le C_2\big(\|\theta_x(t)\|^2 + \|\eta_x(t)\|^2 + \|v_{xx}(t)\|^2 + \|\theta_{xx}(t)\|^2$$
$$+ \|v_{tx}(t)\|^2 + \|\mathbf{b}_x(t)\|^2 + \|\mathbf{b}_{xx}(t)\|^2\big). \quad (1.3.40)$$

Meanwhile,

$$\|v_{xx}(t)\| \le C_1 \left(\|v_t(t)\| + \|v_x(t)\| + \|\eta_x(t)\| + \|\mathbf{b}_x(t)\|\right) \le C_2, \qquad (1.3.41)$$
$$\|\mathbf{w}_{xx}(t)\| \le C_1 \left(\|\mathbf{w}_t(t)\| + \|\eta_x(t)\| + \|\mathbf{b}_x(t)\|\right) \le C_2, \qquad (1.3.42)$$
$$\|\mathbf{b}_{xx}(t)\| \le C_1 \left(\|\mathbf{b}_t(t)\| + \|\eta_x(t)\| + \|\mathbf{w}_x(t)\|\right) \le C_2, \qquad (1.3.43)$$
$$\|\theta_{xx}(t)\| \le C_1 \big(\|\theta_t(t)\| + \|\theta_x(t)\| + \|v_x(t)\|$$
$$+ \|\mathbf{w}_x(t)\| + \|\mathbf{b}_x(t)\| + \|v_{xx}(t)\|\big) \le C_2, \qquad (1.3.44)$$

$$\|v_x(t)\|_{L^\infty}^2 \le C\|v_x(t)\|\,\|v_{xx}(t)\| \le C_2, \quad \|\mathbf{w}_x(t)\|_{L^\infty}^2 \le C\|\mathbf{w}_x(t)\|\,\|\mathbf{w}_{xx}(t)\| \le C_2, \tag{1.3.45}$$

$$\|\mathbf{b}_x(t)\|_{L^\infty}^2 \le C\|\mathbf{b}_x(t)\|\,\|\mathbf{b}_{xx}(t)\| \le C_2, \quad \|\theta_x(t)\|_{L^\infty}^2 \le C\|\theta_x(t)\|\,\|\theta_{xx}(t)\| \le C_2, \tag{1.3.46}$$

$$\|\eta_x(t)\|_{L^\infty}^2 \le C\|\eta_x(t)\|\,\|\eta_{xx}(t)\| \le C_2. \tag{1.3.47}$$

Applying Lemma 1.3.4 to (1.3.36)–(1.3.40) and using Lemmas 1.3.1–1.3.3, we obtain that, as $t \to +\infty$,

$$\|v_t(t)\| \to 0, \ \|\mathbf{w}_t(t)\| \to 0, \ \|\mathbf{b}_t(t)\| \to 0, \ \|\theta_t(t)\| \to 0, \ \|\eta_{xx}(t)\| \to 0 \tag{1.3.48}$$

which together with (1.1.1), (1.1.9) and (1.3.41)–(1.3.47) implies that, as $t \to +\infty$,

$$\|v_{xx}(t)\| + \|\mathbf{w}_{xx}(t)\| + \|\mathbf{b}_{xx}(t)\| + \|\theta_{xx}(t)\| + \|\eta_t(t)\|_{H^1} \to 0, \tag{1.3.49}$$

$$\|\eta_t(t)\|_{L^\infty} + \|(\eta_x(t), v_x(t), \mathbf{w}_x(t), \mathbf{b}_x(t), \theta_x(t))\|_{L^\infty} \to 0. \tag{1.3.50}$$

Thus (1.1.13)–(1.1.14) follows from (1.1.1) and (1.3.48)–(1.3.50). The proof is complete. □

Proof of Theorem 1.1.2. Combining Lemmas 1.3.1–1.3.3 with Lemma 1.3.5, we can complete the proof of Theorem 1.1.2. □

1.4 Global Existence and Asymptotic Behavior in $H^4(\mathbb{R})$

In this section, we shall derive estimates in $H^4(\mathbb{R})$ and complete the proof of Theorem 1.1.3.

Lemma 1.4.1. *Under the assumptions of Theorem 1.1.3, the following estimates hold for any $t > 0$:*

$$\|v_{tx}(x,0)\| + \|\mathbf{w}_{tx}(x,0)\| + \|\mathbf{b}_{tx}(x,0)\| + \|\theta_{tx}(x,0)\| \le C_3, \tag{1.4.1}$$

$$\|v_{tt}(x,0)\| + \|\theta_{tt}(x,0)\| + \|\mathbf{w}_{tt}(x,0) + \|\mathbf{b}_{tt}(x,0)\| + \|v_{txx}(x,0)\|$$
$$+ \|\theta_{txx}(x,0)\| + \|\mathbf{w}_{txx}(x,0)\| + \|\mathbf{b}_{txx}(x,0)\| \le C_4, \tag{1.4.2}$$

$$\|v_{tt}(t)\|^2 + \int_0^t \|v_{ttx}(s)\|^2 ds \le C_4 + C_4 \int_0^t \left(\|\theta_{txx}\|^2 + \|\mathbf{b}_{txx}\|^2\right)(s)ds, \tag{1.4.3}$$

$$\|\mathbf{w}_{tt}(t)\|^2 + \int_0^t \|\mathbf{w}_{ttx}(s)\|^2 ds \le C_4 + C_4 \int_0^t \|\mathbf{b}_{txx}(s)\|^2 ds, \tag{1.4.4}$$

$$\|\mathbf{b}_{tt}(t)\|^2 + \int_0^t \|\mathbf{b}_{ttx}(s)\|^2 ds \le C_4 + C_4 \int_0^t \|\mathbf{w}_{txx}(s)\|^2 ds, \tag{1.4.5}$$

$$\|\theta_{tt}(t)\|^2 + \int_0^t \|\theta_{ttx}(s)\|^2 ds$$

$$\le C_4 + C_4 \int_0^t \left(\|\theta_{txx}\|^2 + \|v_{txx}\|^2 + \|\mathbf{w}_{txx}\|^2 + \|\mathbf{b}_{txx}\|^2\right)(s)ds. \tag{1.4.6}$$

Proof. We easily infer from (1.1.2) and Lemmas 1.3.1–1.3.3 that

$$\|v_t(t)\| \le C_2 \|v_x(t)\|_{H^1} + \|\eta_x(t)\| + \|\theta_x(t)\| + \|\mathbf{b}_x(t)\|). \tag{1.4.7}$$

Differentiating (1.1.2) with respect to x and employing Lemmas 1.3.1–1.2.3, we have

$$\|v_{tx}(t)\| \le C_2 \left(\|v_x(t)\| + \|v_{xxx}(t)\| + \|\theta_x(t)\|_{H^1} + \|\eta_x(t)\|_{H^1} + \|\mathbf{b}_x(t)\|_{H^1} \right) \tag{1.4.8}$$

or

$$\|v_{xxx}(t)\| \le C_2 \left(\|v_x(t)\| + \|\eta_x(t)\|_{H^1} + \|\theta_x(t)\|_{H^1} + \|v_{tx}(t)\| + \|\mathbf{b}_x(t)\|_{H^1} \right). \tag{1.4.9}$$

Differentiating (1.1.2) with respect to x twice and using Lemmas 1.3.1–1.3.3 and the interpolation inequalities, we have

$$\|v_{txx}(t)\| \le C_2 \left(\|\eta_x(t)\|_{H^2} + \|v_x(t)\|_{H^3} + \|\theta_x(t)\|_{H^2} + \|\mathbf{b}_x(t)\|_{H^2} \right) \tag{1.4.10}$$

or

$$\|v_{xxxx}(t)\| \le C_2 \left(\|\eta_x(t)\|_{H^2} + \|v_x(t)\|_{H^2} + \|\theta_x(t)\|_{H^2} + \|v_{txx}(t)\| + \|\mathbf{b}_x(t)\|_{H^2} \right). \tag{1.4.11}$$

In the same manner, we deduce from (1.1.3) and (1.1.4) that

$$\|\mathbf{w}_t(t)\| \le C_2 \left(\|\eta_x(t)\| + \|\mathbf{w}(t)\|_{H^2} + \|\mathbf{b}_x(t)\| \right), \tag{1.4.12}$$

$$\|\mathbf{w}_{tx}(t)\| \le C_2 \left(\|\eta_x(t)\|_{H^1} + \|\mathbf{w}_x(t)\|_{H^2} + \|\mathbf{b}_x(t)\|_{H^1} \right), \tag{1.4.13}$$

$$\|\mathbf{b}_t(t)\| \le C_2 \left(\|\eta_x(t)\| + \|\mathbf{b}(t)\|_{H^2} + \|\mathbf{w}_x(t)\| \right), \tag{1.4.14}$$

$$\|\mathbf{b}_{tx}(t)\| \le C_2 \left(\|\eta_x(t)\|_{H^1} + \|\mathbf{b}_x(t)\|_{H^2} + \|\mathbf{w}_x(t)\|_{H^1} \right), \tag{1.4.15}$$

$$\|\theta_t(t)\| \le C_2 \big(\|\theta_x(t)\|_{H^1} + \|v_x(t)\| + \|\eta_x(t)\|$$
$$+ \|\mathbf{w}_x(t)\| + \|\mathbf{b}_x(t)\| \big), \tag{1.4.16}$$

$$\|\theta_{tx}(t)\| \le C_2 \big(\|\theta_x(t)\|_{H^2} + \|v_x(t)\|_{H^1} + \|\eta_{xx}(t)\|$$
$$+ \|\mathbf{w}_x(t)\|_{H^1} + \|\mathbf{b}_x(t)\|_{H^1} \big), \tag{1.4.17}$$

or

$$\|\mathbf{w}_{xxx}(t)\| \le C_2 \left(\|\mathbf{w}_{tx}(t)\| + \|\eta_x(t)\|_{H^1} + \|\mathbf{b}_x(t)\|_{H^1} \right), \tag{1.4.18}$$

$$\|\mathbf{b}_{xxx}(t)\| \le C_2 \left(\|\mathbf{b}_{tx}(t)\| + \|\eta_x(t)\|_{H^1} + \|\mathbf{w}_x(t)\|_{H^1} \right), \tag{1.4.19}$$

$$\|\theta_{xxx}(t)\| \le C_2 \big(\|\theta_x(t)\|_{H^1} + \|v_x(t)\|_{H^1} + \|\eta_{xx}(t)\|$$
$$+ \|\theta_{tx}(t)\| + \|\mathbf{w}_x(t)\|_{H^1} + \|\mathbf{b}_x(t)\|_{H^1} \big) \tag{1.4.20}$$

and

$$\|\mathbf{w}_{txx}(t)\| \le C_2 \left(\|\mathbf{w}_{xxxx}(t)\| + \|\eta_x(t)\|_{H^2} + \|\mathbf{b}_x(t)\|_{H^2} \right), \tag{1.4.21}$$

$$\|\mathbf{b}_{txx}(t)\| \le C_2 \left(\|\mathbf{b}_{xxxx}(t)\| + \|\eta_x(t)\|_{H^2} + \|\mathbf{w}_x(t)\|_{H^2} \right), \tag{1.4.22}$$

$$\|\theta_{txx}(t)\| \le C_2 \big(\|\eta_x(t)\|_{H^2} + \|v_x(t)\|_{H^2} + \|\theta_x(t)\|_{H^3}$$

$$+ \|\mathbf{w}_x(t)\|_{H^2} + \|\mathbf{b}_x(t)\|_{H^2}), \tag{1.4.23}$$

or

$$\|\mathbf{w}_{xxxx}(t)\| \leq C_2 \left(\|\mathbf{w}_{txx}(t)\| + \|\eta_x(t)\|_{H^2} + \|\mathbf{b}_x(t)\|_{H^2} \right), \tag{1.4.24}$$

$$\|\mathbf{b}_{xxxx}(t)\| \leq C_2 \left(\|\mathbf{b}_{txx}(t)\| + \|\eta_x(t)\|_{H^2} + \|\mathbf{w}_x(t)\|_{H^2} \right), \tag{1.4.25}$$

$$\|\theta_{xxxx}(t)\| \leq C_2 \big(\|\eta_x(t)\|_{H^2} + \|v_x(t)\|_{H^2} + \|\theta_x(t)\|_{H^2}$$
$$+ \|\theta_{txx}(t)\| + \|\mathbf{w}_x(t)\|_{H^2} + \|\mathbf{b}_x(t)\|_{H^2} \big). \tag{1.4.26}$$

Differentiating (1.1.2) with respect to t, and using Lemmas 1.3.1–1.3.3 and (1.1.1), we deduce that

$$\|v_{tt}(t)\| \leq C_2 \big(\|\theta_x(t)\| + \|\eta_x(t)\| + \|v_{xx}(t)\| + \|v_{tx}(t)\|_{H^1}$$
$$+ \|\theta_t(t)\|_{H^1} + \|\mathbf{b}_t(t)\|_{H^1} \big) \tag{1.4.27}$$

which, together with (1.4.8), (1.4.10), (1.4.13), (1.4.15) and (1.4.17), implies

$$\|v_{tt}(t)\| \leq C_2 \left(\|\theta_x(t)\|_{H^2} + \|v_x(t)\|_{H^3} + \|\eta_x(t)\|_{H^2} + \|\mathbf{w}_x(t)\|_{H^2} + \|\mathbf{b}_x(t)\|_{H^2} \right). \tag{1.4.28}$$

Analogously, we derive from (1.1.3)–(1.1.5) that

$$\|\mathbf{w}_{tt}(t)\| \leq C_2 \left(\|\mathbf{w}_{txx}(t)\| + \|v_x(t)\|_{H^2} + \|\eta_x(t)\| + \|\mathbf{b}_x(t)\|_{H^2} \right), \tag{1.4.29}$$

$$\|\mathbf{b}_{tt}(t)\| \leq C_2 \left(\|\mathbf{b}_{txx}(t)\| + \|v_x(t)\|_{H^2} + \|\eta_x(t)\| + \|\mathbf{w}_x(t)\|_{H^2} \right), \tag{1.4.30}$$

$$\|\theta_{tt}(t)\| \leq C_2 \big(\|\theta_{txx}(t)\| + \|v_x(t)\|_{H^2} + \|\eta_x(t)\|_{H^2}$$
$$+ \|\mathbf{w}_x(t)\|_{H^2} + \|\mathbf{b}_x(t)\|_{H^2} \big), \tag{1.4.31}$$

or

$$\|\mathbf{w}_{tt}(t)\| \leq C_2 \left(\|\mathbf{w}_x(t)\|_{H^3} + \|v_x(t)\|_{H^2} + \|\eta_x(t)\| + \|\mathbf{b}_x(t)\|_{H^2} \right), \tag{1.4.32}$$

$$\|\mathbf{b}_{tt}(t)\| \leq C_2 \left(\|\mathbf{b}_x(t)\|_{H^3} + \|v_x(t)\|_{H^2} + \|\eta_x(t)\| + \|\mathbf{w}_x(t)\|_{H^2} \right), \tag{1.4.33}$$

$$\|\theta_{tt}(t)\| \leq C_2 \big(\|\theta_x(t)\|_{H^3} + \|v_x(t)\|_{H^2} + \|\eta_x(t)\|_{H^2}$$
$$+ \|\mathbf{w}_x(t)\|_{H^2} + \|\mathbf{b}_x(t)\|_{H^2} \big). \tag{1.4.34}$$

Thus estimates (1.4.1)–(1.4.2) follow from (1.4.8), (1.4.10), (1.4.13), (1.4.15), (1.4.17), (1.4.21)–(1.4.23) and (1.4.28)–(1.4.34).

Now differentiating (1.1.2) with respect to t twice, multiplying the resulting equation by v_{tt} in $L^2(\mathbb{R})$, and using (1.1.1) and Lemmas 1.3.1–1.3.3, we deduce

$$\frac{1}{2}\frac{d}{dt}\|v_{tt}(t)\|^2 \leq -(2C_1)^{-1}\|v_{ttx}(t)\|^2 + C_2 \big(\|\theta_{tt}(t)\|^2 + \|v_{tx}(t)\|^2$$
$$+ \|\theta_t(t)\|^2 + \|v_x(t)\|^2 + \|\mathbf{b}_t(t)\|^2_{H^1} + \|\mathbf{b}_{tt}(t)\|^2 \big)$$

which, along with (1.4.33)–(1.4.34), implies

$$\frac{d}{dt}\|v_{tt}(t)\|^2 + (2C_1)^{-1}\|v_{ttx}(t)\|^2$$

$$\leq C_2 \left(\|\theta_{txx}(t)\|^2 + \|\theta_{xx}(t)\|^2 + \|\theta_{tx}(t)\|^2 + \|v_x(t)\|_{H^1}^2 \right.$$
$$+ \left. \|v_{tx}(t)\|^2 + \|\theta_t(t)\|^2 + \|\eta_x(t)\|^2 + \|\mathbf{b}_t(t)\|_{H^1}^2 + \|\mathbf{b}_{txx}(t)\|^2 \right). \qquad (1.4.35)$$

Thus estimate (1.4.3) follows from Lemmas 1.3.1–1.3.3 and (1.4.2).

Analogously, we obtain from (1.1.3) that

$$\|\mathbf{w}_{tt}(t)\|^2 + (2C_1)^{-1} \int_0^t \|\mathbf{w}_{ttx}(s)\|^2 ds$$
$$\leq C_4 + C_4 \int_0^t \left(\|v_{tx}(s)\|^2 + \|\mathbf{w}_{tx}(t)\|^2 + \|\mathbf{w}_x\|^2 + \|\mathbf{b}_{txx}(s)\|^2 \right) ds. \qquad (1.4.36)$$

Hence,

$$\|\mathbf{w}_{tt}(t)\|^2 + (2C_1)^{-1} \int_0^t \|\mathbf{w}_{ttx}(s)\|^2 ds \leq C_4 + C_4 \int_0^t \|\mathbf{b}_{txx}(s)\|^2 ds, \qquad (1.4.37)$$

which gives (1.4.4). In the same way, we obtain from (1.1.4) that

$$\|\mathbf{b}_{tt}(t)\|^2 + (2C_1)^{-1} \int_0^t \|\mathbf{b}_{ttx}(s)\|^2 ds \leq C_4 + C_4 \int_0^t \|\mathbf{w}_{txx}(s)\|^2 ds, \qquad (1.4.38)$$

and from (1.1.5), upon combining with (1.4.32), (1.4.34) and (1.4.35) that

$$\|\theta_{tt}(t)\|^2 + (2C_1)^{-1} \int_0^t \|\theta_{ttx}(s)\|^2 ds$$
$$\leq C_4 + C_4 \int_0^t \left(\|v_{txx}\|^2 + \|\theta_{txx}\|^2 + \|\mathbf{w}_{txx}\|^2 + \|\mathbf{b}_{txx}\|^2 \right)(s) ds. \qquad (1.4.39)$$

Therefore, estimates (1.4.5)–(1.4.6) follow from (1.4.35)–(1.4.36). The proof is complete. $\qquad \square$

Lemma 1.4.2. *Under the assumptions of Theorem 1.1.3, the following estimates hold for any $t > 0$:*

$$\|v_{tx}(t)\|^2 + \int_0^t \|v_{txx}(s)\|^2 ds \leq C_3, \qquad (1.4.40)$$

$$\|\mathbf{w}_{tx}(t)\|^2 + \int_0^t \|\mathbf{w}_{txx}(s)\|^2 ds \leq C_3, \qquad (1.4.41)$$

$$\|\mathbf{b}_{tx}(t)\|^2 + \int_0^t \|\mathbf{b}_{txx}(s)\|^2 ds \leq C_3, \qquad (1.4.42)$$

$$\|\theta_{tx}(t)\|^2 + \int_0^t \|\theta_{txx}(s)\|^2 ds \leq C_3, \qquad (1.4.43)$$

$$\|\theta_{tt}(t)\|^2 + \|v_{tt}(t)\|^2 + \|\mathbf{w}_{tt}(t)\|^2 + \|\mathbf{w}_{tt}(t)\|^2$$
$$+ \int_0^t \left(\|v_{ttx}\|^2 + \|\mathbf{w}_{ttx}\|^2 + \|\mathbf{b}_{ttx}\|^2 + \|\theta_{ttx}\|^2 \right)(s) ds \leq C_4. \qquad (1.4.44)$$

Proof. Differentiating (1.1.2) with respect to x and t, multiplying the resulting equation by v_{tx} in $L^2(\mathbb{R})$, and integrating by parts, we deduce that

$$\frac{1}{2}\frac{d}{dt}\|v_{tx}(t)\|^2 \leq -C_1^{-1}\int_{\mathbb{R}}\frac{v_{txx}^2}{\eta}dx + C_2\|v_{txx}(t)\|\big(\|\mathbf{b}_t(t)\|_{H^1} + \|\theta_{tx}(t)\| + \|v_{tx}(t)\|$$
$$+ \|\theta_t(t)\| + \|v_{xx}(t)\| + \|\theta_x(t)\| + \|\eta_x(t)\|\big)$$
$$\leq -(2C_1)^{-1}\|v_{txx}(t)\|^2 + C_2\big(\|\mathbf{b}_{tx}(t)\|^2 + \|\theta_{tx}(t)\|^2 + \|v_{tx}(t)\|^2$$
$$+ \|\theta_t(t)\|^2 + \|v_{xx}(t)\|^2 + \|\theta_x(t)\|^2 + \|\eta_x(t)\|^2\big), \tag{1.4.45}$$

which, combined with Lemmas 1.3.1–1.3.3 and (1.4.2), gives estimate (1.4.40).

In the same way, we infer from (1.1.3) that for any $\delta > 0$,

$$\frac{1}{2}\frac{d}{dt}\|\mathbf{w}_{tx}(t)\|^2 \leq -C_1^{-1}\int_{\mathbb{R}}\frac{\mathbf{w}_{txx}^2}{\eta}dx + \delta\|\mathbf{w}_{txx}(t)\|^2 + C_2\big(\|\mathbf{b}_{tx}(t)\|^2$$
$$+ \|v_x(t)\|_{H^1}^2 + \|\mathbf{w}_t(t)\|_{H^1}^2 + \|\mathbf{w}_x(t)\|^2 + \|\eta_x(t)\|^2\big), \tag{1.4.46}$$

which leads to (1.4.41).

In the same manner, we can prove (4.42)–(4.43). Inserting (1.4.40)–(1.4.43) into (1.4.3)–(1.4.6) yields estimate (1.4.44). The proof is now complete. $\qquad\square$

Lemma 1.4.3. *Under the assumptions of Theorem 1.1.3, the following estimates hold for any $t > 0$:*

$$\|\eta_{xxx}(t)\|_{H^1}^2 + \|\eta_{xx}(t)\|_{W^{1,\infty}}^2 + \int_0^t \big(\|\eta_{xxx}\|_{H^1}^2 + \|\eta_{xx}\|_{W^{1,\infty}}^2\big)(s)ds \leq C_4, \tag{1.4.47}$$

$$\|v_{xxx}(t)\|_{H^1}^2 + \|v_{xx}(t)\|_{W^{1,\infty}}^2 + \|\mathbf{w}_{xxx}(t)\|_{H^1}^2 + \|\mathbf{w}_{xx}(t)\|_{W^{1,\infty}}^2 + \|\mathbf{b}_{xxx}(t)\|_{H^1}^2$$
$$+ \|\mathbf{b}_{xx}(t)\|_{W^{1,\infty}}^2 + \|\theta_{xxx}(t)\|_{H^1}^2 + \|\theta_{xx}(t)\|_{W^{1,\infty}}^2 + \|\eta_{txxx}(t)\|^2 + \|v_{txx}(t)\|^2$$
$$+ \|\mathbf{w}_{txx}(t)\|^2 + \|\mathbf{b}_{txx}\|^2 + \|\theta_{txx}(t)\|^2 + \int_0^t \big(\|v_{tt}\|^2 + \|\mathbf{w}_{tt}\|^2 + \|\mathbf{b}_{tt}\|^2 + \|\theta_{tt}\|^2$$
$$+ \|v_{xx}\|_{W^{2,\infty}}^2 + \|\mathbf{w}_{xx}\|_{W^{2,\infty}}^2 + \|\mathbf{b}_{xx}\|_{W^{2,\infty}}^2 + \|\theta_{xx}\|_{W^{2,\infty}}^2 + \|v_{txx}\|_{H^1}^2$$
$$+ \|\mathbf{w}_{txx}\|_{H^1}^2 + \|\mathbf{b}_{txx}\|_{H^1}^2 + \|\theta_{txx}\|_{H^1}^2 + \|\theta_{tx}\|_{W^{1,\infty}}^2 + \|\mathbf{w}_{tx}\|_{W^{1,\infty}}^2$$
$$+ \|\mathbf{b}_{tx}\|_{W^{1,\infty}}^2 + \|v_{tx}\|_{W^{1,\infty}}^2 + \|\eta_{txxx}\|_{H^1}^2\big)(s)ds \leq C_4, \tag{1.4.48}$$

$$\int_0^t \big(\|v_{xxxx}\|_{H^1}^2 + \|\mathbf{w}_{xxxx}\|_{H^1}^2 + \|\mathbf{b}_{xxxx}\|_{H^1}^2 + \|\theta_{xxxx}\|_{H^1}^2\big)(s)ds \leq C_4. \tag{1.4.49}$$

Proof. Differentiating (1.3.29) with respect to x, using (1.1.1), we arrive at

$$\frac{\partial}{\partial t}\left(\frac{\eta_{xxx}}{\eta}\right) + \frac{R\theta\eta_{xxx}}{\eta^2} = E_1(x, t) \tag{1.4.50}$$

with

$$E_1(x,t) = \left[\frac{v_{xxx}\eta_x + \eta_{xx}v_{xx}}{\eta^2} - \frac{2\eta_x\eta_{xx}v_x}{\eta^3} \right] - \frac{\theta_x\eta_{xx}}{\eta^2} + \frac{2R\theta\eta_x\eta_{xx}}{\eta^3}$$

$$+ v_{txx} + 2\mathbf{b}_x \cdot \mathbf{b}_{xx} + \mathbf{b} \cdot \mathbf{b}_{xxx} + \mathbf{b}_x \cdot \mathbf{b}_{xx} + E_x(x,t),$$

$$E(x,t) = \frac{R\theta_{xx}}{\eta} + \frac{2\mu v_{xx}\eta_x - 2R\theta_x\eta_x}{\eta^2} + \frac{2R\theta\eta_x^2 - 2v_x\eta_x^2}{\eta^3}.$$

An easy calculation with Lemmas 1.3.1–1.3.3 and Lemmas 1.4.1–1.4.2 gives

$$\|E_1(t)\| \le C_2 \left(\|\eta_x(t)\|_{H^1} + \|v_x(t)\|_{H^2} + \|\theta_x(t)\|_{H^2} + \|v_{tx}(t)\|_{H^1} + \|\mathbf{b}_x(t)\|_{H^2} \right) \tag{1.4.51}$$

and

$$\int_0^t \|E_1(s)\|^2 ds \le C_4. \tag{1.4.52}$$

Now multiplying (1.4.50) by η_{xxx}/η in $L^2(\mathbb{R})$, we derive

$$\frac{d}{dt} \left\| \frac{\eta_{xxx}}{\eta}(t) \right\|^2 + C_1^{-1} \left\| \frac{\eta_{xxx}}{\eta}(t) \right\|^2 \le C_1 \|E_1(t)\|^2 \tag{1.4.53}$$

which, combined with (1.4.52), Lemmas 1.3.1–1.3.3 and Lemmas 1.4.1–1.4.2, yields, for any $t > 0$,

$$\|\eta_{xxx}(t)\|^2 + \int_0^t \|\eta_{xxx}(s)\|^2 ds \le C_4. \tag{1.4.54}$$

In view of (1.4.9), (1.4.11), (1.4.18)–(1.4.20), (1.4.24)–(1.4.26), Lemmas 1.3.1–1.3.3, and Lemmas 1.4.1–1.4.2, we get that for any $t > 0$,

$$\|v_{xxx}(t)\|^2 + \|\mathbf{w}_{xxx}(t)\|^2 + \|\mathbf{b}_{xxx}(t)\|^2 + \|\theta_{xxx}(t)\|^2 \tag{1.4.55}$$

$$+ \int_0^t \left(\|v_{xxx}\|_{H^1}^2 + \|\mathbf{w}_{xxx}\|_{H^1}^2 + \|\mathbf{b}_{xxx}\|_{H^1}^2 + \|\theta_{xxx}\|_{H^1}^2 \right)(s) ds \le C_4,$$

$$\|v_{xx}(t)\|_{L^\infty}^2 + \|\mathbf{w}_{xx}(t)\|_{L^\infty}^2 + \|\mathbf{b}_{xx}(t)\|_{L^\infty}^2 + \|\theta_{xx}(t)\|_{L^\infty}^2 \tag{1.4.56}$$

$$+ \int_0^t \left(\|v_{xx}\|_{W^{1,\infty}}^2 + \|\mathbf{w}_{xx}\|_{W^{1,\infty}}^2 + \|\mathbf{b}_{xx}\|_{W^{1,\infty}}^2 + \|\theta_{xx}\|_{W^{1,\infty}}^2 \right)(s) ds \le C_4.$$

Differentiating (1.1.2) with respect to t, we infer that for any $t > 0$,

$$\|v_{txx}(t)\| \le C_1 \|v_{tt}(t)\| + C_2 \left(\|\eta_x(t)\| + \|v_{xx}(t)\| + \|v_{tx}(t)\| + \|\theta_x(t)\| \right.$$

$$\left. + \|\theta_t(t)\| + \|\theta_{tx}(t)\| + \|\mathbf{b}_{tx}(t)\| \right) \le C_4, \tag{1.4.57}$$

which, together with (1.4.11), gives

$$\|v_{xxxx}(t)\|^2 + \int_0^t \left(\|v_{txx}\|^2 + \|v_{xxxx}\|^2 \right)(s) ds \le C_4. \tag{1.4.58}$$

Similarly, we can infer from (1.4.21)–(1.4.26) and (1.4.55)–(1.4.56) that

$$\|\mathbf{w}_{txx}(t)\|^2 + \|\mathbf{w}_{xxxx}(t)\|^2 + \int_0^t \left(\|\mathbf{w}_{txx}\|^2 + \|\mathbf{w}_{xxxx}\|^2\right)(s)ds \leq C_4, \quad (1.4.59)$$

$$\|\mathbf{b}_{txx}(t)\|^2 + \|\mathbf{b}_{xxxx}(t)\|^2 + \int_0^t \left(\|\mathbf{b}_{txx}\|^2 + \|\mathbf{b}_{xxxx}\|^2\right)(s)ds \leq C_4, \quad (1.4.60)$$

$$\|\theta_{txx}(t)\|^2 + \|\theta_{xxxx}(t)\|^2 + \int_0^t \left(\|\theta_{txx}\|^2 + \|\theta_{xxxx}\|^2\right)(s)ds \leq C_4, \quad (1.4.61)$$

which, combined with (1.4.55), (1.4.56) and (4.59)–(4.61), implies

$$\|v_{xxx}(t)\|_{L^\infty}^2 + \|\mathbf{w}_{xxx}(t)\|_{L^\infty}^2 + \|\mathbf{b}_{xxx}(t)\|_{L^\infty}^2 + \|\theta_{xxx}(t)\|_{L^\infty}^2$$
$$+ \int_0^t \left(\|v_{xxx}\|_{L^\infty}^2 + \|\mathbf{w}_{xxx}\|_{L^\infty}^2 + \|\mathbf{b}_{xxx}\|_{L^\infty}^2 + \|\theta_{xxx}\|_{L^\infty}^2\right)(s)ds \leq C_4. \quad (1.4.62)$$

Differentiating (1.4.50) with respect to x, we see that

$$\frac{\partial}{\partial t}\left(\frac{\eta_{xxxx}}{\eta}\right) + \frac{R\theta\eta_{xxxx}}{\eta^2} = E_2(x,t) \quad (1.4.63)$$

with

$$E_2(x,t) = \left[\frac{v_{xx}\eta_{xxx} + \eta_x v_{xxxx}}{\eta^2} - \frac{2\eta_x v_x \eta_{xxx}}{\eta^3}\right] + \frac{2R\theta\eta_x\eta_{xxx}}{\eta^3} - \frac{R\theta_x\eta_{xxx}}{\eta^2}$$
$$+ 3|\mathbf{b}_{xx}|^2 + 4\mathbf{b}_x \cdot \mathbf{b}_{xxx} + \mathbf{b} \cdot \mathbf{b}_{xxxx} + E_{1x}(x,t). \quad (1.4.64)$$

Using Lemmas 1.3.1–1.3.3 and Lemmas 1.4.1–1.4.2, we can deduce that

$$\|E_{xx}(t)\| \leq C_4\big(\|\theta_x(t)\|_{H^3} + \|\eta_x(t)\|_{H^2} + \|v_x(t)\|_{H^3}\big), \quad (1.4.65)$$

$$\|E_{1x}(t)\| \leq C_4\big(\|v_x(t)\|_{H^3} + \|\eta_x(t)\|_{H^2} + \|v_{tx}(t)\|_{H^2}$$
$$+ \|\theta_x(t)\|_{H^3} + \|\mathbf{b}_x(t)\|_{H^3}\big), \quad (1.4.66)$$

$$\|E_2(t)\| \leq C_4\big(\|v_x(t)\|_{H^3} + \|\eta_x(t)\|_{H^2} + \|v_{tx}(t)\|_{H^2}$$
$$+ \|\theta_x(t)\|_{H^3} + \|\mathbf{b}_x(t)\|_{H^3}\big). \quad (1.4.67)$$

On the other hand, differentiating (1.1.2) with respect to t and x, we infer that

$$\|v_{txxx}(t)\| \leq C_1\|v_{ttx}(t)\| + C_2\big(\|v_{xx}\|_{H^1} + \|\theta_x(t)\|_{H^1} + \|\eta_x(t)\|_{H^1}$$
$$+ \|\theta_{tx}(t)\|_{H^1} + \|\theta_t(t)\| + \|v_{tx}(t)\|_{H^1} + \|\mathbf{b}_{txx}(t)\|\big). \quad (1.4.68)$$

Similarly, we have

$$\|\mathbf{w}_{txxx}(t)\| \leq C_1\|\mathbf{w}_{ttx}(t)\| + C_2\left(\|\mathbf{w}_{tx}(t)\|_{H^1} + \|\mathbf{b}_{txx}(t)\| + \|v_x(t)\|_{H^2}\right), \quad (1.4.69)$$

$$\|\mathbf{b}_{txxx}(t)\| \leq C_1\|\mathbf{b}_{ttx}(t)\| + C_2\left(\|\mathbf{b}_{tx}(t)\|_{H^1} + \|\mathbf{w}_{txx}(t)\| + \|v_x(t)\|_{H^2}\right), \quad (1.4.70)$$

$$\|\theta_{txxx}(t)\| \leq C_1\|\theta_{ttx}(t)\| + C_2\big(\|\eta_x(t)\| + \|v_x(t)\|_{H^2} + \|\theta_x(t)\|_{H^2} + \|\theta_{tx}(t)\|_{H^1}$$
$$+ \|\theta_t(t)\| + \|v_{tx}(t)\|_{H^1} + \|\mathbf{w}_{tx}(t)\|_{H^1} + \|\mathbf{b}_{tx}(t)\|_{H^1}\big). \qquad (1.4.71)$$

Thus it follows from Lemmas 1.3.1–1.3.3, Lemmas 1.4.1–1.4.2 and (1.4.69)–(1.4.72) that

$$\int_0^t \big(\|v_{txxx}\|^2 + \|\mathbf{w}_{txxx}\|^2 + \|\mathbf{b}_{txxx}\|^2 + \|\theta_{txxx}\|^2\big)(s)ds \leq C_4, \ \forall t > 0. \qquad (1.4.72)$$

Then we have

$$\int_0^t \|E_2(s)\|^2 ds \leq C_4, \ \ \forall t > 0. \qquad (1.4.73)$$

Multiplying (1.4.63) by η_{xxxx}/η in $L^2(\mathbb{R})$, we get

$$\frac{d}{dt}\left\|\frac{\eta_{xxxx}}{\eta}(t)\right\|^2 + C_1^{-1}\left\|\frac{\eta_{xxxx}}{\eta}(t)\right\|^2 \leq C_1\|E_2(t)\|^2 \qquad (1.4.74)$$

which, combined with (1.4.73), implies

$$\|\eta_{xxxx}(t)\|^2 + \int_0^t \|\eta_{xxxx}(s)\|^2 ds \leq C_4, \ \forall t > 0. \qquad (1.4.75)$$

From (1.4.28)–(1.4.34), Lemmas 1.3.1–1.3.3, Lemmas 1.4.1–1.4.2 and (1.4.55)–(1.4.61), we derive

$$\int_0^t \big(\|v_{tt}\|^2 + \|\mathbf{w}_{tt}\|^2 + \|\mathbf{b}_{tt}\|^2 + \|\theta_{tt}\|^2\big)(s)ds \leq C_4, \ \forall t > 0. \qquad (1.4.76)$$

Differentiating (1.1.2) with respect to x three times, we deduce

$$\|v_{xxxxx}(t)\| \leq C_1\|v_{txxx}(t)\|$$
$$+ C_2\big(\|\eta_x(t)\|_{H^3} + \|v_x(t)\|_{H^3} + \|\theta_x(t)\|_{H^3} + \|\mathbf{b}_x(t)\|_{H^3}\big). \qquad (1.4.77)$$

Thus we conclude from (1.1.1), (1.4.58)–(1.4.61), and Lemmas 1.3.1–1.3.2 and Lemmas 1.4.1–1.4.2 that

$$\int_0^t \big(\|v_{xxxxx}\|^2 + \|\eta_{txxx}\|_{H^1}^2\big)(s)ds \leq C_4, \ \forall t > 0. \qquad (1.4.78)$$

Similarly, we can deduce that, for any $t > 0$,

$$\int_0^t \big(\|\mathbf{w}_{xxxxx}\|^2 + \|\mathbf{b}_{xxxxx}(s)\|^2 + \|\theta_{xxxxx}\|^2\big)(s)ds \leq C_4, \qquad (1.4.79)$$

$$\int_0^t \big(\|v_{xx}\|_{W^{2,\infty}}^2 + \|\mathbf{w}_{xx}\|_{W^{2,\infty}}^2 + \|\mathbf{b}_{xx}\|_{W^{2,\infty}}^2 + \|\theta_{xx}\|_{W^{2,\infty}}^2\big)(s)ds \leq C_4. \qquad (1.4.80)$$

Thus employing (1.1.1), (1.4.54)–(1.4.62), (1.4.73), (1.4.76), (1.4.77), (1.4.79)–(1.4.81) and the interpolation inequality, we derive the desired estimates (1.4.47)–(1.4.49). The proof is complete. $\qquad \square$

Lemma 1.4.4. *Under the assumptions of Theorem 1.1.3, the following estimates hold for any $t > 0$:*

$$\|\eta(t) - \overline{\eta}\|_{H^4}^2 + \|\eta_t(t)\|_{H^3}^2 + \|\eta_{tt}(t)\|_{H^1}^2 + \|v(t)\|_{H^4}^2 + \|v_t(t)\|_{H^2}^2 + \|v_{tt}(t)\|^2$$
$$+ \|\mathbf{w}(t)\|_{H^4}^2 + \|\mathbf{w}_t(t)\|_{H^2}^2 + \|\mathbf{w}_{tt}(t)\|^2 + \|\mathbf{b}(t)\|_{H^4}^2 + \|\mathbf{b}_t(t)\|_{H^2}^2 + \|\mathbf{b}_{tt}(t)\|^2$$
$$+ \|\theta(t) - \overline{\theta}\|_{H^4}^2 + \|\theta_t(t)\|_{H^2}^2 + \|\theta_{tt}(t)\|^2 + \int_0^t \big(\|\eta_x\|_{H^3}^2 + \|v_x\|_{H^4}^2 + \|v_t\|_{H^3}^2$$
$$+ \|v_{tt}\|_{H^1}^2 + \|\mathbf{w}_x\|_{H^4}^2 + \|\mathbf{w}_t\|_{H^3}^2 + \|\mathbf{w}_{tt}\|_{H^1}^2 + \|\mathbf{b}_x\|_{H^4}^2 + \|\mathbf{b}_t\|_{H^3}^2 + \|\mathbf{b}_{tt}\|_{H^1}^2$$
$$+ \|\theta_x\|_{H^4}^2 + \|\theta_t\|_{H^3}^2 + \|\theta_{tt}\|_{H^1}^2 \big)(s)ds \leq C_4, \tag{1.4.81}$$

$$\int_0^t \big(\|\eta_t\|_{H^4}^2 + \|\eta_{tt}\|_{H^2}^2 + \|\eta_{ttt}\|^2 \big)(s)ds \leq C_4. \tag{1.4.82}$$

Proof. These estimates follow from (1.1.1), Lemmas 1.3.1–1.3.3 and Lemmas 1.4.1–1.4.3.   □

We now prove the large-time behavior of the H^4-global solution

$$(\eta(t),\ v(t),\ \mathbf{w}(t),\ \mathbf{b}(t),\ \theta(t)).$$

Lemma 1.4.5. *Under the assumptions in Theorem 1.1.3, the H^4-global solution $(\eta(t), v(t), \mathbf{w}(t), \mathbf{b}(t), \theta(t))$ obtained in Lemmas 1.4.1–1.4.4 to the Cauchy problem (1.1.1)–(1.1.7) satisfies (1.1.22)–(1.1.23).*

Proof. Similarly to the reasoning in Lemmas 1.4.1 and 1.4.2, we derive

$$\frac{d}{dt}\|v_{tt}(t)\|^2 + (2C_1)^{-1}\|v_{ttx}(t)\|^2$$
$$\leq C_2 \big(\|\theta_{xx}(t)\|^2 + \|\theta_{tx}(t)\|_{H^1}^2 + \|v_x(t)\|_{H^1}^2 + \|v_{tx}(t)\|^2 + \|\theta_t(t)\|^2$$
$$+ \|\eta_x(t)\|^2 + \|\mathbf{b}_t(t)\|_{H^2}^2 \big), \tag{1.4.83}$$

$$\frac{d}{dt}\|\mathbf{w}_{tt}(t)\|^2 + (2C_1)^{-1}\|\mathbf{w}_{ttx}(t)\|^2$$
$$\leq C_4 \big(\|v_{tx}(t)\|^2 + \|\mathbf{w}_{tx}(t)\|^2 + \|\mathbf{w}_x(t)\|^2 + \|\mathbf{b}_{txx}(t)\|^2 \big), \tag{1.4.84}$$

$$\frac{d}{dt}\|\mathbf{b}_{tt}(t)\|^2 + (2C_1)^{-1}\|\mathbf{b}_{ttx}(t)\|^2$$
$$\leq C_4 \big(\|v_{tx}(t)\|^2 + \|\mathbf{b}_{tx}(t)\|^2 + \|\mathbf{b}_x(t)\|^2 + \|\mathbf{w}_{txx}(t)\|^2 \big), \tag{1.4.85}$$

$$\frac{d}{dt}\|\theta_{tt}(t)\|^2 + (2C_1)^{-1}\|\theta_{ttx}(t)\|^2$$
$$\leq C_4 \big(\|\theta_{tx}(t)\|^2 + \|v_{tx}(t)\|_{H^1}^2 + \|v_x(t)\|^2 + \|\theta_t(t)\|^2 + \|v_{ttx}(t)\|^2$$
$$+ \|\theta_{tt}(t)\|^2 + \|\mathbf{w}_{tx}(t)\|_{H^1}^2 + \|\mathbf{b}_{tx}(t)\|_{H^1}^2 + \|v_{tx}(t)\|^2 + \|v_x(t)\|^2 \big), \tag{1.4.86}$$

$$\frac{d}{dt}\|v_{tx}(t)\|^2 + (2C_1)^{-1}\|v_{txx}(t)\|^2$$
$$\leq C_2 \big(\|\theta_{tx}(t)\|^2 + \|v_{tx}(t)\|^2 + \|\theta_t(t)\|^2 + \|v_{xx}(t)\|^2 + \|\theta_x(t)\|^2$$

$$+ \|\eta_x(t)\|^2 + \|\mathbf{b}_{tx}(t)\|^2), \tag{1.4.87}$$

$$\frac{d}{dt}\|\mathbf{w}_{tx}(t)\|^2 + (2C_1)^{-1}\|\mathbf{w}_{txx}(t)\|^2$$
$$\leq C_2\big(\|\mathbf{b}_{tx}(t)\|^2 + \|v_x(t)\|_{H^1}^2 + \|\mathbf{w}_t(t)\|_{H^1}^2 + \|\mathbf{w}_x(t)\|^2 + \|\eta_x(t)\|^2\big), \tag{1.4.88}$$

$$\frac{d}{dt}\|\mathbf{b}_{tx}(t)\|^2 + (2C_1)^{-1}\|\mathbf{b}_{txx}(t)\|^2$$
$$\leq C_2\big(\|\mathbf{w}_{tx}(t)\|^2 + \|v_x(t)\|_{H^1}^2 + \|\mathbf{b}_t(t)\|_{H^1}^2 + \|\mathbf{b}_x(t)\|^2 + \|\eta_x(t)\|^2\big), \tag{1.4.89}$$

$$\frac{d}{dt}\|\theta_{tx}(t)\|^2 + (2C_1)^{-1}\|\theta_{txx}(t)\|^2$$
$$\leq C_2\big(\|\theta_{tx}(t)\|^2 + \|\theta_{xx}(t)\|^2 + \|v_{xx}(t)\|^2 + \|\eta_x(t)\|^2 + \|v_{txx}(t)\|^2$$
$$+ \|\mathbf{w}_{txx}(t)\|^2 + \|\mathbf{b}_{txx}(t)\|^2\big), \tag{1.4.90}$$

$$\frac{d}{dt}\left\|\frac{\eta_{xxx}}{\eta}(t)\right\|^2 + (2C_1)^{-1}\left\|\frac{\eta_{xxx}}{\eta}(t)\right\|^2 \leq C_1\|E_1(t)\|^2, \tag{1.4.91}$$

$$\frac{d}{dt}\left\|\frac{\eta_{xxxx}}{\eta}(t)\right\|^2 + (2C_1)^{-1}\left\|\frac{\eta_{xxxx}}{\eta}(t)\right\|^2 \leq C_1\|E_2(t)\|^2 \tag{1.4.92}$$

where, by (1.1.20), (1.4.52) and (1.4.74),

$$\int_0^t \big(\|E_1\|^2 + \|E_2\|^2\big)(s)ds \leq C_4, \quad \forall t > 0. \tag{1.4.93}$$

Applying Lemma 1.3.4 to (1.4.84)–(1.4.93) and using estimates (1.1.20) and (1.4.94), we infer that as $t \to +\infty$,

$$\|v_{tt}(t)\| \to 0, \quad \|\mathbf{w}_{tt}(t)\| \to 0, \quad \|\mathbf{b}_{tt}(t)\| \to 0, \quad \|\theta_{tt}(t)\| \to 0, \tag{1.4.94}$$
$$\|v_{tx}(t)\| \to 0, \quad \|\mathbf{w}_{tx}(t)\| \to 0, \quad \|\mathbf{b}_{tx}(t)\| \to 0, \quad \|\theta_{tx}(t)\| \to 0, \tag{1.4.95}$$
$$\|\eta_{xxx}(t)\| \to 0, \quad \|\eta_{xxxx}(t)\| \to 0. \tag{1.4.96}$$

In the same manner as for (1.4.8), (1.4.10), (1.4.55) and using the interpolation inequality, we deduce that

$$\|v_{xxx}(t)\| \leq C_2\big(\|v_x(t)\| + \|\eta_x(t)\|_{H^1} + \|\theta_x(t)\|_{H^1} + \|v_{tx}(t)\|$$
$$+ \|\mathbf{b}_x(t)\|_{H^1}\big), \tag{1.4.97}$$
$$\|v_{txx}(t)\| \leq C_1\|v_{tt}(t)\| + C_2\big(\|v_{xx}(t)\| + \|u_x(t)\| + \|v_{tx}(t)\| + \|\theta_x(t)\|$$
$$+ \|\theta_t(t)\| + \|\theta_{tx}\| + \|\mathbf{b}_{tx}(t)\|\big), \tag{1.4.98}$$
$$\|v_{xxxx}(t)\| \leq C_2\big(\|v_x(t)\|_{H^2} + \|\eta_x(t)\|_{H^2} + \|\theta_x(t)\|_{H^2} + \|v_{txx}(t)\|$$
$$+ \|\mathbf{b}_{xx}(t)\|_{H^1}\big), \tag{1.4.99}$$
$$\|v_{tx}(t)\|_{L^\infty}^2 \leq C\|v_{tx}(t)\|\|v_{txx}(t)\|, \quad \|v_t(t)\|_{L^\infty}^2 \leq C\|v_t(t)\|\|v_{tx}(t)\|, \tag{1.4.100}$$
$$\|v_{xx}(t)\|_{L^\infty}^2 \leq C\|v_{xx}(t)\|\|v_{xxx}(t)\|, \quad \|v_{xxx}(t)\|_{L^\infty}^2 \leq C\|v_{xxx}(t)\|\|v_{xxxx}(t)\|, \tag{1.4.101}$$

$$\|\eta_{xx}(t)\|_{L^\infty}^2 \le C\|\eta_{xx}(t)\|\|\eta_{xxx}(t)\|, \quad \|\eta_{xxx}(t)\|_{L^\infty}^2 \le C\|\eta_{xxx}(t)\|\|\eta_{xxxx}(t)\|.$$
$$(1.4.102)$$

Thus, it follows from (1.1.1), (1.4.98)–(1.4.103) and Lemma 1.3.5 that, as $t \to +\infty$,

$$\|(\eta_x(t), v_x(t))\|_{H^3} + \|v_t(t)\|_{H^2} + \|\eta_t(t)\|_{H^3} + \|\eta_t(t)\|_{W^{2,\infty}}$$
$$+ \|\eta_{tt}(t)\|_{H^1} + \|(\eta_x(t), v_x(t))\|_{W^{2,\infty}} \to 0. \qquad (1.4.103)$$

In the same manner, we can conclude as $t \to +\infty$

$$\|\mathbf{w}_x(t)\|_{H^3} + \|\mathbf{w}_t(t)\|_{H^2} + \|\mathbf{w}_t(t)\|_{W^{1,\infty}} + \|\mathbf{w}_x(t)\|_{W^{2,\infty}} \to 0, \qquad (1.4.104)$$
$$\|\mathbf{b}_x(t)\|_{H^3} + \|\mathbf{b}_t(t)\|_{H^2} + \|\mathbf{b}_t(t)\|_{W^{1,\infty}} + \|\mathbf{b}_x(t)\|_{W^{2,\infty}} \to 0, \qquad (1.4.105)$$
$$\|\theta_x(t)\|_{H^3} + \|\theta_t(t)\|_{H^2} + \|\theta_t(t)\|_{W^{1,\infty}} + \|\theta_x(t)\|_{W^{2,\infty}} \to 0, \qquad (1.4.106)$$

which, together with Lemma 1.3.5 and (1.4.104), imply (1.1.22)–(1.1.23). The proof is complete. $\qquad\Box$

Proof of Theorem 1.1.3. Lemmas 1.4.1–1.4.5 establish the global existence of H^4-solutions to problem (1.1.1)–(1.1.7). $\qquad\Box$

Proof of Corollary 1.1.1. Employing the Sobolev embedding theorem together with the estimates (1.1.17)–(1.1.18) yield the desired conclusion immediately. $\qquad\Box$

1.5 Bibliographic Comments

Assuming the dependencies $P = P(\eta, \theta)$ and $e = e(\eta, \theta)$, the ideal gas with $P = R\theta/\eta$ and $e = C_V\theta$ was treated earlier, with many good results on global solutions and their exponential stability for various problems (see, e.g., [1, 8, 9, 43, 53–55, 93, 96, 98, 138, 140]). If the Stefan-Bolzmann law, i.e., (1.1.7) is considered, the radiative gas model, of a more general type, was derived and related results were obtained, e.g., in [22, 30, 105, 129]. Ducomet and Zlotnik [30] studied a selfgravitating gas with the mass force g involving only the space variable x in (1.1.2). The subtle introduction of the entropy $S = R\log\eta + C_V\log\theta + 4a\eta\theta^4/3$ leads to the energy-type Lyapunov functional, which is crucial for establishing the global existence and exponential stability of the solutions. Moreover, they employed the technique for estimating $\Theta := \sup_{0\le s\le t}\|\theta(s)\|_{L^\infty}$ for other H^1 norms of the unknown variables in [96, 98]. For the $\overline{\text{MHD}}$ models, this case has been studied in [10, 11, 25, 58, 132] and [145]. Specifically, in Kawashima and Okada [58], assuming $\overline{\mathbf{b}}$ is an arbitrary but fixed two-dimensional vector and letting S denote the entropy, the energy type $E = e - \overline{e} + \overline{P}(1/\eta - 1/\overline{\eta}) - \overline{\theta}(S - \overline{S}) + v^2/2 + |\mathbf{w}|^2/2 + |\mathbf{b} - \overline{\mathbf{b}}|^2/2$, where the steady-state expressions $\overline{N} = N(\overline{\eta}, \overline{\theta})$, $N = e, P$ or S are used, plays an important role in the estimates for $|\mathbf{b} - \overline{\mathbf{b}}|$'s certain uniform spatial norms. In Zhang and Xie [145], another kind of gravitational force is presented as Ψ_x/η^2 where the relation of Ψ and η is governed by Poisson's equation:

$$\Delta \Psi = -\frac{G}{\eta}, \qquad (1.5.1)$$

in which $G > 0$ is the Newtonian gravitational constant. However, the estimates of the lower and upper bounds of η are not uniform in the whole time interval $(0, +\infty)$ and no large-time behavior was obtained when the self-gravitation is present. On the other hand, when the domain is unbounded and the Poincaré inequality is not available, the properties of the initial data for the unknown variables must be specified via some appropriate stationary state data like $\eta_0 - \overline{\eta}$, etc., understood as *small initial data* in [45, 54, 56].

Under the small initial data hypothesis, the global existence and large-time behavior of smooth solutions have been obtained for the Cauchy problem, including the two- or three-dimensional case (see, e.g., [46, 53–55, 58, 74–77, 93]). Although these results are based on the same assumption namely, *small initial data*, the specifics are quite different. In [54], the small data hypothesis in the 1D case appears in weighted form,

$$e_0 := \|\eta_0 - \overline{\eta}\|_{L^\infty}^2 + \int_{\mathbb{R}} (1+x^2)^\gamma \left\{ (\eta_0 - \overline{\eta})^2 + v_0^2 + (\theta_0 - \overline{\theta})^2 + v_0^4 \right\} dx \le \epsilon, \ (1.5.2)$$

where $(1+x^2)^\gamma$ $(\gamma > \frac{1}{2})$ is a weight function and ϵ is a sufficiently small positive constant. In contrast, in Okada and Kawashima [93], the hypothesis takes on the form

$$E_0 E_1 \le \epsilon, \qquad (1.5.3)$$

where

$$E_i = \left\| \left(\log(\eta_0/\overline{\eta}), \log(v_0), \log(\theta_0/\overline{\theta}) \right) \right\|_{H^i} \quad (i = 0,1). \qquad (1.5.4)$$

As another example, Kawashima and Okada [58] use a simpler condition:

$$\|(1/\eta_0 - 1/\overline{\eta}_0, \ v_0, \ \mathbf{w_0}, \ \mathbf{b_0} - \overline{\mathbf{b}}, \ \theta_0 - \overline{\theta})\| \le \epsilon. \qquad (1.5.5)$$

In Qin et al.'s work [110, 112], the former two cases (1.1.9)–(1.1.11) have been summarized together for higher regularity of H^2 and H^4 global solutions. This chapter is ti employ the basic ideas in [54, 110] to obtain the global solutions in H^1 for a 1D ideal gas MHD system (1.1.1)–(1.1.6) similar to that of [58] and then obtain the regularity and asymptotic behavior of H^2 and H^4 global solutions.

Chapter 2

Global Existence and Exponential Stability of 1D Compressible and Radiative Magnetohydrodynamic Flows

2.1 Main Results

In this chapter, we shall study the global existence and exponential stability of solutions to the one-dimensional thermally-radiative magnetohydrodynamic equations.

Magnetohydrodynamics (MHD) is concerned with the study of the interaction between magnetic fields and fluid conductors of electricity. The applications of magnetohydrodynamics covers a very wide range of physical areas, from liquid metals to cosmic plasmas, for example, the intensely heated and ionized fluids in an electromagnetic field in astrophysics, geophysics, high-speed aerodynamics and plasma physics. In addition to these situations, we also take into account effect of the radiation field. The flows mentioned above are described by the following equations in the Lagrangian coordinate system:

$$\tau_t - u_x = 0, \tag{2.1.1}$$

$$u_t + \left(p + \frac{1}{2}|\mathbf{b}|^2\right)_x = \left(\frac{\lambda u_x}{\tau}\right)_x, \tag{2.1.2}$$

$$\mathbf{w}_t - \mathbf{b}_x = \left(\frac{\mu \mathbf{w}_x}{\tau}\right)_x, \tag{2.1.3}$$

$$(\tau \mathbf{b})_t - \mathbf{w}_x = \left(\frac{\nu \mathbf{b}_x}{\tau}\right)_x, \tag{2.1.4}$$

$$E_t + \left(u\left(p + \frac{1}{2}|\mathbf{b}|^2\right) - \mathbf{w}\cdot\mathbf{b}\right)_x = \left(\frac{\lambda u u_x + \mu \mathbf{w}\cdot\mathbf{w}_x + \nu \mathbf{b}\cdot\mathbf{b}_x + \kappa\theta_x}{\tau}\right)_x. \tag{2.1.5}$$

Here $\tau = \frac{1}{\rho}$ denotes the specific volume, $u \in \mathbb{R}$ the longitudinal velocity, $\mathbf{w} \in \mathbb{R}^2$ the transverse velocity, $\mathbf{b} \in \mathbb{R}^2$ the transverse magnetic field, and θ the temperature, $p = p(\tau, \theta)$ the pressure, and $e = e(\tau, \theta)$ the internal energy; λ and μ are the bulk and the shear viscosity coefficients, respectively, ν is the magnetic diffusivity acting as a magnetic diffusion coefficient of the magnetic field, $k = k(\tau, \theta)$ is the heat conductivity, and E is given by

$$E = e + \frac{1}{2}(u^2 + |\mathbf{w}|^2) + \frac{1}{2}\tau|\mathbf{b}|^2.$$

For the constitutive relations, we consider (see, e.g., [22]) the Stefan–Boltzmann model, i.e., the pressure $p(\tau, \theta)$, internal energy $e(\tau, \theta)$ and the thermoradiative flux $Q(\tau, \theta)$ take the following respective forms,

$$p(\tau, \theta) = \frac{R\theta}{\tau} + \frac{a}{3}\theta^4, \quad e(\tau, \theta) = C_v\theta + a\tau\theta^4, \quad Q(\tau, \theta) = Q_F + Q_R = -\kappa\theta_x, \quad (2.1.6)$$

where $R > 0$ is the ideal gas constant, $C_v > 0$ is the specific heat at constant volume, $a > 0$ is a constant and the heat conductivity $\kappa(\tau, \theta) > 0$ is a function of τ and θ. As initial and boundary conditions, we consider

$$(\tau, u, \mathbf{w}, \mathbf{b}, \theta)|_{t=0} = (\tau_0, u_0, \mathbf{w}_0, \mathbf{b}_0, \theta_0)(x), \quad x \in \overline{\Omega} = [0, 1], \quad (2.1.7)$$

$$(u, \mathbf{w}, \mathbf{b}, \theta_x)|_{\partial\Omega} = 0. \quad (2.1.8)$$

In this chapter we shall establish the global existence and exponential stability of solutions in the spaces H_+^i $(i = 1, 2, 4)$ (see below for their definitions) to problem (2.1.1)–(2.1.8). The main difficulty arises from the higher-order nonlinear dependence on the temperature θ of $p(\tau, \theta)$, $e(\tau, \theta)$ and $\kappa(\tau, \theta)$, which makes the upper bound for θ become more complicated. In order to overcome this difficulty, we make use of Corollaries 2.2.1–2.2.2 and interpolation techniques to reduce the higher-order dependence on θ. The estimate $\int_0^1 \tau\theta^4 dx \leq C_1$ in (2.2.3) plays an important part. This will be done by a careful analysis. Another difficulty is that we have to establish uniform estimates independent of time in order to study the large-time behavior.

We define three function classes as follows:

$$H_+^1 = \Big\{(\tau, u, \mathbf{w}, \mathbf{b}, \theta) \in (H^1[0, 1])^7 : \tau(x) > 0, \theta(x) > 0, x \in [0, 1],$$

$$u(0) = u(1) = 0, \mathbf{w}(0) = \mathbf{w}(1) = \mathbf{b}(0) = \mathbf{b}(1) = \mathbf{0}\Big\},$$

$$H_+^i = \Big\{(\tau, u, \mathbf{w}, \mathbf{b}, \theta) \in (H^i[0, 1])^7 : \tau(x) > 0, \theta(x) > 0, x \in [0, 1],$$

$$u(0) = u(1) = 0, \mathbf{w}(0) = \mathbf{w}(1) = \mathbf{b}(0) = \mathbf{b}(1) = \mathbf{0},$$

$$\theta'(0) = \theta'(1) = 0\Big\}, \quad i = 2, 4.$$

In this chapter we will use the following notations:

L^p, $1 \leq p \leq +\infty$, $W^{m,p}$, $m \in N$, $H^1 = W^{1,2}$, $H_0^1 = W_0^{1,2}$ denote the usual (Sobolev) spaces on $[0,1]$. In addition, $\|\cdot\|_B$ denotes the norm in the space

B; we also put $\| \cdot \| = \| \cdot \|_{L^2[0,1]}$. Constants C_i ($i = 1, 2, 3, 4$) are generic constants depending only on the H^i_+ norm of the initial data $(\tau_0, u_0, \mathbf{w}_0, \mathbf{b}_0, \theta_0)$, but independent of time.

Now we are in a position to state our main results.

Theorem 2.1.1. *Assume that the initial data $(\tau_0, u_0, \mathbf{w}_0, \mathbf{b}_0, \theta_0) \in H^1_+$ and compatibility conditions are satisfied. Assume also that the heat conductivity κ is a C^2 function on $0 < \tau < +\infty$ and $0 \leq \theta < +\infty$ and satisfies the growth condition*

$$k_1(1 + \theta^q) \leq \kappa(\tau, \theta) \leq k_2(1 + \theta^q), \quad |\kappa_\tau| + |\kappa_{\tau\tau}| \leq k_2(1 + \theta^q), \quad q > 2, \quad (2.1.9)$$

with positive constants $k_1 \leq k_2$, and there exists a constant $\varepsilon_0 > 0$ such that $\overline{\tau} = \int_0^1 \tau_0 \, dx \leq \varepsilon_0$. Then the problem (2.1.1)–(2.1.8) admits a unique global solution $(\tau, u, \mathbf{w}, \mathbf{b}, \theta) \in H^1_+$ verifying

$$0 < C_1^{-1} \leq \tau(x,t) \leq C_1, \quad 0 < C_1^{-1} \leq \theta(x,t) \leq C_1, \quad \forall (x,t) \in [0,1] \times [0, +\infty) \quad (2.1.10)$$

and for any $t > 0$,

$$\|\tau(t) - \overline{\tau}\|^2_{H^1} + \|u(t)\|^2_{H^1} + \|\mathbf{w}(t)\|^2_{H^1} + \|\mathbf{b}(t)\|^2_{H^1} + \|\theta(t) - \overline{\theta}\|^2_{H^1}$$
$$+ \int_0^t \left(\|\tau - \overline{\tau}\|^2_{H^1} + \|u\|^2_{H^2} + \|\mathbf{w}\|^2_{H^2} + \|\mathbf{b}\|^2_{H^2} + \|\theta - \overline{\theta}\|^2_{H^2} \right.$$
$$\left. + \|u_t\|^2 + \|\mathbf{w}_t\|^2 + \|\mathbf{b}_t\|^2 + \|\theta_t\|^2 \right)(s) ds \leq C_1, \quad (2.1.11)$$

where $\overline{\tau} = \int_0^1 \tau \, dx = \int_0^1 \tau_0 \, dx$ and the constant $\overline{\theta} > 0$ is determined by

$$e(\overline{\tau}, \overline{\theta}) = E_0 \equiv \int_0^1 \left(\frac{1}{2}(u_0^2 + |\mathbf{w}_0|^2 + \tau_0|\mathbf{b}_0|^2) + e(\tau_0, \theta_0) \right) dx.$$

Moreover, there are constants $C_1 > 0$ and $\gamma_1 = \gamma_1(C_1) > 0$ such that, for any fixed $\gamma \in (0, \gamma_1]$, the following estimate holds for any $t > 0$:

$$e^{\gamma t} \left(\|\tau(t) - \overline{\tau}\|^2_{H^1} + \|u(t)\|^2_{H^1} + \|\mathbf{w}(t)\|^2_{H^1} + \|\mathbf{b}(t)\|^2_{H^1} + \|\theta(t) - \overline{\theta}\|^2_{H^1} \right)$$
$$+ \int_0^t e^{\gamma s} \left(\|\tau - \overline{\tau}\|^2_{H^1} + \|u\|^2_{H^2} + \|\mathbf{w}\|^2_{H^2} + \|\mathbf{b}\|^2_{H^2} + \|\theta - \overline{\theta}\|^2_{H^2} \right.$$
$$\left. + \|u_t\|^2 + \|\mathbf{w}_t\|^2 + \|\mathbf{b}_t\|^2 + \|\theta_t\|^2 \right)(s) ds \leq C_1. \quad (2.1.12)$$

Theorem 2.1.2. *Assume that the initial data $(\tau_0, u_0, \mathbf{w}_0, \mathbf{b}_0, \theta_0) \in H^2_+$ and compatibility conditions are satisfied. Assume also that the heat conductivity κ is a C^3 function satisfying (2.1.13) on $0 < \tau < +\infty$ and $0 \leq \theta < +\infty$, and there exists a constant $\varepsilon_0 > 0$ such that $\overline{\tau} = \int_0^1 \tau_0 \, dx \leq \varepsilon_0$. Then the problem (2.1.1)–(2.1.8) admits a unique global solution $(\tau, u, \mathbf{w}, \mathbf{b}, \theta) \in H^2_+$ verifying that, for any $t > 0$,*

$$\|\tau(t) - \overline{\tau}\|^2_{H^2} + \|u(t)\|^2_{H^2} + \|\mathbf{w}(t)\|^2_{H^2} + \|\mathbf{b}(t)\|^2_{H^2} + \|\theta(t) - \overline{\theta}\|^2_{H^2}$$

$$+ \|u_t(t)\|^2 + \|\mathbf{w}_t(t)\|^2 + \|\mathbf{b}_t(t)\|^2 + \|\theta_t(t)\|^2 + \int_0^t \Big(\|\tau - \overline{\tau}\|_{H^2}^2$$

$$+ \|u\|_{H^3}^2 + \|\mathbf{w}\|_{H^3}^2 + \|\mathbf{b}\|_{H^3}^2 + \|\theta - \overline{\theta}\|_{H^3}^2 + \|u_{tx}\|^2 + \|\mathbf{w}_{tx}\|^2$$

$$+ \|\mathbf{b}_{tx}\|^2 + \|\theta_{tx}\|^2 \Big)(s)ds \le C_2. \tag{2.1.13}$$

Moreover, there are constants $C_2 > 0$ and $\gamma_2 = \gamma_2(C_2) > 0$ such that, for any fixed $\gamma \in (0, \gamma_2]$, the following estimate holds for any $t > 0$:

$$e^{\gamma t} \Big(\|\tau(t) - \overline{\tau}\|_{H^2}^2 + \|u(t)\|_{H^2}^2 + \|\mathbf{w}(t)\|_{H^2}^2 + \|\mathbf{b}(t)\|_{H^2}^2 + \|\theta(t) - \overline{\theta}\|_{H^2}^2$$

$$+ \|u_t(t)\|^2 + \|\mathbf{w}_t(t)\|^2 + \|\mathbf{b}_t(t)\|^2 + \|\theta_t(t)\|^2 \Big) + \int_0^t e^{\gamma s} \Big(\|\tau - \overline{\tau}\|_{H^2}^2$$

$$+ \|u\|_{H^3}^2 + \|\mathbf{w}\|_{H^3}^2 + \|\mathbf{b}\|_{H^3}^2 + \|\theta - \overline{\theta}\|_{H^3}^2 + \|u_{tx}\|^2 + \|\mathbf{w}_{tx}\|^2$$

$$+ \|\mathbf{b}_{tx}\|^2 + \|\theta_{tx}\|^2 \Big)(s)ds \le C_2. \tag{2.1.14}$$

Theorem 2.1.3. *Assume that the initial data $(\tau_0, u_0, \mathbf{w}_0, \mathbf{b}_0, \theta_0) \in H_+^4$ and compatibility conditions are satisfied. Assume that the heat conductivity κ is a C^5 function satisfying (2.1.13) on $0 < \tau < +\infty$ and $0 \le \theta < +\infty$, and there exists a constant $\varepsilon_0 > 0$ such that $\overline{\tau} = \int_0^1 \tau_0 \, dx \le \varepsilon_0$. Then the problem (2.1.1)–(2.1.8) admits a unique global solution $(\tau, u, \mathbf{w}, \mathbf{b}, \theta) \in H_+^4$ verifying that, for any $t > 0$,*

$$\|\tau(t) - \overline{\tau}\|_{H^4}^2 + \|u(t)\|_{H^4}^2 + \|\mathbf{w}(t)\|_{H^4}^2 + \|\mathbf{b}(t)\|_{H^4}^2 + \|\theta(t) - \overline{\theta}\|_{H^4}^2 + \|u_{tt}(t)\|^2$$

$$+ \|\mathbf{w}_{tt}(t)\|^2 + \|\mathbf{b}_{tt}(t)\|^2 + \|u_t(t)\|_{H^2}^2 + \|\mathbf{w}_t(t)\|_{H^2}^2 + \|\mathbf{b}_t(t)\|_{H^2}^2 + \|\theta_t(t)\|_{H^2}^2$$

$$+ \|\theta_{tt}(t)\|^2 + \int_0^t \Big(\|\tau - \overline{\tau}\|_{H^4}^2 + \|u\|_{H^5}^2 + \|\mathbf{w}\|_{H^5}^2 + \|\mathbf{b}\|_{H^5}^2 + \|\theta - \overline{\theta}\|_{H^5}^2$$

$$+ \|u_t\|_{H^3}^2 + \|\mathbf{w}_t\|_{H^3}^2 + \|\mathbf{b}_t\|_{H^3}^2 + \|\theta_t\|_{H^3}^2 + \|u_{tt}\|_{H^1}^2 + \|\mathbf{w}_{tt}\|_{H^1}^2$$

$$+ \|\mathbf{b}_{tt}\|_{H^1}^2 + \|\theta_{tt}\|_{H^1}^2 \Big)(s)ds \le C_4. \tag{2.1.15}$$

Moreover, there are constants $C_4 > 0$ and $\gamma_4 = \gamma_4(C_4) > 0$ such that, for any fixed $\gamma \in (0, \gamma_4]$, the following estimate holds for any $t > 0$:

$$e^{\gamma t} \Big(\|\tau(t) - \overline{\tau}\|_{H^4}^2 + \|u(t)\|_{H^4}^2 + \|\mathbf{w}(t)\|_{H^4}^2 + \|\mathbf{b}(t)\|_{H^4}^2 + \|\theta(t) - \overline{\theta}\|_{H^4}^2$$

$$+ \|u_{tt}(t)\|^2 + \|\mathbf{w}_{tt}(t)\|^2 + \|\mathbf{b}_{tt}(t)\|^2 + \|u_t(t)\|_{H^2}^2 + \|\mathbf{w}_t(t)\|_{H^2}^2 + \|\mathbf{b}_t(t)\|_{H^2}^2$$

$$+ \|\theta_t(t)\|_{H^2}^2 + \|\theta_{tt}(t)\|^2 \Big) + \int_0^t e^{\gamma s} \Big(\|\tau - \overline{\tau}\|_{H^4}^2 + \|u\|_{H^5}^2 + \|\mathbf{w}\|_{H^5}^2 + \|\mathbf{b}\|_{H^5}^2$$

$$+ \|\theta - \overline{\theta}\|_{H^5}^2 + \|u_t\|_{H^3}^2 + \|\mathbf{w}_t\|_{H^3}^2 + \|\mathbf{b}_t\|_{H^3}^2 + \|\theta_t\|_{H^3}^2 + \|u_{tt}\|_{H^1}^2 + \|\mathbf{w}_{tt}\|_{H^1}^2$$

$$+ \|\mathbf{b}_{tt}\|_{H^1}^2 + \|\theta_{tt}\|_{H^1}^2 \Big)(s)ds \le C_4. \tag{2.1.16}$$

Corollary 2.1.1. *The global solution in Theorem 2.1.3 is in fact a classical solution* $(\tau, u, \mathbf{w}, \mathbf{b}, \theta) \in \left(C^{3+1/2}(\Omega)\right)^7$ *that obeys for any* $\gamma \in (0, \gamma_4]$ *the estimate*

$$\left\| \left(\tau(t) - \overline{\tau}, u(t), \mathbf{w}(t), \mathbf{b}(t), \theta(t) - \overline{\theta} \right) \right\|_{(C^{3+1/2}(\Omega))^7} \leq C_4 e^{-\gamma t}.$$

2.2 Global Existence and Exponential Stability in H^1

In this section, we shall study the global existence and exponential stability of problem solutions to (2.1.1)–(2.1.8) in H^1 under the assumptions of Theorem 2.1.1. We begin with the following lemmas.

Lemma 2.2.1. *Under the assumptions of Theorem 2.1.1, the following relations and estimates hold for any* $t > 0$,

$$\int_0^1 \tau(x,t)dx = \int_0^1 \tau_0(x)dx, \tag{2.2.1}$$

$$\int_0^1 E(x,t)dx = \int_0^1 E(x,0)dx \equiv E_0, \tag{2.2.2}$$

$$\int_0^1 \left(\theta + \tau\theta^4 + u^2 + |\mathbf{w}|^2 + \tau|\mathbf{b}|^2 \right)(x,t)dx \leq C_1, \tag{2.2.3}$$

$$\Phi(t) + \int_0^t V(s)ds \leq C_1, \tag{2.2.4}$$

where

$$\Phi(t) = \int_0^1 \left[C_v(\theta - \log\theta - 1) + R(\tau - \log\tau - 1) \right](x,t)dx,$$

$$V(t) = \int_0^1 \left(\frac{\kappa\theta_x^2}{\tau\theta^2} + \frac{\lambda u_x^2 + \mu|\mathbf{w}_x|^2 + \nu|\mathbf{b}_x|^2}{\tau\theta} \right)(x,t)dx,$$

$$E(x,t) = e(\tau,\theta) + \frac{1}{2}(u^2 + |\mathbf{w}|^2) + \frac{1}{2}\tau|\mathbf{b}|^2$$

$$= C_v\theta + a\tau\theta^4 + \frac{1}{2}(u^2 + |\mathbf{w}|^2) + \frac{1}{2}\tau|\mathbf{b}|^2.$$

Proof. By (2.1.1) and (2.1.8), we get (2.2.1). Integrating (2.1.5) over $Q_t := (0,1) \times (0,t)$ and using (2.1.8), we get (2.2.2), the conservation law of total energy. Estimate (2.2.3) follows directly from (2.2.2) and (2.1.6).

Equation (2.1.5) can be rewritten as

$$e_t + pu_x = \left(\frac{\kappa\theta_x}{\tau} \right)_x + \frac{\lambda u_x^2 + \mu|\mathbf{w}_x|^2 + \nu|\mathbf{b}_x|^2}{\tau}, \tag{2.2.5}$$

i.e.,

$$C_v\theta_t + 4a\tau\theta^3\theta_t + \frac{R\theta\tau_t}{\tau} + \frac{4a}{3}\tau_t\theta^4 = \left(\frac{\kappa\theta_x}{\tau}\right)_x + \frac{\lambda u_x^2 + \mu|\mathbf{w}_x|^2 + \nu|\mathbf{b}_x|^2}{\tau}. \quad (2.2.6)$$

Multiplying (2.2.6) by θ^{-1} and integrating the resulting equation over Q_t, we get (2.2.4).   □

Lemma 2.2.2. *For any $t \geq 0$, there exists a point $x_1 = x_1(t) \in [0,1]$ such that specific volume $\tau(x,t)$ in problem (2.1.1)–(2.1.8) can be expressed for any $\delta \geq 0$ as*

$$\tau(x,t) = D(x,t)Z(t)\left\{1 + \lambda^{-1}\int_0^t D^{-1}(x,s)Z^{-1}(s)\tau(x,s)\Big[\sigma_1(x,s) - \delta\Big]ds\right\},$$
$$(2.2.7)$$

where

$$D(x,t) = \tau_0(x)\exp\left\{\lambda^{-1}\left(\int_{x_1(t)}^x u(y,t)dy\right.\right.$$
$$\left.\left. -\int_0^x u_0(y)dy + \bar\tau_0{}^{-1}\int_0^1 \tau_0\int_0^x u_0\,dydx\right)\right\},$$

$$Z(t) = \exp\left\{-(\lambda\bar\tau_0)^{-1}\int_0^t\int_0^1 (u^2 + \tau\sigma_1)dyds + \frac{\delta t}{\lambda}\right\},$$

$$\sigma_1 = p + \frac{1}{2}|\mathbf{b}|^2, \quad \bar\tau_0 = \int_0^1 \tau_0 dx.$$

Proof. For any $\delta \geq 0$, we can rewrite (2.1.1) as

$$\tau_t = u_x = \frac{1}{\lambda}(\delta + \sigma)\tau + \frac{1}{\lambda}(\sigma_1 - \delta)\tau,$$

i.e.,

$$\tau_t - \frac{1}{\lambda}(\delta + \sigma)\tau = \frac{1}{\lambda}(\sigma_1 - \delta)\tau, \quad (2.2.8)$$

where $\sigma = \frac{\lambda u_x}{\tau} - \sigma_1$.

Multiplying (2.2.8) by $\exp\left\{-\frac{1}{\lambda}\int_0^t(\sigma + \delta)ds\right\}$, we obtain the expression

$$\tau(x,t) = \exp\left(\frac{1}{\lambda}\int_0^t(\sigma+\delta)ds\right)\left(\tau_0 + \frac{1}{\lambda}\int_0^t\tau(\sigma_1-\delta)\exp\left(-\frac{1}{\lambda}\int_0^s(\sigma+\delta)dr\right)ds\right).$$
$$(2.2.9)$$

Let

$$h(x,t) = \int_0^t \sigma(x,s)ds + \int_0^x u_0(y)dy.$$

Then from (2.1.2) it follows that $h(x,t)$ satisfies

$$h_x = u, \quad h_t = \sigma = \frac{\lambda u_x}{\tau} - \sigma_1. \quad (2.2.10)$$

Hence, we have

$$(\tau h)_t = \tau_t h + \tau h_t = (uh)_x - u^2 + \lambda u_x - \tau \left(p + \frac{1}{2}|\mathbf{b}|^2 \right). \qquad (2.2.11)$$

Integrating (2.2.11) over Q_t, we get

$$\int_0^1 \tau h\, dx = \int_0^1 \tau_0 h_0\, dx - \int_0^t \int_0^1 \left(u^2 + \tau(p + \frac{1}{2}|\mathbf{b}|^2) \right) dx ds$$

$$= \int_0^1 \tau_0 \left(\int_0^x u_0\, dy \right) dx - \int_0^t \int_0^1 (u^2 + \tau \sigma_1)\, dx ds \equiv \Gamma(t). \qquad (2.2.12)$$

Then for any $t \geq 0$, there exists a point $x_1 = x_1(t) \in [0, 1]$ such that

$$\Gamma(t) = \int_0^1 \tau h\, dx = \int_0^1 \tau dx \cdot h(x_1(t), t) = \overline{\tau}_0 \cdot h(x_1(t), t),$$

i.e.,

$$h(x_1(t), t) = \frac{1}{\overline{\tau}_0} \Gamma(t). \qquad (2.2.13)$$

Thus from (2.2.10)–(2.2.13) we deduce that

$$\int_0^t \sigma(x_1(t), s)ds = h(x_1(t), t) - \int_0^{x_1(t)} u_0\, dy = \frac{1}{\overline{\tau}_0} \Gamma(t) - \int_0^{x_1(t)} u_0\, dy. \qquad (2.2.14)$$

Integrating (2.1.2) over $[x_1(t), x] \times [0, t]$, we get

$$\int_0^t \sigma(x, s)ds = \int_0^t \sigma(x_1(t), s)ds + \int_{x_1(t)}^x (u - u_0)dy. \qquad (2.2.15)$$

Then we infer from (2.2.14) and (2.2.15) that

$$\int_0^t \sigma(x, s)ds = \frac{1}{\overline{\tau}_0} \Gamma(t) - \int_0^{x_1(t)} u_0\, dy + \int_{x_1(t)}^x (u - u_0)dy$$

$$= \frac{1}{\overline{\tau}_0} \Gamma(t) - \int_0^x u_0\, dy + \int_{x_1(t)}^x u(y, t)dy,$$

which, together with (2.2.9), gives (2.2.7). The proof is complete. $\qquad \square$

Lemma 2.2.3. *Under the assumptions of Theorem 2.1.1, the following estimate holds:*

$$0 < C_1^{-1} \leq \tau(x, t) \leq C_1, \quad \forall(x, t) \in [0, 1] \times [0, +\infty). \qquad (2.2.16)$$

Proof. It follows from (2.2.4) and the convexity of the function $-\log y$ that

$$\int_0^1 \theta\, dx - \log \int_0^1 \theta\, dx - 1 \le \int_0^1 (\theta - \log\theta - 1)dx \le C_1/C_v,$$

which implies that there exist a point $b(t) \in [0,1]$ and two positive constants r_1, r_2 such that

$$0 < r_1 \le \int_0^1 \theta\, dx = \theta(b(t), t) \le r_2,$$

with r_1, r_2 being the two positive roots of the equation $y - \log y - 1 = C_1/C_v$. It follows that

$$\int_0^1 \tau p\, dx \le \max(R, a/3) \int_0^1 (\theta + \tau\theta^4) dx$$

$$\le \frac{\max(R, a/3)}{\min(C_v, a)} \int_0^1 e\, dx \le \frac{\max(R, a/3)}{\min(C_v, a)} E_0$$

and

$$\int_0^1 \left(u^2 + \frac{\tau}{2}|\mathbf{b}|^2\right) dx \le 2E_0,$$

whence

$$0 < a_1 \le \frac{1}{\lambda\bar{\tau}_0} \int_0^1 \left(u^2 + \tau(p + \frac{1}{2}|\mathbf{b}|^2)\right)(x, s)dx \le a_2 \qquad (2.2.17)$$

with

$$a_1 = \frac{Rr_1}{\lambda\bar{\tau}_0}, \quad a_2 = \frac{1}{\lambda\bar{\tau}_0}\left[Rr_2 + \left(2 + \frac{\max(R, a/3)}{\min(C_v, a)}\right)E_0\right].$$

By Lemmas 2.2.1 and 2.2.2, we derive

$$0 < C_1^{-1} \le D(x, t) \le C_1, \quad \forall (x, t) \in [0, 1] \times [0, +\infty). \qquad (2.2.18)$$

On the other hand, for $0 \le m \le (q+4)/2$, we infer from Lemma 2.2.1 that

$$|\theta^m(x, t) - \theta^m(b(t), t)| \le C_1 \left|\int_{b(t)}^x \theta^{m-1}\theta_y\, dy\right|$$

$$\le C_1 \left(\int_0^1 \frac{\kappa\theta_y^2}{\tau\theta^2}dy\right)^{\frac{1}{2}} \left(\int_0^1 \frac{\tau\theta^{2m}}{\kappa}dy\right)^{\frac{1}{2}}$$

$$\le C_1 V(t)^{\frac{1}{2}} \left(\int_0^1 \tau(1 + \theta)^{2m-q}dy\right)^{\frac{1}{2}}$$

$$\le C_1 V(t)^{\frac{1}{2}}.$$

Thus,

$$\frac{1}{2}r_1^{2m} - C_1V(t) \le \theta^{2m}(x,t) \le 2r_2^{2m} + C_1V(t), \quad 0 \le m \le \frac{q+4}{2}. \tag{2.2.19}$$

Obviously, from Lemmas 2.2.1–2.2.2 and (2.2.17), for $\delta \ge 0$ and $0 \le s \le t$, we get

$$e^{-(a_2-\delta/\lambda)(t-s)} \le Z(t)Z^{-1}(s) \le e^{-(a_1-\delta/\lambda)(t-s)}, \quad 0 \le s \le t, \ 0 < a_1 < a_2. \tag{2.2.20}$$

Noting that $\sigma_1\tau = \tau p + \frac{\tau}{2}|\mathbf{b}|^2 = R\theta + \frac{a}{3}\tau\theta^4 + \frac{\tau}{2}|\mathbf{b}|^2 \ge R\theta$, we use (2.2.17)–(2.2.20) to derive that there exists a large time $t_0 > 0$ such that as $t \ge t_0 > 0$, for $\delta = 0$ in Lemma 2.2.2,

$$\tau(x,t) \ge \lambda^{-1}\int_0^t D(x,t)Z(t)D^{-1}(x,s)Z^{-1}(s)\sigma_1(x,s)\tau(x,s)ds$$

$$\ge C_1^{-1}\int_0^t \theta(x,s)e^{-a_2(t-s)}ds \ge C_1^{-1}\int_0^t \left(\frac{1}{2}r_1^2 - C_1V(s)\right)e^{-a_2(t-s)}ds$$

$$\ge \frac{r_1^2}{2a_2C_1}(1 - e^{-a_2t}) - C_1^{-1}\int_0^t V(s)e^{-a_2(t-s)}ds$$

$$\ge \frac{r_1^2}{4a_2C_1} > 0, \tag{2.2.21}$$

where we have used the fact that, as $t \to +\infty$,

$$\int_0^t V(s)e^{-a_2(t-s)}ds \le e^{-a_2t/2}\int_0^{+\infty} V(s)ds + \int_{\frac{t}{2}}^t V(s)ds \to 0.$$

On the other hand, we can also derive from Lemma 2.2.2 with $\delta = 0$ and (2.2.20) that for any $t \in [0, t_0]$,

$$\tau(x,t) \ge \frac{D(x,t)}{Z(t)} \ge C_1^{-1}e^{-a_2t} \ge C_1^{-1}e^{-a_2t_0} > 0,$$

which, together with (2.2.21), gives

$$\tau(x,t) \ge C_1^{-1} > 0, \quad \forall(x,t) \in [0,1] \times [0,+\infty). \tag{2.2.22}$$

Now let us prove that $\tau(x,t) \le C_1$. If we choose $0 < \varepsilon_0 < \frac{3Rr_1}{ar_2^4}$, then as $\overline{\tau}_0 \le \varepsilon_0$, we get

$$\frac{Rr_1}{\overline{\tau}_0} \ge \frac{Rr_1}{\varepsilon_0} > \frac{ar_2^4}{3} \ge \frac{a\theta_1^4}{3}. \tag{2.2.23}$$

Thanks to (2.2.23), we can pick $\delta > 0$ such that

$$\frac{Rr_1}{\overline{\tau}_0} \ge \frac{Rr_1}{\varepsilon_0} > \delta > \frac{ar_2^4}{3} \ge \frac{a\theta_1^4}{3},$$

which gives

$$\zeta \equiv a_1 - \frac{\delta}{\lambda} = \frac{1}{\lambda}[\lambda a_1 - \delta] = \frac{1}{\lambda}\left[\frac{Rr_1}{\bar{\tau}_0} - \delta\right] > 0, \qquad (2.2.24)$$

$$\delta_1 \equiv \frac{3}{a}\theta_1^{-4}\delta - 1 \geq \frac{3}{a}r_2^{-4}\delta - 1 \equiv \delta_0 > 0. \qquad (2.2.25)$$

Noting that for any ε, c_1, $c_2 > 0$, $c \in \mathbb{R}$, we have, by the Young inequality, that $c_1 c_2 \leq \varepsilon c_1^2 + c_2^2/(2\varepsilon)$ and also inequality $c^4 - 1 - \varepsilon \leq (c^2 - 1)^2(1 + \varepsilon^{-1})$, we derive from (2.2.24)–(2.2.25) that

$$\tau(\sigma_1 - \delta) = R\theta + \frac{\tau}{2}|\mathbf{b}|^2 + \tau\left(\frac{a}{3}\theta^4 - \delta\right), \qquad (2.2.26)$$

and

$$\tau\left(\frac{a}{3}\theta^4 - \delta\right) = \frac{a}{3}\theta_1^4\tau\left(\tilde{\theta}^4 - \frac{3}{a}\theta_1^{-4}\delta\right)$$

$$\leq \frac{a}{3}\theta_1^4\tau \max[\tilde{\theta}^4 - 1 - \delta_1, 0]$$

$$\leq \frac{a}{3}\theta_1^4\tau \max[\tilde{\theta}^4 - 1 - \delta_0, 0]$$

$$\leq \frac{a}{3}\theta_1^4\tau(\tilde{\theta}^2 - 1)^2(1 + \delta_0^{-1})$$

$$\leq C_1\tau(\tilde{\theta}^2 - 1)^2, \qquad (2.2.27)$$

where

$$\tilde{\theta} = \theta(x,t)/\theta_1, \quad \theta_1 \equiv \theta(b(t),t) = \int_0^1 \theta(x,t)dx \in [r_1, r_2].$$

Noting that $\tilde{\theta}(b(t),t)$ and using the Poincaré inequality, we have

$$(\tilde{\theta}^2 - 1)^2 \leq \left(\int_0^1 2|\tilde{\theta}\tilde{\theta}_x|dx\right)^2 \leq C_1\left(\int_0^1 |\theta\theta_x|dx\right)^2$$

$$\leq C_1\left(\int_0^1 \frac{\kappa\theta_x^2}{\tau\theta^2}dx\right)\left(\int_0^1 \frac{\tau\theta^4}{\kappa}dx\right)$$

$$\leq C_1 V(t)\int_0^1 \tau(1 + \theta^4)dx \leq C_1 V(t). \qquad (2.2.28)$$

By Lemma 2.2.1, we get

$$\|\mathbf{b}(t)\|_{L^\infty}^2 \leq C_1\left(\int_0^1 |\mathbf{b}_x|dx\right)^2 \leq C_1\left(\int_0^1 \frac{|\mathbf{b}_x|^2}{\tau\theta}dx\right)\left(\int_0^1 \tau\theta dx\right)$$

$$\leq C_1 V(t)\int_0^1 \tau(1 + \theta^4)dx \leq C_1 V(t),$$

which, together with (2.2.26)–(2.2.28), yields

$$\tau(\sigma_1 - \delta) \le R\theta + C_1 V(t)\tau, \quad \int_0^{+\infty} \|\mathbf{b}(t)\|_{L^\infty}^2 dt \le C_1. \tag{2.2.29}$$

Thus it follows from Lemma 2.2.2 that

$$\tau(x,t) \le C_1 + C_1 \int_0^t [1 + V(s) + \tau V(s)] \exp[-\zeta(t-s)]ds$$
$$\le C_1 + C_1 \int_0^t \tau V(s) \exp[-\zeta(t-s)]ds,$$

i.e.,

$$M(t) \le C_1 \exp(\zeta t) + C_1 \int_0^t V(s)M(s)ds, \tag{2.2.30}$$

with $M(t) = e^{\zeta t} \max_{x \in [0,1]} \tau(x,t) \equiv e^{\zeta t} M_\tau(t)$. Thus, by Gronwall's inequality,

$$M(t) \le C_1 \exp(\zeta t) \exp\left[C_1 \int_0^t V(s)ds\right] \le C_1 \exp(\zeta t),$$

i.e.,

$$M_\tau(t) \le C_1,$$

which, together with (2.2.22), gives (2.2.16). $\qquad\square$

Corollary 2.2.1. *Under the assumptions of Theorem 2.1.1, the following estimate holds:*

$$C_1^{-1} - C_1 V(t) \le \theta^{2m}(x,t) \le C_1 + C_1 V(t), \ 0 \le m \le \frac{q+4}{2}, \ \forall(x,t) \in [0,1] \times [0, +\infty). \tag{2.2.31}$$

Lemma 2.2.4. *Under the assumptions in Theorem 2.1.1, the following estimate holds for any $t > 0$:*

$$\|\mathbf{w}(t)\|_{H^1}^2 + \|\mathbf{b}(t)\|_{H^1}^2 + \int_0^t \left(\|\mathbf{w}_x\|^2 + \|\mathbf{b}_x\|^2\right.$$
$$\left. + \|\mathbf{w}_t\|^2 + \|\mathbf{b}_t\|^2 + \|\mathbf{b}\|_{L^\infty}^2 + \|\mathbf{w}\|_{L^\infty}^2\right)(s)ds \le C_1. \tag{2.2.32}$$

Proof. Multiplying (2.1.3) by $\mathbf{w}$, $\mathbf{w}_t$ and (2.1.4) by $\mathbf{b}$, $\mathbf{b}_t$, respectively, and then adding the resulting equalities, integrating over Q_t, and using Lemmas 2.2.1–2.2.3, we get

$$\|\mathbf{w}(t)\|_{H^1}^2 + \|\mathbf{b}(t)\|_{H^1}^2 + C_1 \int_0^t \left(\|\mathbf{w}_x\|^2 + \|\mathbf{b}_x\|^2 + \|\mathbf{w}_t\|^2 + \|\mathbf{b}_t\|^2\right)(s)ds$$
$$\le C_1 \int_0^t \int_0^1 \left|(\mathbf{w} \cdot \mathbf{b})_x + \mathbf{w}_x \cdot \mathbf{b}_t + \mathbf{b}_x \cdot \mathbf{w}_t + \left(\mathbf{b} \cdot \mathbf{b}_x + \mathbf{b}_t \cdot \mathbf{b}_x + \mathbf{b} \cdot \mathbf{b}_{tx}\right)u\right|(x,s)dxds$$

$$\leq \frac{C_1}{4} \int_0^t \left(\|\mathbf{w}_x\|^2 + \|\mathbf{b}_x\|^2 + \|\mathbf{w}_t\|^2 + \|\mathbf{b}_t\|^2 \right)(s)ds$$

$$+ C_1 \int_0^t \|u(s)\|_{L^\infty} \left| \int_0^1 \left(\mathbf{b} \cdot \mathbf{b}_x + \mathbf{b}_t \cdot \mathbf{b}_x + \mathbf{b} \cdot \mathbf{b}_{tx} \right) dx \right| ds$$

$$= \frac{C_1}{4} \int_0^t \left(\|\mathbf{w}_x\|^2 + \|\mathbf{b}_x\|^2 + \|\mathbf{w}_t\|^2 + \|\mathbf{b}_t\|^2 \right)(s)ds$$

$$+ C_1 \int_0^t \|u(s)\|_{L^\infty} \left| \int_0^1 \left(\mathbf{b} \cdot \mathbf{b}_x + \mathbf{b}_t \cdot \mathbf{b}_x - \mathbf{b}_t \cdot \mathbf{b}_x \right) dx \right| ds$$

$$\leq \frac{C_1}{2} \int_0^t \left(\|\mathbf{w}_x\|^2 + \|\mathbf{b}_x\|^2 + \|\mathbf{w}_t\|^2 + \|\mathbf{b}_t\|^2 \right)(s)ds$$

$$+ C_1 \int_0^t \|u(s)\|_{L^\infty} \left| \int_0^1 \mathbf{b} \cdot \mathbf{b}_x dx \right| ds$$

$$\leq \frac{C_1}{2} \int_0^t \left(\|\mathbf{w}_x\|^2 + \|\mathbf{b}_x\|^2 + \|\mathbf{w}_t\|^2 + \|\mathbf{b}_t\|^2 \right)(s)ds$$

$$+ C_1 \int_0^t \|\mathbf{b}(s)\|_{L^\infty}^2 \left(\|\mathbf{w}\|_{H^1}^2 + \|\mathbf{b}\|_{H^1}^2 \right)(s)ds,$$

i.e.,

$$\|\mathbf{w}(t)\|_{H^1}^2 + \|\mathbf{b}(t)\|_{H^1}^2 + \int_0^t \left(\|\mathbf{w}_x\|^2 + \|\mathbf{b}_x\|^2 + \|\mathbf{w}_t\|^2 + \|\mathbf{b}_t\|^2 \right)(s)ds$$

$$\leq C_1 + C_1 \int_0^t \|\mathbf{b}(s)\|_{L^\infty}^2 \left(\|\mathbf{w}\|_{H^1}^2 + \|\mathbf{b}\|_{H^1}^2 \right)(s)ds. \tag{2.2.33}$$

Applying the Gronwall inequality to (2.2.33) and using (2.2.29) and the Poincaré inequality, we get (2.2.32). $\qquad \square$

Lemma 2.2.5. *Under the assumptions of Theorem 2.1.1, the following estimate holds for any $t > 0$:*

$$\|\tau_x(t)\|^2 + \int_0^t \int_0^1 \left(\tau_x^2 + \theta \tau_x^2 \right)(x,s)dxds \leq C_1 + C_1 A^{q_0}, \tag{2.2.34}$$

with $A = \sup\limits_{0 \leq s \leq t} \|\theta(s)\|_{L^\infty}$ and $q_0 = \max(4 - q, 0)$.

Proof. Equation (2.1.2) can be rewritten as

$$\left(u - \frac{\lambda \tau_x}{\tau} \right)_t = \frac{R\theta \tau_x}{\tau^2} - \frac{R\tau_x}{\tau} - \frac{4a}{3}\theta^3 \theta_x - \mathbf{b} \cdot \mathbf{b}_x. \tag{2.2.35}$$

Multiplying (2.2.35) by $u - \frac{\lambda \tau_x}{\tau}$, and integrating the resulting equation over Q_t, we have

$$\frac{1}{2} \left\| u(t) - \frac{\lambda \tau_x}{\tau}(t) \right\|^2 + R\lambda \int_0^t \int_0^1 \frac{\theta \tau_x^2}{\tau^3}(x,s)dxds = \frac{1}{2} \left\| u_0 - \frac{\lambda \tau_{0x}}{\tau_0} \right\|^2 \tag{2.2.36}$$

$$+ \int_0^t \int_0^1 \left\{ R\frac{\theta\tau_x^2 u}{\tau^2} - \left[\left(\frac{R}{\tau} + \frac{4a}{3}\theta^3 \right) \theta_x - \mathbf{b} \cdot \mathbf{b}_x \right] \left(u - \frac{\lambda\tau_x}{\tau} \right) \right\} (x,s)dxds,$$

which gives

$$\|\tau_x(t)\|^2 + \int_0^t \int_0^1 \theta\tau_x^2(x,s)dxds$$

$$\leq C_1 + C_1 \int_0^t \int_0^1 \Big[|\theta\tau_x u| + |u\theta_x| + |\theta^3\theta_x u| + |\theta_x\tau_x| + |\theta^3\theta_x\tau_x|$$

$$+ |\mathbf{b} \cdot \mathbf{b}_x u| + |\mathbf{b} \cdot \mathbf{b}_x\tau_x| \Big](x,s)dxds. \tag{2.2.37}$$

Using Lemmas 2.2.1–2.2.4, we easily derive that for any $\varepsilon > 0$,

$$\int_0^t \int_0^1 |\theta\tau_x u|dxds \leq \varepsilon \int_0^t \int_0^1 \theta\tau_x^2 dxds + C_1(\varepsilon) \int_0^t \|u(s)\|_{L^\infty}^2 \left(\int_0^1 \theta dx \right) ds$$

$$\leq C_1(\varepsilon) + \varepsilon \int_0^t \int_0^1 \theta\tau_x^2 dxds, \tag{2.2.38}$$

$$\int_0^t \int_0^1 |\theta_x u|dxds \leq \left(\int_0^t \int_0^1 \frac{\kappa\theta_x^2}{\tau\theta^2}dxds \right)^{\frac{1}{2}} \left(\int_0^t \int_0^1 \frac{\tau\theta^2 u^2}{\kappa}dxds \right)^{\frac{1}{2}}$$

$$\leq C_1 \left(\int_0^t \|u(s)\|_{L^\infty}^2 \left(\int_0^1 \tau(1+\theta)^{2-q}dx \right) ds \right)^{\frac{1}{2}} \leq C_1, \tag{2.2.39}$$

$$\int_0^t \int_0^1 |\theta^3\theta_x u|dxds \leq \left(\int_0^t \int_0^1 \frac{\kappa\theta_x^2}{\tau\theta^2}dxds \right)^{\frac{1}{2}} \left(\int_0^t \int_0^1 \frac{\tau\theta^8 u^2}{\kappa}dxds \right)^{\frac{1}{2}}$$

$$\leq C_1 + A^{q_0}, \tag{2.2.40}$$

$$\int_0^t \int_0^1 |\theta_x\tau_x|dxds \leq \left(\int_0^t V(s)ds \right)^{\frac{1}{2}} \left(\int_0^t \int_0^1 \frac{\tau\theta^2\tau_x^2}{\kappa}dxds \right)^{\frac{1}{2}}$$

$$\leq C_1 \sup_{0\leq s\leq t} \left\| \frac{\tau\theta}{\kappa}(s) \right\|_{L^\infty}^{\frac{1}{2}} \left(\int_0^t \int_0^1 \theta\tau_x^2 dxds \right)^{\frac{1}{2}}$$

$$\leq \varepsilon \int_0^t \int_0^1 \theta\tau_x^2 dxds + C_1(\varepsilon) + C_1(\varepsilon)A^{\max(1-q,0)}, \tag{2.2.41}$$

$$\int_0^t \int_0^1 |(\theta^2 - \theta_1^2)\theta\theta_x\tau_x|dxds \leq C_1 \int_0^t \int_0^1 |\theta\theta_x|dxds \int_0^t \int_0^1 |\theta\theta_x\tau_x|dxds$$

$$\leq C_1 \sup_{0 \leq s \leq t} \left[\left(\int_0^1 \frac{\tau \theta^4}{\kappa} \, dx \right)^{\frac{1}{2}} \left(\int_0^1 \frac{\tau \theta^4 \tau_x^2}{\kappa} \, dx \right)^{\frac{1}{2}} \right] \int_0^t V(s) \, ds$$

$$\leq C_1 + C_1 \sup_{0 \leq s \leq t} \left[\left(\int_0^1 (1 + \theta^4) \tau dx \right)^{\frac{1}{2}} \left(\left\| \frac{\tau \theta^4}{\kappa}(s) \right\|_{L^\infty}^{\frac{1}{2}} \| \tau_x(s) \| \right) \right]$$

$$\leq \varepsilon \sup_{0 \leq s \leq t} \| \tau_x(s) \|^2 + C_1(\varepsilon) + C_1(\varepsilon) A^{q_0}, \qquad (2.2.42)$$

which gives

$$\int_0^t \int_0^1 |\theta^3 \theta_x \tau_x| dx ds \leq \int_0^t \int_0^1 |(\theta^2 - \theta_1^2) \theta \theta_x \tau_x| \, dx ds + \int_0^t \int_0^1 |\theta_1^2 \theta \theta_x \tau_x| dx ds$$

$$\leq \varepsilon \sup_{0 \leq s \leq t} \| \tau_x(s) \|^2 + C_1 + C_1 A^{q_0} + C_1 \left(\int_0^t V(s) \, ds \right)^{\frac{1}{2}} \left(\int_0^t \int_0^1 \frac{\tau \theta^4 \tau_x^2}{\kappa} \, dx ds \right)^{\frac{1}{2}}$$

$$\leq \varepsilon \sup_{0 \leq s \leq t} \| \tau_x(s) \|^2 + C_1 + C_1 A^{q_0} + C_1 \sup_{0 \leq s \leq t} \left\| \frac{\tau \theta^3}{\kappa}(s) \right\|_{L^\infty}^{\frac{1}{2}} \left(\int_0^t \int_0^1 \theta \tau_x^2 \, dx ds \right)^{\frac{1}{2}}$$

$$\leq \varepsilon \sup_{0 \leq s \leq t} \| \tau_x(s) \|^2 + C_1(\varepsilon) + C_1(\varepsilon) A^{q_0} + \varepsilon \int_0^t \int_0^1 \theta \tau_x^2 \, dx ds, \qquad (2.2.43)$$

$$\int_0^t \int_0^1 |\mathbf{b} \cdot \mathbf{b}_x u| dx ds \leq \int_0^t \| u(s) \|_{L^\infty} \left(\int_0^1 |\mathbf{b} \cdot \mathbf{b}_x| dx \right) ds$$

$$\leq C_1 + C_1 \int_0^t \int_0^1 |\mathbf{b}|^2 |\mathbf{b}_x|^2 dx ds$$

$$\leq C_1 + C_1 \int_0^t \| \mathbf{b}(s) \|_{L^\infty}^2 \| \mathbf{b}_x(s) \|^2 ds \leq C_1, \qquad (2.2.44)$$

$$\int_0^t \int_0^1 |\mathbf{b} \cdot \mathbf{b}_x \tau_x| dx ds \leq C_1(\varepsilon) + \varepsilon \int_0^t \int_0^1 \tau_x^2 \, dx ds. \qquad (2.2.45)$$

Inserting (2.2.38)–(2.2.45) into (2.2.37), we easily obtain (2.2.34). $\qquad \square$

Lemma 2.2.6. *Under the assumptions of Theorem 2.1.1, the following estimates hold for any $t > 0$:*

$$\int_0^t \int_0^1 (1 + \theta)^{2m} \tau_x^2 \, dx ds \leq C_1 + C_1 A^{q_0}, \qquad (2.2.46)$$

$$\int_0^t \int_0^1 (1 + \theta)^{2m} u^2 dx ds \leq C_1, \qquad (2.2.47)$$

$$\int_0^t \int_0^1 (1 + \theta)^{2m} \left(|\mathbf{w}_x|^2 + |\mathbf{b}_x|^2 \right) dx ds \leq C_1, \qquad (2.2.48)$$

where $0 \leq m \leq \frac{q+4}{2}$.

Proof. It follows from (2.2.33) and Lemma 2.2.5 that

$$\int_0^t \int_0^1 (1+\theta)^{2m}\tau_x^2 \, dxds \leq C_1 \int_0^t \|\tau_x(s)\|^2 ds + C_1 \int_0^t V(s)\|\tau_x(s)\|^2 ds \leq C_1 + C_1 A^{q_0}.$$

The proof of (2.2.47) and (2.2.48) is similar to that of (2.2.46). $\qquad\square$

Lemma 2.2.7. *Under the assumptions of Theorem 2.1.1, the following estimates hold for any $t > 0$:*

$$\int_0^t \|u_x(s)\|^2 ds \leq C_1 + C_1 A^{q_0}, \tag{2.2.49}$$

$$\|u_x(t)\|^2 + \int_0^t \|u_t(s)\|^2 ds \leq C_1 + C_1 A^{q_1}, \tag{2.2.50}$$

$$\int_0^t \|u_{xx}(s)\|^2 ds \leq C_1 + C_1 A^{q_2}, \tag{2.2.51}$$

where $q_1 = \max(8 - q, 0)$, $q_2 = \max(3q_0, q_1)$.

Proof. Multiplying (2.1.2) by u, u_t, u_{xx}, respectively, and then integrating the result over Q_t and using Lemmas 2.2.1–2.2.6, we get

$$\|u(t)\|^2 + \int_0^t \|u_x(s)\|^2 ds \leq C_1 + C_1 \left| \int_0^t \int_0^1 \left(\tau_x \theta u + (1+\theta^3)\theta_x u + |\mathbf{b} \cdot \mathbf{b}_x u| \right) dxds \right|$$

$$\leq C_1 + C_1 \int_0^t V(s)ds + C_1 \int_0^t \int_0^1 \left[\theta u^2 + \theta \tau_x^2 + \frac{\tau(1+\theta^3)^2\theta^2 u^2}{\kappa} + \mathbf{b} \cdot \mathbf{b}_x u \right] dxds$$

$$\leq C_1 + C_1 A^{q_0},$$

$$\|u_x(t)\|^2 + \int_0^t \|u_t(s)\|^2 ds \leq C_1 \int_0^t \int_0^1 \left(|p_x u_t| + |\mathbf{b} \cdot \mathbf{b}_x u_t| \right) dxds$$

$$\leq \frac{1}{2} \int_0^t \|u_t(s)\|^2 ds + C_1 \int_0^t \int_0^1 \left[(1+\theta^3)^2\theta_x^2 + \theta^2\tau_x^2 + |\mathbf{b}|^2|\mathbf{b}_x|^2 \right] dxds$$

$$\leq \frac{1}{2} \int_0^t \|u_t(s)\|^2 ds + C_1 + C_1 A^{q_1} + C_1 A^{q_0}$$

$$\leq \frac{1}{2} \int_0^t \|u_t(s)\|^2 ds + C_1 + C_1 A^{q_1},$$

which gives

$$\|u_x(t)\|^2 + \int_0^t \|u_t(s)\|^2 ds \leq C_1 + C_1 A^{q_1},$$

$$\|u_x(t)\|^2 + \int_0^t \|u_{xx}(s)\|^2 ds$$

$$\leq C_1 \int_0^t \int_0^1 \left(|(1+\theta^3)\theta_x u_{xx}| + |\theta \tau_x u_{xx}| + |u_x \tau_x u_{xx}| + |\mathbf{b} \cdot \mathbf{b}_x u_{xx}| \right) dx ds$$

$$\leq C_1(\varepsilon) + \varepsilon \int_0^t \|u_{xx}(s)\|^2 ds + C_1(\varepsilon) A^{q_1} + C_1(\varepsilon) A^{q_0}$$

$$+ C_1 A^{q_0} \left(\int_0^t \|u_x(s)\|^2 ds \right)^{\frac{1}{2}} \left(\int_0^t \|u_{xx}(s)\|^2 ds \right)^{\frac{1}{2}}$$

$$\leq C_1(\varepsilon) + \varepsilon \int_0^t \|u_{xx}(s)\|^2 ds + C_1(\varepsilon) A^{q_1} + C_1(\varepsilon) A^{3q_0}$$

$$\leq \varepsilon C_1(\varepsilon) + \int_0^t \|u_{xx}(s)\|^2 ds + C_1(\varepsilon) A^{q_2},$$

i.e., for ε small enough,

$$\|u_x(t)\|^2 + \int_0^t \|u_{xx}(s)\|^2 ds \leq C_1 + C_1 A^{q_2}. \qquad \square$$

Corollary 2.2.2. *Under the assumptions of Theorem 2.1.1, the following estimate holds:*

$$\int_0^t \int_0^1 (1+\theta)^{2m} u_x^2 \, dx ds \leq C_1(1+A)^{q_1}, \quad \forall t > 0, \ 0 \leq m \leq \frac{q+4}{2}. \qquad (2.2.52)$$

Proof. We easily derive from (2.2.33), (2.2.49) and (2.2.50) that

$$\int_0^t \int_0^1 (1+\theta)^{2m} u_x^2 \, dx ds \leq C_1 \int_0^t \int_0^1 u_x^2 \, dx ds + C_1 \int_0^t \int_0^1 V(s) u_x^2 \, dx ds$$

$$\leq C_1 + C_1 A^{q_0} + C_1 A^{q_1} \leq C_1 + C_1 A^{q_1},$$

which gives (2.2.52). $\qquad \square$

Lemma 2.2.8. *Under the assumptions of Theorem 2.1.1, the following estimates hold for any $t > 0$:*

$$\int_0^t \left(\|\mathbf{w}_{xx}\|^2 + \|\mathbf{b}_{xx}\|^2 \right)(s) ds \leq C_1 + C_1 A^{q_1}. \qquad (2.2.53)$$

Proof. Multiplying (2.1.3), (2.1.4) by $\mathbf{w}_{xx}$, $\mathbf{b}_{xx}$, respectively, and then integrating the results over Q_t and using Lemmas 2.2.1–2.2.7, we get for any $\varepsilon > 0$

$$\|\mathbf{w}_x(t)\|^2 + \|\mathbf{b}_x(t)\|^2 + \int_0^t \left(\|\mathbf{w}_{xx}\|^2 + \mathbf{b}_{xx}\|^2 \right)(s) ds$$

$$\leq C_1 + C_1 \left| \int_0^t \int_0^1 \left(\mathbf{w}_{xx} \cdot \mathbf{b}_x + \mathbf{w}_{xx} \cdot \mathbf{w}_x \tau_x + \mathbf{b}_{xx} \cdot \mathbf{w}_x \right. \right.$$

$$\left. \left. + \mathbf{b}_{xx} \cdot \mathbf{b}_x \tau_x + u_x \mathbf{b}_{xx} \cdot \mathbf{b} \right) dx ds \right|$$

$$\leq C_1 + \varepsilon \int_0^t \left(\|\mathbf{w}_{xx}\|^2 + \|\mathbf{b}_{xx}\|^2 \right)(s)ds + C_1(\varepsilon) \int_0^t \left(\|\mathbf{w}_x\|^2 + \|\mathbf{b}_x\|^2 \right)(s)ds$$

$$+ C_1(\varepsilon) \int_0^t \left(\|\tau_x \mathbf{w}_x\|^2 + \|\tau_x \mathbf{b}_x\|^2 + \|u_x \mathbf{b}\|^2 \right)(s)ds$$

$$\leq C_1 + \varepsilon \int_0^t \left(\|\mathbf{w}_{xx}\|^2 + \|\mathbf{b}_{xx}\|^2 \right)(s)ds$$

$$+ C_1(\varepsilon) \int_0^t \left(\left[\|\mathbf{w}_x\|_{L^\infty}^2 + \|\mathbf{b}_x\|_{L^\infty}^2 \right] \|\tau_x\|^2 + \|\mathbf{b}\|_{L^\infty}^2 \|u_x\|^2 \right)(s)ds$$

$$\leq C_1 + \varepsilon \int_0^t \left(\|\mathbf{w}_{xx}\|^2 + \|\mathbf{b}_{xx}\|^2 \right)(s)ds + C_1(\varepsilon)A^{2q_0}$$

$$+ C_1(\varepsilon)(1 + A^{q_1}) \int_0^t \|\mathbf{b}\|_{L^\infty}^2(s)ds$$

$$\leq C_1 + \varepsilon \int_0^t \left(\|\mathbf{w}_{xx}\|^2 + \|\mathbf{b}_{xx}\|^2 \right)(s)ds + C_1(\varepsilon)A^{2q_0} + C_1(\varepsilon)A^{q_1}$$

$$\leq C_1 + \varepsilon \int_0^t \left(\|\mathbf{w}_{xx}\|^2 + \|\mathbf{b}_{xx}\|^2 \right)(s)ds + C_1(\varepsilon)A^{q_1}.$$

Therefore, for $\varepsilon > 0$ small enough,

$$\int_0^t \left(\|\mathbf{w}_{xx}\|^2 + \|\mathbf{w}_{xx}\|^2 \right)(s)ds \leq C_1 + C_1 A^{q_1}. \qquad \square$$

Lemma 2.2.9. *Under the assumptions of Theorem 2.1.1, the following estimate holds for any $t > 0$:*

$$\|\theta + \theta^4\|^2 + \int_0^t \int_0^1 (1 + \theta)^{q+3} \theta_x^2 dxds \leq C_1 + C_1 A^{q_4}, \qquad (2.2.54)$$

where $q_4 = \max(q_3, 2q_0, q_1)$ and $q_3 = \max(7 - 2q, 0)$.

Proof. Multiplying (2.2.5) by e and integrating the resulting equation over Q_t, we have

$$\|\theta + \theta^4\|^2 + \int_0^t \int_0^1 (1 + \theta)^{q+3} \theta_x^2 dxds$$

$$\leq C_1 \int_0^t \int_0^1 \left| (pe)_x u + (u_x^2 + |\mathbf{w}_x|^2 + |\mathbf{b}_x|^2)e \right| dxds \qquad (2.2.55)$$

$$\leq C_1 \int_0^t \int_0^1 \left\{ (1 + \theta^7)|\theta_x u| + (1 + \theta^8)|\tau_x u| + (1 + \theta^4)\left[u_x^2 + |\mathbf{w}_x|^2 + |\mathbf{b}_x|^2 \right] \right\} dxds.$$

By Lemmas 2.2.1–2.2.8 and Corollaries 2.2.1–2.2.2, it holds that for any $\varepsilon > 0$,

$$\int_0^t \int_0^1 (1 + \theta^7)|\theta_x u| dxds$$

$$\leq \varepsilon \int_0^t \int_0^1 (1+\theta)^{q+3}\theta_x^2 \, dxds + C_1(\varepsilon) \int_0^t \int_0^1 (1+\theta)^{11-q} u^2 dxds$$

$$\leq \varepsilon \int_0^t \int_0^1 (1+\theta)^{q+3}\theta_x^2 \, dxds + C_1(\varepsilon) + C_1(\varepsilon)A^{q_3}, \tag{2.2.56}$$

$$\int_0^t \int_0^1 (1+\theta^8)|\tau_x u| dxds$$

$$\leq C_1 \int_0^t \int_0^1 (1+\theta)^8 \tau_x^2 \, dxds + C_1 \int_0^t \int_0^1 (1+\theta)^8 u^2 dxds$$

$$\leq C_1 A^{q_0} \int_0^t \int_0^1 (1+\theta)^{q+4}\tau_x^2 \, dxds + C_1 A^{q_0} \int_0^t \int_0^1 (1+\theta)^{q+4} u^2 dxds$$

$$\leq C_1 + C_1 A^{2q_0}, \tag{2.2.57}$$

$$\int_0^t \int_0^1 (1+\theta^4)\left|u_x^2 + |\mathbf{w}_x|^2 + |\mathbf{b}_x|^2\right| dxds$$

$$\leq \int_0^t \int_0^1 (1+\theta)^4 u_x^2 \, dxds + \int_0^t \int_0^1 (1+\theta)^4 \left(|\mathbf{w}_x|^2 + |\mathbf{b}_x|^2\right) dxds$$

$$\leq C_1 + C_1 A^{q_1}. \tag{2.2.58}$$

Substituting the estimates (2.2.56) to (2.2.58) into (2.2.55), we get (2.2.54).   □

Lemma 2.2.10. *Under the assumptions of Theorem 2.1.1, the following estimates hold for any $t > 0$:*

$$\int_0^t \int_0^1 (1+\theta)^{q+3}\theta_t^2 \, dxds + \int_0^1 (1+\theta)^{2q}\theta_x^2 \, dx \leq C_1 + C_1 A^{q_{12}}, \tag{2.2.59}$$

where

$$q_5 = \max(3q - 1, 0), \quad q_6 = \max(3 - q, 0),$$

$$q_7 = \max(q - 3, 0), \quad q_9 = \max(q - 2, 0),$$

$$q_8 = \max\Big\{ q_0 + \frac{3q_7 + q_4}{2} + \frac{q_0 + 2q_1 + q_2}{4},$$

$$q_0 + \frac{3q_7 + q_4}{2} + \frac{q_0 + q_1}{2}, 2q_0 + 3q_7 + q_4 + q_6 \Big\},$$

$$q_{10} = \max\Big\{ q_1, q_9 + \frac{q_0 + 2q_1 + q_2}{2}, q_9 + q_1 \Big\},$$

$$q_{11} = \max\Big\{ q_7 + \frac{q_0 + 2q_1 + q_2}{2}, q_7 + q_1 \Big\},$$

$$q_{12} = \max\Big\{ q_1 + 1, \frac{q_2 + q_4 + q_5}{2}, \frac{q_4 + q_5}{2} + \frac{q_1 + 3q_0}{4}, q_8, q_{10}, q_{11} \Big\}.$$

Proof. Let

$$H(x,t) = H(\tau, \theta) = \int_0^\theta \frac{\kappa(\tau, s)}{\tau} ds.$$

Then it is easy to verify that

$$H_t = H_\tau u_x + \frac{\kappa(x, \theta)\theta_t}{\tau},$$

$$H_{xt} = \left[\frac{\kappa(x, \theta)\theta_x}{\tau} \right]_t + H_\tau u_{xx} + H_{\tau\tau} u_x \tau_x + \left(\frac{\kappa}{\tau} \right)_\tau \tau_x \theta_t,$$

$$|H_\tau| + |H_{\tau\tau}| \le C_1 (1 + \theta)^{q+1}. \tag{2.2.60}$$

We rewrite (2.1.5) as

$$e_\theta \theta_t + \theta p_\theta u_x = \left(\frac{\kappa \theta_x}{\tau} \right)_x + \frac{\lambda u_x^2 + \mu |\mathbf{w}_x|^2 + \nu |\mathbf{b}_x|^2}{\tau}. \tag{2.2.61}$$

Multiplying (2.2.61) by H_t and integrating the resulting equation over Q_t, we obtain

$$\int_0^t \int_0^1 (e_\theta \theta_t + \theta p_\theta u_x) H_t dx ds + \int_0^t \int_0^1 \left(\frac{\kappa \theta_x}{\tau} \right) H_{tx} dx ds$$

$$= \int_0^t \int_0^1 \left(\frac{\lambda u_x^2 + \mu |\mathbf{w}_x|^2 + \nu |\mathbf{b}_x|^2}{\tau} \right) H_t dx ds. \tag{2.2.62}$$

Now we estimate each term in (2.2.62) by using Lemmas 2.2.1–2.2.9.

First we have, for any $\varepsilon > 0$, that

$$\int_0^t \int_0^1 e_\theta \theta_t H_\tau u_x dx ds \le C_1 \int_0^t \int_0^1 (1 + \theta)^{q+4} |\theta_t u_x| dx ds$$

$$\le \varepsilon \int_0^t \int_0^1 (1 + \theta)^{q+3} \theta_t^2 dx ds + C_1 \int_0^t \int_0^1 (1 + \theta)^{q+5} u_x^2 dx ds$$

$$\le \varepsilon \int_0^t \int_0^1 (1 + \theta)^{q+3} \theta_t^2 dx ds + C_1(\varepsilon) + C_1(\varepsilon) A^{q_1+1}, \tag{2.2.63}$$

$$\int_0^t \int_0^1 e_\theta \theta_t \frac{\kappa \theta_t}{\tau} dx ds \ge C_0 \int_0^t \int_0^1 (1 + \theta)^{q+3} \theta_t^2 dx ds, \tag{2.2.64}$$

$$\int_0^t \int_0^1 \theta p_\theta u_x H_\tau u_x dx ds \le C_1 \int_0^t \int_0^1 (1 + \theta)^{q+5} u_x^2 dx ds$$

$$\le C_1 + C_1 A^{q_1+1}, \tag{2.2.65}$$

$$\int_0^t \int_0^1 \theta p_\theta u_x \frac{\kappa \theta_t}{\tau} dx ds \le C_1 \int_0^t \int_0^1 (1 + \theta)^{q+4} |\theta_t u_x| dx ds \tag{2.2.66}$$

$$\le \varepsilon \int_0^t \int_0^1 (1+\theta)^{q+3}\theta_t^2 dxds + C_1(\varepsilon) + C_1(\varepsilon)A^{q_1+1},$$

$$\int_0^t \int_0^1 \frac{\kappa\theta_x}{\tau}\left(\frac{\kappa\theta_x}{\tau}\right)_t dxds = \frac{1}{2}\int_0^1 \left(\frac{\kappa\theta_x}{\tau}\right)^2 dx\Big|_0^t \ge C_1^{-1}\int_0^1 (1+\theta)^{2q}\theta_x^2 dx - C_1,$$

$$(2.2.67)$$

$$\int_0^t \int_0^1 \frac{\kappa\theta_t}{\tau}H_\tau u_{xx}dxds \le C_1 \int_0^t \int_0^1 (1+\theta)^{2q+1}|\theta_x u_{xx}|dxds$$

$$\le C_1 \left(\int_0^t \int_0^1 (1+\theta)^{q+3}\theta_x^2 dxds\right)^{\frac{1}{2}}\left(\int_0^t \int_0^1 (1+\theta)^{3q-1}u_{xx}^2 dxds\right)^{\frac{1}{2}}$$

$$\le C_1 + C_1 A^{\frac{q_2+q_4+q_5}{2}},$$

$$(2.2.68)$$

$$\int_0^t \int_0^1 \frac{\kappa\theta_x}{\tau}H_{\tau\tau}u_x\tau_x dxds \le C_1 \int_0^t \int_0^1 (1+\theta)^{2q+1}|\theta_x u_x\tau_x|dxds$$

$$\le C_1 \left(\int_0^t \int_0^1 (1+\theta)^{q+3}\theta_x^2 dxds\right)^{\frac{1}{2}}\left(\int_0^t \int_0^1 (1+\theta)^{3q-1}u_x^2\tau_x^2 dxds\right)^{\frac{1}{2}}$$

$$\le C_1 A^{\frac{q_4+q_5}{2}}\left(\int_0^t \|u_x(s)\|_{L^\infty}^2 \|\tau_x(s)\|^2 ds\right)^{\frac{1}{2}}$$

$$\le C_1 A^{\frac{q_4+q_5}{2}+\frac{3q_0+q_2}{4}},$$

$$(2.2.69)$$

$$\int_0^t \left\|\left(\frac{\kappa\theta_x}{\tau}\right)_x\right\|^2 ds \le C_1 \int_0^t \left(\|e_\theta\theta_t\|^2 + \|\theta p_\theta u_x\|^2 + \|u_x^2 + |\mathbf{w}_x|^2 + |\mathbf{b}_x|^2\|^2\right) ds$$

$$\le C_1 \int_0^t \int_0^1 \left[(1+\theta)^6\theta_t^2 + (1+\theta)^8 u_x^2 + u_x^4 + |\mathbf{w}_x|^4 + |\mathbf{b}_x|^4\right] dxds$$

$$\le C_1(1+A)^{3-q}\int_0^t \int_0^1 (1+\theta)^{q+3}\theta_t^2 dxds + (1+A)^{q_0}\int_0^t \int_0^1 (1+\theta)^{q+4}u_x^2 dxds$$

$$+ \int_0^t \|u_x(s)\|^3\|u_{xx}(s)\|ds + \int_0^t \|\mathbf{w}_x(s)\|^3\|\mathbf{w}_{xx}(s)\|ds$$

$$+ \int_0^t \|\mathbf{b}_x(s)\|^3\|\mathbf{b}_{xx}(s)\|ds$$

$$\le C_1(1+A)^{q_6}\int_0^t \int_0^1 (1+\theta)^{q+3}\theta_t^2 dxds + C_1(1+A)^{q_0+q_1}$$

$$+ C_1(1+A)^{\frac{q_0+2q_1+q_2}{2}} + C_1(1+A)^{q_1},$$

$$(2.2.70)$$

$$\int_0^t \int_0^1 \frac{\kappa\theta_x}{\tau}\left(\frac{\kappa}{\tau}\right)_\tau \tau_x\theta_t dxds \le C_1 \int_0^t \int_0^1 (1+\theta)^{2q}|\theta_x\theta_t\tau_x|dxds$$

$$\leq \varepsilon \int_0^t \int_0^1 (1+\theta)^{q+3}\theta_t^2 \, dxds + C_1 \int_0^t \int_0^1 \frac{(1+\theta)^{3q-3}}{\kappa^2}\left(\frac{\kappa\theta_x}{\tau}\right)^2 \tau_x^2 \, dxds$$

$$\leq \varepsilon \int_0^t \int_0^1 (1+\theta)^{q+3}\theta_t^2 \, dxds + C_1(1+A)^{q_7}\int_0^t \left\|\frac{\kappa\theta_x}{\tau}(s)\right\|_{L^\infty}^2 \|\tau_x(s)\|^2 \, ds$$

$$\leq \varepsilon \int_0^t \int_0^1 (1+\theta)^{q+3}\theta_t^2 \, dxds + C_1(1+A)^{q_0+q_7}\int_0^t \left\|\frac{\kappa\theta_x}{\tau}(s)\right\|\left\|\left(\frac{\kappa\theta_x}{\tau}\right)_x(s)\right\| \, ds$$

$$\leq \varepsilon \int_0^t \int_0^1 (1+\theta)^{q+3}\theta_t^2 \, dxds + C_1(1+A)^{q_0+q_7}\left(\int_0^t \left\|\frac{\kappa\theta_x}{\tau}(s)\right\|^2 \, ds\right)^{\frac{1}{2}}$$

$$\times \left(\int_0^t \left\|\left(\frac{\kappa\theta_x}{\tau}\right)_x(s)\right\|^2 \, ds\right)^{\frac{1}{2}}$$

$$\leq \varepsilon \int_0^t \int_0^1 (1+\theta)^{q+3}\theta_t^2 \, dxds + C_1(1+A)^{q_0+\frac{3q_7}{2}}\left(\int_0^t \int_0^1 (1+\theta)^{q+3}\theta_x^2 \, dxds\right)^{\frac{1}{2}}$$

$$\times \left(\int_0^t \left\|\left(\frac{\kappa\theta_x}{\tau}\right)_x(s)\right\|^2 \, ds\right)^{\frac{1}{2}}$$

$$\leq \varepsilon \int_0^t \int_0^1 (1+\theta)^{q+3}\theta_t^2 \, dxds + C_1 + C_1 A^{q_0+\frac{3q_7+q_4}{2}+\frac{q_0+2q_1+q_2}{4}}$$

$$+ C_1 A^{q_0+\frac{3q_7+q_4}{2}+\frac{q_1+q_0}{2}} + C_1 A^{2q_0+3q_7+q_4+q_6}$$

$$\leq \varepsilon \int_0^t \int_0^1 (1+\theta)^{q+3}\theta_t^2 \, dxds + C_1(\varepsilon) + C_1 A^{q_8}, \tag{2.2.71}$$

$$\int_0^t \int_0^1 \frac{\lambda u_x^2 + \mu|\mathbf{w}_x|^2 + \nu|\mathbf{b}_x|^2}{\tau} H_\tau u_x \, dxds$$

$$\leq C_1 \int_0^t \int_0^1 (1+\theta)^{q+1}|u_x|\frac{\lambda u_x^2 + \mu|\mathbf{w}_x|^2 + \nu|\mathbf{b}_x|^2}{\tau} \, dxds$$

$$\leq C_1 \int_0^t \int_0^1 (1+\theta)^{q+4}u_x^2 \, dxds$$

$$+ C_1 \int_0^t \int_0^1 (1+\theta)^{q-2}\left(\lambda u_x^4 + \mu|\mathbf{w}_x|^4 + \nu|\mathbf{w}_x|^4\right) \, dxds$$

$$\leq C_1(1+A)^{q_1} + C_1(1+A)^{q_9+\frac{q_0+2q_1+q_2}{2}} + C_1(1+A)^{q_9+q_1}$$

$$\leq C_1 + C_1(1+A)^{q_{10}}, \tag{2.2.72}$$

$$\int_0^t \int_0^1 \frac{\lambda u_x^2 + \mu|\mathbf{w}_x|^2 + \nu|\mathbf{b}_x|^2}{\tau}\frac{\kappa\theta_t}{\tau} \, dxds$$

$$\leq \varepsilon \int_0^t \int_0^1 (1+\theta)^{q+3}\theta_t^2 \, dxds$$

$$+ C_1 \int_0^t \int_0^1 (1 + \theta)^{q-3} \left(\lambda u_x^4 + \mu |\mathbf{w}_x|^4 + \nu |\mathbf{b}_x|^4 \right) dx ds$$

$$\leq \varepsilon \int_0^t \int_0^1 (1 + \theta)^{q+3} \theta_t^2 dx ds + C_1 (1 + A)^{q_7 + \frac{q_0 + 2q_1 + q_2}{2}} + C_1 (1 + A)^{q_7 + q_1}$$

$$\leq \varepsilon \int_0^t \int_0^1 (1 + \theta)^{q+3} \theta_t^2 dx ds + C_1(\varepsilon) + C_1(\varepsilon)(1 + A)^{q_{11}}. \tag{2.2.73}$$

Therefore estimate (2.2.59) follows from (2.2.62)–(2.2.73) for $\varepsilon > 0$ small enough. $\qquad\square$

Lemma 2.2.11. *Under the assumptions of Theorem 2.1.1, the following estimate holds for any $t > 0$:*

$$\|\theta(t)\|_{L^\infty} \leq C_1. \tag{2.2.74}$$

Proof. By Lemmas 2.2.1–2.2.10,

$$|\theta^{q+3} - \theta_1^{q+3}| \leq C_1 \int_0^1 |\theta^{q+2} \theta_x| dx$$

$$\leq C_1 \left(\int_0^1 |\theta^{2q} \theta_x^2| dx \right)^{\frac{1}{2}} \left(\int_0^1 |\theta^4| dx \right)^{\frac{1}{2}}$$

$$\leq C_1 (1 + A)^{\frac{q_{12}}{2}},$$

which gives

$$A^{q+3} \leq C_1 + C_1 A^{\frac{q_{12}}{2}}.$$

It is easy to verify that $q_{12} < 2q + 6$ if $q > 2$. Therefore, by the Young inequality, we obtain $A \leq C_1$, i.e., $\|\theta(t)\|_{L^\infty} \leq C_1$. $\qquad\square$

Lemma 2.2.12. *Under the assumptions of Theorem 2.1.1, the following estimates hold for any $t > 0$:*

$$\frac{d}{dt} \|\tau_x(t)\|^2 \leq C_1 \left(\|\tau_x(t)\|^2 + \|u_{xx}(t)\|^2 \right), \tag{2.2.75}$$

$$\frac{d}{dt} \|u_x(t)\|^2 \leq C_1 \left(\|\theta_x(t)\|^2 + \|\tau_x(t)\|^2 + \|\mathbf{b}(t) \cdot \mathbf{b}_x(t)\|^2 + \|u_x(t)\|^2 + \|u_{xx}(t)\|^2 \right), \tag{2.2.76}$$

$$\frac{d}{dt} \|\mathbf{w}_x(t)\|^2 \leq C_1 \left(\|\tau_x(t)\|^2 + \|\mathbf{b}_x(t)\|^2 + \|\mathbf{w}_{xx}(t)\|^2 \right), \tag{2.2.77}$$

$$\frac{d}{dt} \|\mathbf{b}_x(t)\|^2 \leq C_1 \left(\|\tau_x(t)\|^2 + \|\mathbf{w}_x(t)\|^2 + \|\mathbf{b}_{xx}(t)\|^2 \right), \tag{2.2.78}$$

$$\frac{d}{dt} \|\theta_x(t)\|^2 + \int_0^1 (1 + \theta)^{q+3} \theta_{xx}^2 dx$$

$$\leq C_1 \left(\|\theta_x(t)\|^2 + \|u_{xx}(t)\|^2 + \|\mathbf{w}_{xx}(t)\|^2 + \|\mathbf{b}_{xx}(t)\|^2 \right). \tag{2.2.79}$$

Proof. Differentiating (2.1.1) with respect to x and multiplying the resulting equation by τ_x, we obtain (2.2.75).

Similarly, multiplying (2.1.2)–(2.1.4) by u_{xx}, $\mathbf{w}_{xx}$, $\mathbf{b}_{xx}$, respectively, and then integrating the resulting equation over $[0, 1]$, we get estimates (2.2.76)–(2.2.78).

Multiplying (2.2.61) by $e_\theta^{-1} \theta_{xx}$ and integrating the resulting equation on $[0, 1]$, we deduce

$$\frac{d}{dt}\|\theta_x(t)\|^2 + 2\int_0^1 \frac{\kappa\theta_{xx}^2}{\tau e_\theta} dx$$

$$= \int_0^1 \left(\frac{\theta p_\theta u_x}{e_\theta} - \frac{\lambda u_x^2}{\tau e_\theta} - \frac{\kappa_x \theta_x}{\tau e_\theta} + \frac{\kappa\theta_x \tau_x \theta_{xx}}{\tau^2 e_\theta} - \frac{\mu \mathbf{w}_x^2}{\tau e_\theta} - \frac{\nu \mathbf{b}_x^2}{\tau e_\theta} \right) \theta_{xx} dx$$

$$\leq \varepsilon \|\theta_{xx}(t)\|^2 + C_1(\varepsilon)\Big(\|u_x(t)\|^2 + \|u_x(t)\|_{L^4}^4 + \|\theta_x(t)\|_{L^4}^4 + \|\mathbf{w}_x(t)\|_{L^4}^4$$

$$+ \|\mathbf{b}_x(t)\|_{L^4}^4 + \|\tau_x \theta_x(t)\|^2 \Big)$$

$$\leq 2\varepsilon\|\theta_{xx}(t)\|^2 + C_1(\varepsilon)\Big(\|u_x(t)\|^2 + \|u_{xx}(t)\|^2 + \|\theta_x(t)\|^2$$

$$+ \|\mathbf{w}_{xx}(t)\|^2 + \|\mathbf{b}_{xx}(t)\|^2 \Big),$$

which, by taking $\varepsilon > 0$ small enough, gives (2.2.79). $\qquad\square$

Lemma 2.2.13. *Under the assumptions of Theorem 2.1.1, the following estimates hold as $t \to +\infty$:*

$$\|\tau(t) - \bar\tau\|_{H^1} \to 0, \quad \|u(t)\| \to 0, \quad \|u(t)\|_{L^\infty} \to 0, \qquad (2.2.80)$$

$$\|\mathbf{w}(t)\| \to 0, \quad \|\mathbf{w}(t)\|_{L^\infty} \to 0, \quad \|\mathbf{b}(t)\| \to 0, \quad \|\mathbf{b}(t)\|_{L^\infty} \to 0, \qquad (2.2.81)$$

$$\|\theta(t) - \bar\theta\|_{H^1} \to 0, \quad \|\theta_x(t)\| \to 0, \quad \|\theta(t) - \bar\theta\|_{L^\infty} \to 0, \qquad (2.2.82)$$

$$0 < C_1^{-1} \leq \theta(x, t) \leq C_1, \quad \forall(x, t) \in [0, 1] \times [0, +\infty). \qquad (2.2.83)$$

Proof. By using Lemmas 2.2.1–2.2.12, we have

$$\int_0^t \left(\|u_x\|^2 + \|\mathbf{w}_x\|^2 + \|\mathbf{b}_x\|^2 + \|\tau_x\|^2 + \|\theta_x\|^2 \right)(s)ds \leq C_1 \qquad (2.2.84)$$

and

$$\int_0^t \left(|\frac{d}{dt}\|u_x\|^2| + |\frac{d}{dt}\|\mathbf{w}_x\|^2| + |\frac{d}{dt}\|\mathbf{b}_x\|^2| + |\frac{d}{dt}\|\tau_x\|^2| + |\frac{d}{dt}\|\theta_x\|^2| \right)(s)ds \leq C_1 \qquad (2.2.85)$$

which yield (2.2.80)–(2.2.82).

We derive from (2.2.82) that there exists a large time $t_0 > 0$ such that

$$\theta(x, t) \geq \frac{1}{2}\bar\theta > 0, \quad \forall t \geq t_0. \qquad (2.2.86)$$

On the other hand, if we put $\omega := \frac{1}{\theta}$, (2.2.61) becomes

$$e_\theta \omega_t = \left(\frac{\kappa \omega_x}{\tau} \right)_x + \frac{\tau p_\theta^2}{4\lambda} - \left[\frac{2\kappa \omega_x^2}{\tau \omega} + \frac{\omega^2}{\tau} \left(\mu |\mathbf{w}_x|^2 + \nu |\mathbf{b}_x|^2 \right) + \frac{\lambda \omega^2}{\tau} \left(u_x - \frac{\tau p_\theta}{2\lambda \omega} \right)^2 \right]$$

which, along with (2.2.16) and (2.2.74), implies that there exists a positive constant C_1 such that

$$\omega_t \leq \frac{1}{e_\theta} \left(\frac{\kappa \omega_x}{\tau} \right)_x + C_1.$$

Defining $\tilde{\omega}(x,t) := C_1 t + \max_{[0,1]} \frac{1}{\theta_0(x)} - \omega(x,t)$ and introducing the parabolic operator $\mathcal{L} := -\frac{\partial}{\partial t} + \frac{1}{e_\theta} \frac{\partial}{\partial x} \left(\frac{\kappa}{\tau} \frac{\partial}{\partial x} \right)$, we have a system

$$\mathcal{L}\tilde{\omega} \leq 0, \quad \text{on} \quad Q_T = [0,1] \times [0, t_0 + 1],$$
$$\tilde{\omega}|_{t=0} \geq 0, \quad \text{on} \quad [0,1],$$
$$\tilde{\omega}_x|_{x=0,1} = 0, \quad \text{on} \quad [0, t_0 + 1].$$

The standard comparison argument implies

$$\min_{(x,t) \in \overline{Q}_T} \tilde{\omega}(x,t) \geq 0$$

which gives for any $(x,t) \in \overline{Q}_T$,

$$\theta(x,t) \geq \left(C_1 t + \max_{x \in [0,1]} \frac{1}{\theta_0(x)} \right)^{-1}.$$

Thus,

$$\theta(x,t) \geq \left(C_1 t_0 + \max_{x \in [0,1]} \frac{1}{\theta_0(x)} \right)^{-1} \geq C_1^{-1}, \quad 0 \leq t \leq t_0$$

which together with (2.2.86) and (2.2.74) gives (2.2.83).   □

In what follows we shall prove the exponential stability of the solution in H_+^1. If we now introduce the flow density $\rho = \frac{1}{\tau}$, then we easily get that the specific entropy

$$\eta = \eta(\tau, \theta) = \eta(\rho, \theta) = R \log \tau + \frac{4a}{3} \tau \theta^3 + C_v \log \theta, \qquad (2.2.87)$$

satisfies

$$\frac{\partial \eta}{\partial \rho} = -\frac{p_\theta}{\rho^2}, \quad \frac{\partial \eta}{\partial \theta} = \frac{e_\theta}{\theta}, \qquad (2.2.88)$$

with $p = R\rho\theta + \frac{a}{3}\theta^4$ and $e = C_v \theta + \frac{a\theta^4}{\rho}$.

We consider the transform

$$\mathcal{A} : \mathcal{D}_{\rho,\theta} = \{(\rho,\theta) : \rho > 0, \ \theta > 0\} \to \mathcal{A}\mathcal{D}_{\rho,\theta}, \quad (\rho,\theta) \mapsto (\tau,\eta),$$

where $\tau = 1/\rho$ and $\eta = \eta(1/\rho, \theta)$. Since the Jacobian

$$\frac{\partial(\tau, \eta)}{\partial(\rho, \theta)} = -\frac{e_\theta}{\rho^2 \theta} = -\frac{1}{\rho^2}\left(C_v \theta^{-1} + 4a\rho^{-1}\theta^2\right) < 0,$$

there is a unique inverse function $(\tau, \eta) \mapsto (\rho, \theta)$. Thus e and p can be regarded as the smooth functions of (τ, η). We denote by

$$e = e(\tau, \eta) := e(\tau, \theta(\tau, \eta)) = e(\rho^{-1}, \theta),$$
$$p = p(\tau, \eta) := p(\tau, \theta(\tau, \eta)) = p(\rho^{-1}, \theta).$$

Let

$$\mathcal{E} = \frac{1}{2}(u^2 + |\mathbf{w}|^2 + \tau|\mathbf{b}|^2) + e(\tau, \eta) - e(\overline{\tau}, \overline{\eta}) - \frac{\partial e}{\partial \tau}(\overline{\tau}, \overline{\eta})(\tau - \overline{\tau}) - \frac{\partial e}{\partial \eta}(\overline{\tau}, \overline{\eta})(\eta - \overline{\eta}),$$

$$\text{(2.2.89)}$$

where $\overline{\tau} = \int_0^1 \tau_0 \, dx$ and $\overline{\theta} > 0$ is determined by

$$e(\overline{\tau}, \overline{\theta}) = e(\overline{\tau}, \overline{\eta}) = \int_0^1 \left(\frac{1}{2}(u_0^2 + |\mathbf{w}_0|^2 + \tau_0|\mathbf{b}_0|^2) + e(\tau_0, \theta_0)\right) dx \equiv E_0, \quad \text{(2.2.90)}$$

and

$$\overline{\eta} = \eta(\overline{\tau}, \overline{\theta}).$$

Lemma 2.2.14. *Under the assumptions of Theorem 2.1.1, the following estimate holds:*

$$\frac{1}{2}(u^2 + |\mathbf{w}|^2 + \tau|\mathbf{b}|^2) + C_1^{-1}(|\tau - \overline{\tau}|^2 + |\eta - \overline{\eta}|^2) \leq \mathcal{E}(\tau, u, \mathbf{w}, \mathbf{b}, \theta)$$

$$\leq \frac{1}{2}(u^2 + |\mathbf{w}|^2 + \tau|\mathbf{b}|^2) + C_1(|\tau - \overline{\tau}|^2 + |\eta - \overline{\eta}|^2). \quad \text{(2.2.91)}$$

Proof. By the mean value theorem, there exists a point $(\tilde{\tau}, \tilde{\eta})$ between (τ, η) and $(\overline{\tau}, \overline{\eta})$ such that

$$\mathcal{E}(\tau, u, \mathbf{w}, \mathbf{b}, \theta) = \frac{1}{2}(u^2 + |\mathbf{w}|^2 + \tau|\mathbf{b}|^2) + \frac{1}{2}\left[\frac{\partial^2 e}{\partial \tau^2}(\tilde{\tau}, \tilde{\eta})(\tau - \overline{\tau})^2 + \frac{\partial^2 e}{\partial \eta^2}(\tilde{\tau}, \tilde{\eta})(\eta - \overline{\eta})^2\right.$$

$$\left. + \frac{\partial^2 e}{\partial \tau \partial \eta}(\tilde{\tau}, \tilde{\eta})(\tau - \overline{\tau})(\eta - \overline{\eta})\right], \quad \text{(2.2.92)}$$

where $\tilde{\tau} = \lambda_0 \overline{\tau} + (1 - \lambda_0)\tau$, $\tilde{\eta} = \lambda_0 \overline{\eta} + (1 - \lambda_0)\eta$, $0 \leq \lambda_0 \leq 1$.

It follows from Lemmas 2.2.1–2.2.13 that

$$\left|\frac{\partial^2 e}{\partial \tau^2}(\tilde{\tau}, \tilde{\eta})\right|^2 + \left|\frac{\partial^2 e}{\partial \tau \partial \eta}(\tilde{\tau}, \tilde{\eta})\right|^2 + \left|\frac{\partial^2 e}{\partial \eta^2}(\tilde{\tau}, \tilde{\eta})\right|^2 \leq C_1. \quad \text{(2.2.93)}$$

Thus, by (2.2.92), (2.2.93) and the Cauchy inequality, we get

$$\mathcal{E}(\tau, u, \mathbf{w}, \mathbf{b}, \theta) \leq \frac{1}{2}(u^2 + |\mathbf{w}|^2 + \tau|\mathbf{b}|^2) + C_1(|\tau - \overline{\tau}|^2 + |\eta - \overline{\eta}|^2). \quad \text{(2.2.94)}$$

On the other hand, from (2.1.6) it follows that

$$e_{\tau\tau} = -p_\tau = \rho^2 p_\rho + \frac{\theta p_\theta^2}{e_\theta}, \quad e_{\tau\eta} = -p_\eta = \theta_\tau = \frac{\theta p_\theta}{e_\theta}, \quad e_{\eta\eta} = \theta_\eta = \frac{\theta}{e_\theta},$$

which implies that the Hessian of $e(\tau, \theta)$ is positive definite for any $\tau > 0$ and $\theta > 0$. Thus we infer from (2.2.93) that

$$\mathcal{E}(\tau, u, \mathbf{w}, \mathbf{b}, \theta) \geq \frac{1}{2}(u^2 + |\mathbf{w}|^2 + \tau|\mathbf{b}|^2) + C_1^{-1}(|\tau - \overline{\tau}|^2 + |\eta - \overline{\eta}|^2),$$

which together with (2.2.94) gives (2.2.91).   □

The next lemma is crucial in proving the exponential stability of solutions in H_+^i ($i = 1, 2, 4$).

Lemma 2.2.15. *Under assumptions of Theorem 2.1.1, there are positive constants $C_1 > 0$ and $\gamma_1' = \gamma_1'(C_1) < \gamma_0/2$ such that, for any fixed $\gamma \in (0, \gamma_1']$, there holds for any $t > 0$,*

$$e^{\gamma t}\left(\|\tau(t) - \overline{\tau}\|^2 + \|u(t)\|^2 + \|\mathbf{w}(t)\|^2 + \|\mathbf{b}(t)\|^2 + \|\theta(t) - \overline{\theta}\|^2 + \|\tau_x(t)\|^2 + \|\rho_x(t)\|^2\right)$$

$$+ \int_0^t e^{\gamma s}\left(\|\rho_x\|^2 + \|u_x\|^2 + \|\mathbf{w}_x\|^2 + \|\mathbf{b}_x\|^2 + \|\theta_x\|^2 + \|\tau_x\|^2\right)(s)ds \leq C_1.$$

$$(2.2.95)$$

Proof. Using equations (2.1.1)–(2.1.5), it is easy to verify that the following relations:

$$\left(e + \frac{1}{2}(u^2 + |\mathbf{w}|^2 + \tau|\mathbf{b}|^2)\right)_t$$

$$= \left(\lambda \rho u u_x + \mu \rho \mathbf{w} \cdot \mathbf{w}_x + \nu \rho \mathbf{b} \cdot \mathbf{b}_x + \mathbf{w} \cdot \mathbf{b} - u\left(p + \frac{1}{2}|\mathbf{b}|^2\right)\right)_x, \quad (2.2.96)$$

$$\eta_t = \left(\frac{\kappa \rho \theta_x}{\theta}\right)_x + \kappa \rho \left(\frac{\theta_x}{\theta}\right)_x + \frac{\lambda \rho u_x^2 + \mu \rho |\mathbf{w}_x|^2 + \nu \rho |\mathbf{b}_x|^2}{\theta}. \quad (2.2.97)$$

Since $\overline{\tau}_t = 0$, $\overline{\theta}_t = 0$, we infer from (2.2.96) and (2.2.97) that

$$\mathcal{E}_t + \frac{\overline{\theta}}{\theta}\left[\lambda \rho u_x^2 + \mu \rho |\mathbf{w}_x|^2 + \nu \rho |\mathbf{b}_x|^2 + \kappa \rho \theta_x^2/\theta\right] \quad (2.2.98)$$

$$= \left[\lambda \rho u u_x + \mu \rho \mathbf{w} \cdot \mathbf{w}_x + \nu \rho \mathbf{b} \cdot \mathbf{b}_x + \kappa\left(1 - \frac{\overline{\theta}}{\theta}\right)\rho \theta_x + \mathbf{w} \cdot \mathbf{b} - u(p - \overline{p}) - \frac{1}{2}|\mathbf{b}|^2\right]_x,$$

$$\left[\frac{\lambda^2}{2}\left(\frac{\rho_x}{\rho}\right)^2 + \frac{\lambda \rho_x u}{\rho} - \frac{|\mathbf{w}|^2}{2} - \frac{\tau}{2}|\mathbf{b}|^2\right]_t + \frac{\lambda \rho_x^2 p_\rho}{\rho} + (\mathbf{w} \cdot \mathbf{b})_x$$

$$= -\lambda(\rho u u_x)_x + \lambda\rho u_x^2 - \mu(\rho\mathbf{w}\cdot\mathbf{w}_x)_x + \mu\rho|\mathbf{w}_x|^2 - \nu(\rho\,\mathbf{b}\cdot\,\mathbf{b}_x)_x + \nu\rho|\mathbf{b}_x|^2$$
$$- \frac{\lambda p_\theta \rho_x \theta_x}{\rho} - \frac{\lambda\rho_x\,\mathbf{b}\cdot\,\mathbf{b}_x}{\rho} + \frac{1}{2}u_x|\mathbf{b}|^2. \tag{2.2.99}$$

Let

$$G(t) = e^{\gamma t}\left[\mathcal{E}(t) + \beta\left(\frac{\lambda^2\rho_x^2}{2\rho^2} + \frac{\lambda\rho_x u}{\rho} - \frac{|\mathbf{w}|^2}{2} - \frac{\tau}{2}|\mathbf{b}|^2\right)\right].$$

Multiplying (2.2.98) and (2.2.99) by $e^{\gamma t}$ and $\beta e^{\gamma t}$, respectively, and adding the results, we get

$$\frac{\partial}{\partial t}G(t) + e^{\gamma t}\left[\frac{\overline{\theta}}{\theta}(\lambda\rho u_x^2 + \mu\rho|\mathbf{w}_x|^2 + \nu\rho|\mathbf{b}_x|^2 + \kappa\rho\theta_x^2/\theta)\right.$$
$$+ \beta\left(\frac{\lambda\rho_x^2 p_\rho}{\rho} - \lambda\rho u_x^2 - \mu\rho|\mathbf{w}_x|^2 - \nu\rho|\mathbf{b}_x|^2 + \frac{\lambda p_\theta\rho_x\theta_x}{\rho} - \frac{1}{2}u_x|\mathbf{b}|^2 + \frac{\lambda\rho_x\mathbf{b}\cdot\,\mathbf{b}_x}{\rho}\right)\right]$$
$$= \gamma e^{\gamma t}\left[\mathcal{E} + \frac{\beta}{2}\left(\frac{\lambda^2\rho_x^2}{\rho^2} - |\mathbf{w}|^2 - \tau|\mathbf{b}|^2\right) + \frac{\beta\lambda u\rho_x}{\rho}\right]$$
$$+ e^{\gamma t}\left[(1-\beta)(\lambda\rho u u_x + \mu\rho\mathbf{w}\cdot\mathbf{w}_x + \nu\rho\mathbf{b}\cdot\mathbf{b}_x)\right.$$
$$\left.+ \kappa\left(1 - \frac{\overline{\theta}}{\theta}\right)\rho\theta_x - u(p - \overline{p}) - \frac{1}{2}|\mathbf{b}|^2 + (1-\beta)\mathbf{w}\cdot\mathbf{b}\right]_x. \tag{2.2.100}$$

For the boundary conditions (2.1.8), integrating (2.1.5) over $(0,1)$, we have

$$\int_0^1\left[e(\tau,\theta) + \frac{1}{2}(u^2 + |\mathbf{w}|^2 + \tau|\mathbf{b}|^2)\right]dx$$
$$= \int_0^1\left[e_0(\tau_0,\theta_0) + \frac{1}{2}(u_0^2 + |\mathbf{w}_0|^2 + \tau_0|\mathbf{b}_0|^2)\right]dx := e(\overline{\tau},\overline{\theta}),$$

which, together with the Poincaré inequality and the mean value theorem, implies

$$\|e(\tau,\theta) - e(\overline{\tau},\overline{\theta})\| \leq \|e(\tau,\theta) - \int_0^1 e(\tau,\theta)dx\| + \frac{\|u(t)\|^2 + \|\mathbf{w}(t)\|^2 + \|\sqrt{\tau}\mathbf{b}(t)\|^2}{2}$$
$$\leq C_1(\|e_x(t)\| + \|u_x(t)\| + \|\mathbf{w}_x(t)\| + \|\mathbf{b}_x(t)\|)$$
$$\leq C_1(\|\theta_x(t)\| + \|u_x(t)\| + \|\mathbf{w}_x(t)\| + \|\mathbf{b}_x(t)\| + \|\rho_x(t)\|). \tag{2.2.101}$$

On the other hand, by (2.1.1), Lemmas 2.2.1–2.2.14, the mean value theorem and the Poincaré inequality, we have

$$\|\tau(t) - \overline{\tau}\| \leq C_1\|\tau_x(t)\|,$$
$$\|\theta(t) - \overline{\theta}\| \leq C_1(\|e(\tau,\theta) - e(\overline{\tau},\overline{\theta})\| + \|\tau(t) - \overline{\tau}\|),$$

$$\leq C_1(\|e(\tau,\theta) - e(\overline{\tau},\overline{\theta})\| + \|\tau_x(t)\|),$$

which combined with (2.2.101) gives

$$\|\theta(t) - \overline{\theta}\| \leq C_1(\|\theta_x(t)\| + \|u_x(t)\| + \|\mathbf{w}_x(t)\| + \|\mathbf{b}_x(t)\| + \|\rho_x(t)\| + \|\tau_x(t)\|). \quad (2.2.102)$$

By the mean value theorem, Lemmas 2.2.1–2.2.14 and (2.2.102), we get

$$\|\eta(t) - \overline{\eta}\| = C_1\|\eta(\tau,\theta) - \eta(\overline{\tau},\overline{\theta})\| \leq C_1(\|\tau(t) - \overline{\tau}\| + \|\theta(t) - \overline{\theta}\|)$$
$$\leq C_1(\|\theta(t) - \overline{\theta}\| + \|\tau_x(t)\|)$$
$$\leq C_1(\|\theta_x(t)\| + \|u_x(t)\| + \|\mathbf{w}_x(t)\| + \|\mathbf{b}_x(t)\| + \|\rho_x(t)\| + \|\tau_x(t)\|). \quad (2.2.103)$$

Integrating (2.2.100) over Q_t and using Lemmas 2.2.1–2.2.14, the Cauchy and Poincaré inequalities and (2.2.102), we deduce that for small $\beta > 0$ and for any $\gamma > 0$,

$$\int_0^1 G(t)dx + \int_0^t \int_0^1 e^{\gamma s} \left[\frac{\overline{\theta}}{\theta}(\lambda \rho u_x^2 + \mu \rho |\mathbf{w}_x|^2 + \nu \rho |\mathbf{b}_x|^2 + \kappa \rho \theta_x^2/\theta) \right](x,s)dxds$$

$$+ \beta \int_0^t \int_0^1 e^{\gamma s} \left(\frac{\lambda \rho_x^2 p_\rho}{\rho} - \lambda \rho u_x^2 - \mu \rho |\mathbf{w}_x|^2 - \nu \rho |\mathbf{b}_x|^2 \right.$$

$$+ \frac{\lambda p_\theta \rho_x \theta_x}{\rho} - \frac{1}{2}u_x|\mathbf{b}|^2 + \frac{\lambda \rho_x \mathbf{b} \cdot \mathbf{b}_x}{\rho} \Big) dxds$$

$$= \int_0^1 G(0)dx$$

$$+ C_1\gamma \int_0^t \int_0^1 e^{\gamma s} \left(\mathcal{E} + \frac{\beta}{2} \left(\frac{\lambda^2 \rho_x^2}{\rho^2} - |\mathbf{w}|^2 - \tau |\mathbf{b}|^2 \right) + \frac{\beta \lambda u \rho_x}{\rho} \right)(x,s)dxds$$

$$\leq C_1 + C_1\gamma \int_0^t e^{\gamma s} \left(\|\tau - \overline{\tau}\|^2 + \|\eta - \overline{\eta}\|^2 + \|u\|^2 + \|\mathbf{w}\|^2 + \|\mathbf{b}\|^2 + \|\rho_x\|^2 \right)(s)ds$$

$$\leq C_1 + C_1\gamma \int_0^t e^{\gamma s} \left(\|u_x\|^2 + \|\mathbf{w}_x\|^2 + \|\mathbf{b}_x\|^2 + \|\theta_x\|^2 + \|\rho_x\|^2 + \|\tau_x\|^2 \right)(s)ds.$$
$$(2.2.104)$$

Using Lemmas 2.2.1–2.2.14 we easily infer that the following estimate holds for small $\beta > 0$:

$$\int_0^1 G(t)dx \geq e^{\gamma t} \Big\{ C_1^{-1}(\|\tau(t) - \overline{\tau}\|^2 + \|\eta(t) - \overline{\eta}\|^2 + \|u(t)\|^2 + \|\mathbf{w}(t)\|^2 + \|\mathbf{b}(t)\|^2)$$

$$+ \beta \int_0^1 \left(\frac{\lambda^2}{2} \left(\frac{\rho_x}{\rho} \right)^2 + \frac{\lambda \rho_x u}{\rho} \right) dx \Big\}$$

$$\geq e^{\gamma t} \Big\{ C_1^{-1}(\|\tau(t) - \overline{\tau}\|^2 + \|\eta(t) - \overline{\eta}\|^2 + \|u(t)\|^2 + \|\mathbf{w}(t)\|^2 + \|\mathbf{b}(t)\|^2)$$

$$+ \beta C_1^{-1} \|\rho_x(t)\|^2 - C_1 \beta \|u(t)\|^2 \Big\}$$
$$\geq C_1^{-1} e^{\gamma t} \Big(\|\tau(t) - \overline{\tau}\|^2 + \|\eta(t) - \overline{\eta}\|^2 + \|u(t)\|^2$$
$$+ \|\mathbf{w}(t)\|^2 + \|\mathbf{b}(t)\|^2 + \beta \|\rho_x(t)\|^2 \Big). \qquad (2.2.105)$$

Finally, the Young inequality gives

$$\frac{\lambda p_\theta \rho_x \theta_x}{\rho} \geq -\frac{1}{2} \frac{\lambda \rho_x^2 p_\rho}{\rho} - C_1 \theta_x^2 \qquad (2.2.106)$$

with $p_\rho = R\theta > 0$. It follows from (2.2.104)–(2.2.106) that there exists a constant $\gamma_1' = \gamma_1'(C_1) > 0$ such that for any fixed $\gamma \in (0, \gamma_1']$, (2.2.95) holds. $\qquad \square$

Lemma 2.2.16. *There exists a positive constant $\gamma_1 = \gamma_1(C_1) \leq \gamma_1'$ such that for any fixed $\gamma \in (0, \gamma_1']$, the following estimate holds for any $t > 0$:*

$$e^{\gamma t} \Big(\|u_x(t)\|^2 + \|\mathbf{w}_x(t)\|^2 + \|\mathbf{b}_x(t)\|^2 + \|\theta_x(t)\|^2 \Big) + \int_0^t e^{\gamma s} \Big(\|u_{xx}\|^2 + \|\mathbf{w}_{xx}\|^2$$
$$+ \|\mathbf{b}_{xx}\|^2 + \|\theta_{xx}\|^2 + \|u_t\|^2 + \|\mathbf{w}_t\|^2 + \|\mathbf{b}_t\|^2 + \|\theta_t\|^2 \Big)(s)ds \leq C_1. \quad (2.2.107)$$

Proof. By (2.1.2)–(2.1.5), Lemmas 2.2.1–2.2.15 and the Poincaré inequality, we have

$$\|u_t(t)\| \leq C_1 (\|\tau_x(t)\| + \|\mathbf{b}_x(t)\| + \|\theta_x(t)\| + \|u_{xx}(t)\|), \quad \|u_x(t)\| \leq C_1 \|u_{xx}(t)\|, \tag{2.2.108}$$
$$\|\mathbf{w}_t(t)\| \leq C_1 (\|\tau_x(t)\| + \|\mathbf{b}_x(t)\| + \|\mathbf{w}_{xx}(t)\|), \quad \|\mathbf{w}_x(t)\| \leq C_1 \|\mathbf{w}_{xx}(t)\|, \tag{2.2.109}$$
$$\|\mathbf{b}_t(t)\| \leq C_1 (\|\tau_x(t)\| + \|\mathbf{w}_x(t)\| + \|\mathbf{b}_{xx}(t)\|), \quad \|\mathbf{b}_x(t)\| \leq C_1 \|\mathbf{b}_{xx}(t)\|, \tag{2.2.110}$$
$$\|\theta_t(t)\| \leq C_1 (\|\mathbf{w}_{xx}(t)\| + \|\mathbf{b}_{xx}(t)\| + \|\theta_{xx}(t)\| + \|u_{xx}(t)\|), \quad \|\theta_x(t)\| \leq C_1 \|\theta_{xx}(t)\|. \tag{2.2.111}$$

Multiplying (2.1.2) by $-e^{\gamma t} u_{xx}$, integrating the result over $(0, 1)$, using Young's inequality, the embedding theorem, Lemmas 2.2.1–2.2.15, we deduce for any $\varepsilon > 0$,

$$\frac{1}{2} \frac{d}{dt} (e^{\gamma t} \|u_x(t)\|^2) + \lambda e^{\gamma t} \int_0^1 \frac{u_{xx}^2}{\tau} dx$$
$$= e^{\gamma t} \int_0^1 \Big[\Big(p + \frac{1}{2} |\mathbf{b}|^2 \Big)_x + \frac{\lambda u_x \tau_x}{\tau^2} \Big] u_{xx} dx + \frac{\gamma}{2} e^{\gamma t} \|u_x(t)\|^2$$
$$\leq \frac{\gamma}{2} e^{\gamma t} \|u_x(t)\|^2 + \varepsilon e^{\gamma t} \|u_{xx}(t)\|^2 + C_1(\varepsilon) e^{\gamma t} \Big(\|\tau_x(t)\|^2 + \|\theta_x(t)\|^2 + \|\mathbf{b}(t) \cdot \mathbf{b}_x(t)\|^2$$
$$+ \|\tau_x(t)\| \|u_x(t)\|_{L^\infty} \Big)$$
$$\leq \frac{\gamma}{2} e^{\gamma t} \|u_x(t)\|^2 + \varepsilon e^{\gamma t} \|u_{xx}(t)\|^2 + C_1(\varepsilon) e^{\gamma t} \Big(\|\tau_x(t)\|^2 + \|\theta_x(t)\|^2 + \|\mathbf{b}(t) \cdot \mathbf{b}_x(t)\|^2$$

$$+ \|u_x\|^{1/2}\|u_{xx}(t)\|^{1/2}\|\tau_x(t)\|\Big)$$
$$\leq C_1\gamma e^{\gamma t}\|u_{xx}(t)\|^2 + C_1(\varepsilon)e^{\gamma t}\Big(\|\tau_x(t)\|^2 + \|\theta_x(t)\|^2 + \|\mathbf{b}(t)\cdot\mathbf{b}_x(t)\|^2 + \|u_x(t)\|^2\Big),$$
$$\text{(2.2.112)}$$

i.e., for $\gamma > 0$ small enough,

$$\frac{1}{2}\frac{d}{dt}(e^{\gamma t}\|u_x(t)\|^2) + \frac{1}{2C_1}e^{\gamma t}\|u_{xx}(t)\|^2$$
$$\leq C_1\gamma^{-1}e^{\gamma t}(\|\tau_x(t)\|^2 + \|\theta_x(t)\|^2 + \|\mathbf{b}(t)\cdot\mathbf{b}_x(t)\|^2 + \|u_x(t)\|^2). \quad \text{(2.2.113)}$$

Similarly, we can get

$$\frac{1}{2}\frac{d}{dt}\Big(e^{\gamma t}\|\mathbf{w}_x(t)\|^2\Big) + \frac{1}{2C_1}e^{\gamma t}\|\mathbf{w}_{xx}(t)\|^2$$
$$\leq C_1\gamma^{-1}e^{\gamma t}\Big(\|\tau_x(t)\|^2 + \|\mathbf{b}_x(t)\|^2 + \|\mathbf{w}_x(t)\|^2\Big), \quad \text{(2.2.114)}$$

$$\frac{1}{2}\frac{d}{dt}\Big(e^{\gamma t}\|\mathbf{b}_x(t)\|^2\Big) + \frac{1}{2C_1}e^{\gamma t}\|\mathbf{b}_{xx}(t)\|^2$$
$$\leq C_1\gamma^{-1}e^{\gamma t}\Big(\|\tau_x(t)\|^2 + \|\mathbf{b}_x(t)\|^2 + \|\mathbf{w}_x(t)\|^2\Big), \quad \text{(2.2.115)}$$

$$\frac{1}{2}\frac{d}{dt}\Big(e^{\gamma t}\|\theta_x(t)\|^2\Big) + \frac{1}{2C_1}e^{\gamma t}\|\theta_{xx}(t)\|^2$$
$$\leq \gamma e^{\gamma t}\Big(\|\theta_{xx}(t)\|^2 + \|\mathbf{w}_{xx}(t)\|^2 + \|\mathbf{b}_{xx}(t)\|^2 + \|u_{xx}(t)\|^2\Big). \quad \text{(2.2.116)}$$

Adding the relations (2.2.113) through (2.2.116), integrating the result with respect to t and using Lemma 2.2.15, we obtain (2.2.107) for $\gamma \in (0, \gamma_1']$ small enough. $\qquad\square$

Thus now we have completed the proof of Theorem 2.1.1. $\qquad\square$

2.3   Global Existence and Exponential Stability in H^2

In this section, we shall study the global existence and exponential stability of solutions to problem (2.1.1)–(2.1.8) in H_+^2. We begin with the following lemma.

Lemma 2.3.1. *Under the assumptions of Theorem 2.1.2, the following estimate holds for any $t > 0$:*

$$\|u_t(t)\|^2 + \|\mathbf{w}_t(t)\|^2 + \|\mathbf{b}_t(t)\|^2 + \|\theta_t(t)\|^2$$
$$+ \int_0^t \Big(\|u_{tx}\|^2 + \|\mathbf{w}_{tx}\|^2 + \|\mathbf{b}_{tx}\|^2 + \|\theta_{tx}\|^2\Big)(s)\,ds \leq C_2. \quad \text{(2.3.1)}$$

Proof. Differentiating (2.1.2) with respect to t, multiplying the result by u_t, and integrating over $(0, 1)$, we infer that

$$\frac{d}{dt}\|u_t(t)\|^2 + \|u_{tx}(t)\|^2$$

$$\leq \frac{1}{2}\|u_{tx}(t)\|^2 + C_1\Big(\|u_x(t)\|^2 + \|\mathbf{b}(t)\|_{L^\infty}^2\|\mathbf{b}_t(t)\|^2 + \|(1+\theta^3)\theta_t(t)\|^2 + \|u_x(t)\|_{L^4}^4\Big)$$

$$\leq \frac{1}{2}\|u_{tx}(t)\|^2 + C_2\Big(\|u_{xx}(t)\|^2 + \|\theta_t(t)\|^2 + \|\tau_x(t)\|^2 + \|\mathbf{w}_x(t)\|^2 + \|\mathbf{b}_{xx}(t)\|^2\Big),$$

which, together with Theorem 2.1.1, gives

$$\|u_t(t)\|^2 + \int_0^t \|u_{tx}(s)\|^2 ds \leq C_2. \tag{2.3.2}$$

Analogously, we have

$$\|\theta_t(t)\|^2 + \int_0^t \|\theta_{tx}(s)\|^2 ds \leq C_2, \tag{2.3.3}$$

$$\|\mathbf{w}_t(t)\|^2 + \int_0^t \|\mathbf{w}_{tx}(s)\|^2 ds \leq C_2, \tag{2.3.4}$$

$$\|\mathbf{b}_t(t)\|^2 + \int_0^t \|\mathbf{b}_{tx}(s)\|^2 ds \leq C_2. \tag{2.3.5}$$

Thus (2.3.1) follows from (2.3.2)–(2.3.5).   □

Lemma 2.3.2. *Under the assumptions of Theorem 2.1.2, the following estimate holds for any $t > 0$:*

$$\|u_{xx}(t)\|^2 + \|\mathbf{w}_{xx}(t)\|^2 + \|\mathbf{b}_{xx}(t)\|^2 + \|\theta_{xx}(t)\|^2$$
$$+ \|u_x(t)\|_{L^\infty}^2 + \|\mathbf{w}_x(t)\|_{L^\infty}^2 + \|\mathbf{b}_x(t)\|_{L^\infty}^2 + \|\theta_x(t)\|_{L^\infty}^2$$
$$+ \int_0^t \Big(\|u_{xxx}\|^2 + \|\mathbf{w}_{xxx}\|^2 + \|\mathbf{b}_{xxx}\|^2 + \|\theta_{xxx}\|^2\Big)(s)ds \leq C_2. \tag{2.3.6}$$

Proof. Equation (2.1.2) can be rewritten as

$$u_t = -\left(p + \frac{1}{2}|\mathbf{b}|^2\right)_x + \left(\frac{\lambda u_x}{\tau}\right)_x$$
$$= \frac{R\theta\tau_x - \lambda u_x \tau_x}{\tau^2} + \frac{\lambda u_{xx} - R\theta_x}{\tau} - \mathbf{b} \cdot \mathbf{b}_x. \tag{2.3.7}$$

Using (2.3.7), Theorem 2.1.1, Lemma 2.3.1, the Sobolev embedding theorem and Young's inequality, we have

$$\|u_{xx}(t)\| \leq C_2\big(\|u_t(t)\| + \|\theta_x(t)\| + \|\tau_x(t)\theta(t)\| + \|\mathbf{b}(t) \cdot \mathbf{b}_x(t)\|$$

$$+ \|\tau_x(t)\| + \|\tau_x(t)u_x(t)\|)$$

$$\leq C_2(\|u_t(t)\| + \|\theta_x(t)\| + \|\tau_x(t)\| + \|\mathbf{b}_x(t)\|^2 + \|u_x(t)\|^{\frac{1}{2}}\|u_{xx}(t)\|^{\frac{1}{2}})$$

$$\leq \frac{1}{2}\|u_{xx}(t)\| + C_2(\|u_t(t)\| + 1), \tag{2.3.8}$$

$$\int_0^t \|u_{xxx}(s)\|^2 ds \leq C_2 + C_2 \int_0^t \|u_{tx}(s)\|^2 ds \leq C_2, \tag{2.3.9}$$

which leads to

$$\|u_{xx}(t)\| \leq C_2, \quad \|u_x(t)\|_{L^\infty} \leq C_2.$$

Similarly, we have

$$\int_0^t \left(\|\mathbf{w}_{xxx}\|^2 + \|\mathbf{b}_{xxx}\|^2 + \|\theta_{xxx}\|^2 \right)(s)ds \leq C_2, \tag{2.3.10}$$

$$\|\mathbf{w}_{xx}(t)\| + \|\mathbf{b}_{xx}(t)\| + \|\theta_{xx}(t)\|$$
$$\leq C_2(\|\mathbf{w}_t(t)\| + \|\mathbf{b}_t(t)\| + \|\theta_t(t)\| + 1) \leq C_2, \tag{2.3.11}$$

$$\|\mathbf{w}_x(t)\|_{L^\infty} + \|\mathbf{b}_x(t)\|_{L^\infty} + \|\theta_x(t)\|_{L^\infty} \leq C_2. \tag{2.3.12}$$

Thus (2.3.6) follows from (2.3.8)–(2.3.12). □

Lemma 2.3.3. *Under the assumptions of Theorem 2.1.2, the following estimate holds for any $t > 0$:*

$$\|\tau_{xx}(t)\|^2 + \int_0^t \|\tau_{xx}(s)\|^2 ds \leq C_2. \tag{2.3.13}$$

Proof. Differentiating (2.1.2) with respect to x, we obtain

$$\lambda \frac{d}{dt}\left(\frac{\tau_{xx}}{\tau}\right) + \frac{R\theta}{\tau}\frac{\tau_{xx}}{\tau} = u_{tx} + E(x,t), \tag{2.3.14}$$

where

$$E(x,t) = \left(\frac{R}{\tau} + \frac{4a}{3}\theta^3\right)\theta_{xx} - \frac{2\tau_x(R\theta_x - \lambda u_{xx})}{\tau^2} + \frac{2\tau_x^2(R\theta - \lambda u_x)}{\tau^3}$$
$$+ 4a\theta^2\theta_x + \mathbf{b}\cdot\mathbf{b}_{xx} + |\mathbf{b}_x|^2.$$

Multiplying (2.3.14) by $\frac{\tau_{xx}}{\tau}$ and using the Young inequality and Theorem 2.1.1, we conclude that, for any $\varepsilon > 0$,

$$\frac{d}{dt}\left\|\frac{\tau_{xx}}{\tau}(t)\right\|^2 + C_1\left\|\frac{\tau_{xx}}{\tau}(t)\right\|^2 \leq \varepsilon\left\|\frac{\tau_{xx}}{\tau}(t)\right\|^2 + C_2\Big(\|u_{tx}(t)\|^2 + \|\theta_{xx}(t)\|^2$$

$$+ \|\mathbf{b}_x(t)\|_{L^4}^4 + \|\theta_x(t)\|_{L^4}^4 + \|\tau_x(t)\|_{L^4}^4 + \|u_x(t)\tau_x^2(t)\|^2 + \|\theta(t)\tau_x^2(t)\|^2\Big)$$

$$\leq \varepsilon\left\|\frac{\tau_{xx}}{\tau}(t)\right\|^2 + C_2(\varepsilon)\Big(\|u_{tx}(t)\|^2 + \|\theta_{xx}(t)\|^2 + \|\mathbf{b}_{xx}(t)\|^2 + \|\tau_x(t)\|^2\Big),$$

which, combined with Lemma 2.3.1 and Theorem 2.1.1 gives, for $\varepsilon > 0$ small enough,

$$\|\tau_{xx}(t)\|^2 + \int_0^t \|\tau_{xx}(s)\|^2 ds \leq C_2. \tag{2.3.15}$$

Thus (2.3.13) follows from Theorem 2.1.1 and (2.3.15). $\qquad\square$

Lemma 2.3.4. *Under assumptions of Theorem 2.1.2, for any* $(\tau_0, u_0, \theta_0, \mathbf{w}_0, \mathbf{b}_0) \in H_+^2$, *there exists a positive constant* $\gamma_2' = \gamma_2'(C_2) \leq \gamma_1$ *such that, for any fixed* $\gamma \in (0, \gamma_2']$, *the following estimate holds for any* $t > 0$:

$$e^{\gamma t}\Big(\|u_t(t)\|^2 + \|\mathbf{w}_t(t)\|^2 + \|\mathbf{b}_t(t)\|^2 + \|\theta_t(t)\|^2 + \|u_{xx}(t)\|^2$$

$$+ \|\mathbf{w}_{xx}(t)\|^2 + \|\mathbf{b}_{xx}(t)\|^2 + \|\theta_{xx}(t)\|^2\Big)$$

$$+ \int_0^t e^{\gamma s}\Big(\|u_{tx}\|^2 + \|\mathbf{w}_{tx}\|^2 + \|\mathbf{b}_{tx}\|^2 + \|\theta_{tx}\|^2\Big)(s)ds \leq C_2. \tag{2.3.16}$$

Proof. Differentiating (2.1.2) with respect to t, multiplying the result by $u_t e^{\gamma t}$ and integrating the resulting equation, we conclude that

$$\frac{1}{2}e^{\gamma t}\|u_t(t)\|^2 + \lambda \int_0^t e^{\gamma s}\left\|\frac{u_{tx}}{\sqrt{\tau}}(s)\right\|^2 ds$$

$$\leq C_2 + \frac{\gamma}{2}\int_0^t e^{\gamma s}\left(\|u_t\|^2 + \left\|\frac{u_{tx}}{\sqrt{\tau}}\right\|^2\right)(s)d\tau$$

$$+ C_2 \int_0^t e^{\gamma s}\Big(\|u_x\|^2 + \|\theta_t\|^2 + \|\mathbf{b}_t\|^2 + \|\mathbf{b}_x\|^2 + \|u_x\|_{L^4}^4\Big)(s)ds$$

$$\leq C_2 + \left(C_2\gamma + \frac{\gamma}{2}\right)\int_0^t e^{\gamma s}\left\|\frac{u_{tx}}{\sqrt{\tau}}(s)\right\|^2 ds$$

$$+ C_2 \int_0^t e^{\gamma s}\left(\|\theta_t\|^2 + \|\mathbf{b}_t\|^2 + \|\mathbf{b}_x\|^2 + \|u_{xx}\|^2\right)(s)ds. \tag{2.3.17}$$

Integrating (2.3.17) over $(0, t)$ and using Lemmas 2.3.1–2.3.3 and (2.3.2), we get

$$e^{\gamma t}(\|u_t(t)\|^2 + \|u_{xx}(t)\|^2) + \int_0^t e^{\gamma s}\|u_{tx}\|^2 ds \leq C_2, \quad \forall t > 0. \tag{2.3.18}$$

Analogously, we have

$$e^{\gamma t}\Big(\|\mathbf{w}_t(t)\|^2 + \|\mathbf{w}_{xx}(t)\|^2 + \|\mathbf{b}_t(t)\|^2 + \|\mathbf{b}_{xx}(t)\|^2 + \|\theta_t(t)\|^2 + \|\theta_{xx}(t)\|^2\Big)$$

$$+ \int_0^t e^{\gamma s}(\|\mathbf{w}_{tx}\|^2 + \|\mathbf{b}_{tx}\|^2 + \|\theta_{tx}\|^2)ds \leq C_2, \quad \forall t > 0. \tag{2.3.19}$$

By (2.3.18) and (2.3.19), we get (2.3.16). $\qquad\square$

Lemma 2.3.5. *There exists a positive constant* $\gamma_2 = \gamma_2(C_2) \leq \gamma_2'$ *such that, for any fixed* $\gamma \in (0, \gamma_2]$, *the following estimate holds:*

$$\|\tau(t) - \overline{\tau}\|_{H^2} \leq C_2 e^{-\gamma t}, \quad \forall t > 0. \tag{2.3.20}$$

Proof. Multiplying (2.1.1) by $e^{t/2C_1}$, choosing γ so small that $\gamma \leq \gamma_2(C_2)$, and using Lemma 2.3.4, we have

$$\|\tau_{xx}(t)\|^2 \leq C_2 e^{-t/2C_1} + C_2 e^{-\gamma t} \leq C_2 e^{-\gamma t}, \tag{2.3.21}$$

which, together with Lemmas 2.1.14 and 2.1.15, gives (2.3.20). □

Thus we have completed the proof of Theorem 2.1.2. □

2.4   Global Existence and Exponential Stability in H^4

In this section, we shall study the global existence and exponential stability of solutions to problem (2.1.1)–(2.1.8) in H_+^4. We begin with the following lemma.

Lemma 2.4.1. *Under the assumptions of Theorem 2.1.3, the following estimates hold:*

$$\|u_{tx}(x,0)\| + \|\mathbf{w}_{tx}(x,0)\| + \|\mathbf{b}_{tx}(x,0)\| + \|\theta_{tx}(x,0)\| \leq C_3, \tag{2.4.1}$$

$$\|u_{tt}(x,0)\| + \|\mathbf{w}_{tt}(x,0)\| + \|\mathbf{b}_{tt}(x,0)\| + \|\theta_{tt}(x,0)\|$$
$$+ \|u_{txx}(x,0)\| + \|\mathbf{w}_{txx}(x,0)\| + \|\mathbf{b}_{txx}(x,0)\| + \|\theta_{txx}(x,0)\| \leq C_3. \tag{2.4.2}$$

Proof. We easily infer from (2.1.2) and Theorems 2.1.1–2.1.2 that

$$\|u_t(t)\| \leq C_3 \Big(\|\tau_x(t)\| + \|\theta_x(t)\| + \|u_{xx}(t)\| + \|u_x(t)\|_{L^\infty} \|\tau_x(t)\| + \|\mathbf{b}(t)\|_{L^\infty} \|\mathbf{b}_x(t)\| \Big)$$

$$\leq C_3 \Big(\|\tau_x(t)\| + \|\theta_x(t)\| + \|u_{xx}(t)\| + \|\mathbf{b}_x(t)\| \Big).$$

Differentiating (2.1.2) with respect to x and using Theorems 2.1.1–2.1.2 we get

$$\|u_{tx}(t)\| \leq C_3 \Big(\|\tau_x(t)\|_{H^1} + \|\theta_x(t)\|_{H^1} + \|u_x(t)\|_{H^2} + \|\mathbf{b}_x(t)\|_{H^1} \Big), \tag{2.4.3}$$

and

$$\|u_{xxx}(t)\| \leq C_3 \Big(\|\tau_x(t)\|_{H^1} + \|\theta_x(t)\|_{H^1} + \|u(t)\|_{H^2} + \|\mathbf{b}_x(t)\|_{H^1} + \|u_{tx}(t)\| \Big). \tag{2.4.4}$$

Differentiating (2.1.2) with respect to x twice, using the embedding theorem and Theorems 2.1.1–2.1.2, we conclude that

$$\|u_{txx}(t)\| \leq C_3 \Big(\|\tau_x(t)\|_{H^2} + \|\theta_x(t)\|_{H^2} + \|u_x(t)\|_{H^3} + \|\mathbf{b}_x(t)\|_{H^2} \Big), \tag{2.4.5}$$

and

$$\|u_{xxxx}(t)\| \leq C_3 \Big(\|\tau_x(t)\|_{H^2} + \|\theta_x(t)\|_{H^2} + \|u(t)\|_{H^3} + \|\mathbf{b}_x(t)\|_{H^2} + \|u_{txx}(t)\| \Big).$$
$$(2.4.6)$$

Similarly, we have

$$\|\mathbf{w}_t(t)\| \leq C_3 \Big(\|\mathbf{w}_x(t)\|_{H^1} + \|\mathbf{b}_x(t)\| + \|\tau_x(t)\| \Big),$$

$$\|\mathbf{w}_{tx}(t)\| \leq C_3 \Big(\|\mathbf{w}_x(t)\|_{H^2} + \|\mathbf{b}_x(t)\|_{H^1} + \|\tau_x(t)\|_{H^1} \Big), \qquad (2.4.7)$$

$$\|\mathbf{w}_{xxx}(t)\| \leq C_3 \Big(\|\mathbf{w}_x(t)\|_{H^1} + \|\mathbf{b}_x(t)\|_{H^1} + \|\tau_x(t)\|_{H^1} + \|\mathbf{w}_{tx}(t)\| \Big), \qquad (2.4.8)$$

$$\|\mathbf{w}_{txx}(t)\| \leq C_3 \Big(\|\mathbf{w}_x(t)\|_{H^3} + \|\mathbf{b}_x(t)\|_{H^2} + \|\tau_x(t)\|_{H^2} \Big), \qquad (2.4.9)$$

$$\|\mathbf{w}_{xxxx}(t)\| \leq C_3 \Big(\|\mathbf{w}_x(t)\|_{H^2} + \|\mathbf{b}_x(t)\|_{H^2} + \|\tau_x(t)\|_{H^2} + \|\mathbf{w}_{txx}(t)\| \Big), \qquad (2.4.10)$$

$$\|\mathbf{b}_t(t)\| \leq C_3 \Big(\|\mathbf{b}_x(t)\|_{H^1} + \|\mathbf{w}_x(t)\| + \|\tau_x(t)\| \Big),$$

$$\|\mathbf{b}_{tx}(t)\| \leq C_3 \Big(\|\mathbf{b}_x(t)\|_{H^2} + \|\mathbf{w}_x(t)\|_{H^1} + \|\tau_x(t)\|_{H^1} \Big), \qquad (2.4.11)$$

$$\|\mathbf{b}_{xxx}(t)\| \leq C_3 \Big(\|\mathbf{b}_x(t)\|_{H^1} + \|\mathbf{w}_x(t)\|_{H^1} + \|\tau_x(t)\|_{H^1} + \|\mathbf{b}_{tx}(t)\| \Big), \qquad (2.4.12)$$

$$\|\mathbf{b}_{txx}(t)\| \leq C_3 \Big(\|\mathbf{b}_x(t)\|_{H^3} + \|\mathbf{w}_x(t)\|_{H^2} + \|\tau_x(t)\|_{H^2} \Big), \qquad (2.4.13)$$

$$\|\mathbf{b}_{xxxx}(t)\| \leq C_3 \Big(\|\mathbf{b}_x(t)\|_{H^2} + \|\mathbf{w}_x(t)\|_{H^2} + \|\tau_x(t)\|_{H^2} + \|\mathbf{b}_{txx}(t)\| \Big), \qquad (2.4.14)$$

$$\|\theta_t(t)\| \leq C_3 \Big(\|u_x(t)\| + \|\tau_x(t)\| + \|\theta_{xx}(t)\| + \|u_x(t)\|_{L^\infty} \|u_x(t)\|$$
$$+ \|\mathbf{w}_x(t)\|_{L^\infty} \|\mathbf{w}_x(t)\| + \|\mathbf{b}_x(t)\|_{L^\infty} \|\mathbf{b}_x(t)\| + \|\theta_x(t)\|_{L^\infty} \|\theta_x(t)\| \Big),$$
$$\leq C_3 \Big(\|\theta_{xx}(t)\| + \|u_{xx}(t)\| + \|\mathbf{w}_{xx}(t)\| + \|\mathbf{b}_{xx}(t)\| \Big) \qquad (2.4.15)$$

$$\|\theta_{tx}(t)\| \leq C_3 \Big(\|\theta_t(t)\| + \|\theta_x(t)\|_{H^2} + \|\tau_x(t)\|_{H^1} + \|u_x(t)\|_{H^1}$$
$$+ \|\mathbf{w}_x(t)\|_{H^1} + \|\mathbf{b}_x(t)\|_{H^1} \Big), \qquad (2.4.16)$$

$$\|\theta_{xxx}(t)\| \leq C_3 \Big(\|\theta_x(t)\|_{H^1} + \|\tau_x(t)\|_{H^1} + \|u_x(t)\|_{H^1} + \|\mathbf{w}_x(t)\|_{H^1}$$
$$+ \|\mathbf{b}_x(t)\|_{H^1} + \|\theta_{tx}(t)\| \Big), \qquad (2.4.17)$$

$$\|\theta_{txx}(t)\| \leq C_3 \Big(\|\theta_x(t)\|_{H^3} + \|\tau_x(t)\|_{H^2} + \|u_x(t)\|_{H^2} + \|\mathbf{w}_x(t)\|_{H^2} + \|\mathbf{b}_x(t)\|_{H^2} \Big), \qquad (2.4.18)$$

$$\|\theta_{xxxx}(t)\| \leq C_3 \Big(\|\theta_x(t)\|_{H^2} + \|\tau_x(t)\|_{H^2} + \|u_x(t)\|_{H^2} + \|\mathbf{w}_x(t)\|_{H^2}$$
$$+ \|\mathbf{b}_x(t)\|_{H^2} + \|\theta_{txx}(t)\| \Big). \qquad (2.4.19)$$

Differentiating (2.1.2) with respect to t, and using Theorems 2.1.1–2.1.2 and relations (2.4.3), (2.4.5), (2.4.11)–(2.4.12) and (2.4.16), we derive that

$$\|u_{tt}(t)\| \le C_3\Big(\|\tau_x(t)\|_{H^2} + \|u_x(t)\|_{H^3} + \|\mathbf{b}_x(t)\|_{H^2} + \|\mathbf{w}_x(t)\|_{H^1} + \|\theta_x(t)\|_{H^2}\Big).$$
$$(2.4.20)$$

Similarly, we obtain

$$\|\mathbf{w}_{tt}(t)\| \le C_3\Big(\|\tau_x(t)\|_{H^2} + \|\mathbf{b}_x(t)\|_{H^2} + \|\mathbf{w}_x(t)\|_{H^3} + \|u_x(t)\|_{H^1}\Big), \qquad (2.4.21)$$

$$\|\mathbf{b}_{tt}(t)\| \le C_3\Big(\|\tau_x(t)\|_{H^2} + \|\mathbf{b}_x(t)\|_{H^3} + \|\mathbf{w}_x(t)\|_{H^2} + \|u_x(t)\|_{H^1}\Big), \qquad (2.4.22)$$

$$\|\theta_{tt}(t)\| \le C_3\Big(\|\tau_x(t)\|_{H^2} + \|u_x(t)\|_{H^2} + \|\mathbf{b}_x(t)\|_{H^2} + \|\mathbf{w}_x(t)\|_{H^2} + \|\theta_x(t)\|_{H^3}\Big).$$
$$(2.4.23)$$

Thus (2.4.1) and (2.4.2) follow from (2.4.3), (2.4.7), (2.4.11), (2.4.16) and from (2.4.5), (2.4.9), (2.4.13), (2.4.18) and (2.4.20)–(2.4.23), respectively. □

Lemma 2.4.2. *Under the assumptions of Theorem 2.1.3, the following estimates hold for any $t > 0$ and $\varepsilon > 0$:*

$$\|u_{tt}(t)\|^2 + \int_0^t \|u_{ttx}(s)\|^2 ds \le C_3 + C_3 \int_0^t (\|\mathbf{b}_{txx}\|^2 + \|\theta_{txx}\|^2)(s)ds, \quad (2.4.24)$$

$$\|\mathbf{w}_{tt}(t)\|^2 + \int_0^t \|\mathbf{w}_{ttx}(s)\|^2 ds \le C_3 + C_3 \int_0^t \|\mathbf{b}_{txx}(s)\|^2 ds, \qquad (2.4.25)$$

$$\|\mathbf{b}_{tt}(t)\|^2 + \int_0^t \|\mathbf{b}_{ttx}(s)\|^2 ds \le C_3 + C_3 \int_0^t \|\mathbf{w}_{txx}(s)\|^2 ds, \qquad (2.4.26)$$

$$\|\theta_{tt}(t)\|^2 + \int_0^t \|u_{ttx}(s)\|^2 ds \le C_3 + C_2\varepsilon^{-1} \int_0^t \|\theta_{txx}(s)\|^2 ds \qquad (2.4.27)$$

$$+ C_1\varepsilon \int_0^t \Big(\|u_{txx}\|^2 + \|u_{ttx}\|^2 + \|\mathbf{w}_{txx}\|^2 + \|\mathbf{w}_{ttx}\|^2 + \|\mathbf{b}_{txx}\|^2 + \|\mathbf{b}_{ttx}\|^2\Big)(s)ds.$$

Proof. Differentiating (2.1.2) with respect to t twice, multiplying the resulting equation by u_{tt}, performing an integration by parts, and using Theorems 2.1.1–2.1.2 and Lemma 2.4.1, we have

$$\frac{1}{2}\frac{d}{dt}\int_0^1 u_{tt}^2 dx = -\int_0^1 \left[-(p + \frac{1}{2}|\mathbf{b}|^2) + \frac{\lambda u_x}{\tau}\right]_{tt} u_{ttx} dx$$

$$\le -C_1^{-1}\|u_{ttx}(t)\|^2 + C_2\Big(\|\theta_{tt}(t)\|^2 + \|\mathbf{b}_{tt}(t)\|^2 + \|u_{tt}(t)\|^2 + \|\mathbf{b}_t(t)\|^4\Big)$$

$$+ C_1\Big(\|u_x(t)\|^2 + \|u_{tx}(t)\|^2\Big). \qquad (2.4.28)$$

Thus, using Theorems 2.1.1–2.1.2 and Lemma 2.4.1, we get

$$\|u_{tt}(t)\|^2 + \int_0^t \|u_{ttx}(s)\|^2 ds \leq C_3 + C_3 \int_0^t \left(\|\mathbf{b}_{txx}\|^2 + \|\theta_{txx}\|^2 \right)(s)ds.$$

Analogously, we can obtain (2.4.25)–(2.4.27). The proof is now complete. □

Lemma 2.4.3. *Under the assumptions of Theorem 2.1.3, the following estimates hold for any $t > 0$ and $\varepsilon > 0$:*

$$\|u_{tx}(t)\|^2 + \int_0^t \|u_{txx}\|^2(s)ds$$

$$\leq C_3\varepsilon^{-6} + C_2\varepsilon^2 \int_0^t \left(\|\mathbf{b}_{txx}\|^2 + \|\theta_{txx}\|^2 + \|u_{ttx}\|^2 \right)(s)ds, \qquad (2.4.29)$$

$$\|\mathbf{w}_{tx}(t)\|^2 + \int_0^t \|\mathbf{w}_{txx}\|^2(s)ds$$

$$\leq C_3\varepsilon^{-6} + C_2\varepsilon^2 \int_0^t \left(\|\mathbf{w}_{txx}\|^2 + \|\mathbf{b}_{ttx}\|^2 \right)(s)ds, \qquad (2.4.30)$$

$$\|\mathbf{b}_{tx}(t)\|^2 + \int_0^t \|\mathbf{b}_{txx}\|^2(s)ds$$

$$\leq C_3\varepsilon^{-6} + C_2\varepsilon^2 \int_0^t \left(\|\mathbf{b}_{txx}\|^2 + \|\mathbf{w}_{ttx}\|^2 \right)(s)ds, \qquad (2.4.31)$$

$$\|\theta_{tx}(t)\|^2 + \int_0^t \|\theta_{txx}\|^2(s)ds$$

$$\leq C_3\varepsilon^{-6} + C_2\varepsilon^2 \int_0^t \left(\|\mathbf{b}_{txx}\|^2 + \|u_{txx}\|^2 + \|\mathbf{w}_{txx}\|^2 \right.$$

$$\left. + \|\theta_{ttx}\|^2 + \|\theta_{xxx}\| \|\theta_{tx}\| \right)(s)ds. \qquad (2.4.32)$$

Proof. Differentiating (2.1.2) with respect to x and t, multiplying the resulting equation by u_{tx}, and integrating by parts, we arrive at

$$\frac{1}{2}\frac{d}{dt}\|u_{tx}(t)\|^2 = B_0(x,t) + B_1(t), \qquad (2.4.33)$$

where

$$B_0(x,t) = \sigma_{tx} u_{tx} \Big|_{x=0}^{x=1}, \quad B_1(t) = -\int_0^1 \sigma_{tx} u_{txx} dx.$$

Using Theorems 2.1.1 and 2.1.2, the interpolation inequality and Poincaré's inequality, we obtain

$$B_0 \leq C_1 \Big[(\|u_x(t)\|_{L^\infty} + \|\theta_t(t)\|_{L^\infty})(\|\tau_x(t)\|_{L^\infty} + \|\theta_x(t)\|_{L^\infty}) + \|\mathbf{b}_t(t)\|_{L^\infty} \|\mathbf{b}_x(t)\|_{L^\infty}$$

$$+ \|\mathbf{b}_{tx}(t)\|_{L^\infty} \|\mathbf{b}(t)\|_{L^\infty} + \|\theta_{tx}(t)\|_{L^\infty} + \|\theta(t)\|_{L^\infty} \|\tau_{tx}(t)\|_{L^\infty}$$

$$+ \|u_x(t)\|_{L^\infty} \|u_{xx}(t)\|_{L^\infty} + \|u_{tx}(t)\|_{L^\infty} \|\tau_x(t)\|_{L^\infty} + \|u_{txx}(t)\|_{L^\infty} \Big] \|u_{tx}(t)\|_{L^\infty}$$

$$\leq C_3 (B_{01} + B_{01}) \|u_{tx}(t)\|^{\frac{1}{2}} \|u_{txx}(t)\|^{\frac{1}{2}} \tag{2.4.34}$$

where

$$B_{01} = \|u_x(t)\|_{H^2} + \|\theta_t(t)\| + \|\theta_{tx}(t)\| + \|\mathbf{b}_t(t)\| + \|\mathbf{b}_{tx}(t)\|,$$

$$B_{02} = \|\theta_{tx}(t)\|^{\frac{1}{2}} \|\theta_{txx}(t)\|^{\frac{1}{2}} + \|u_{txx}(t)\|^{\frac{1}{2}} \|u_{txxx}(t)\|^{\frac{1}{2}} + \|u_{txx}(t)\|$$

$$+ \|u_{tx}(t)\|^{\frac{1}{2}} \|u_{txx}(t)\|^{\frac{1}{2}} + \|\mathbf{b}_{tx}(t)\|^{\frac{1}{2}} \|\mathbf{b}_{txx}(t)\|^{\frac{1}{2}}.$$

Now, using the Young inequality several times, we see that, for any $\varepsilon > 0$,

$$C_3 B_{01} \|u_{tx}(t)\|^{\frac{1}{2}} \|u_{txx}(t)\|^{\frac{1}{2}} \leq C_3 \left(2^{1/4} \varepsilon^{1/2} \|u_{txx}(t)\|^{\frac{1}{2}} \right) \left(\frac{B_{01} \|u_{tx}(t)\|^{\frac{1}{2}}}{2^{1/4} \varepsilon^{1/2}} \right)$$

$$\leq \frac{\varepsilon^2}{2} \|u_{txx}(t)\|^2 + C_3 \varepsilon^{-\frac{4}{3}} B_{01}^{-\frac{4}{3}} \|u_{tx}(t)\|^{-\frac{2}{3}}$$

$$\leq \frac{\varepsilon^2}{2} \|u_{txx}(t)\|^2 + C_3 \varepsilon^{-\frac{2}{3}} (B_{01}^2 + \|u_{tx}(t)\|^2)$$

$$\leq \frac{\varepsilon^2}{2} \|u_{txx}(t)\|^2 + C_3 \varepsilon^{-\frac{2}{3}} \Big(\|u_{tx}(t)\|^2 + \|u_x(t)\|_{H^2}^2 + \|\theta_t(t)\|^2$$

$$+ \|\theta_{tx}(t)\|^2 + \|\mathbf{b}_t(t)\|^2 + \|\mathbf{b}_{tx}(t)\|^2 \Big), \tag{2.4.35}$$

and

$$C_3 B_{02} \|u_{tx}(t)\|^{\frac{1}{2}} \|u_{txx}(t)\|^{\frac{1}{2}}$$

$$\leq \frac{\varepsilon^2}{2} \|u_{txx}(t)\|^2 + \varepsilon^2 \Big(\|u_{txxx}(t)\|^2 + \|\mathbf{b}_{txx}(t)\|^2 + \|\theta_{txx}(t)\|^2 \Big)$$

$$+ C_3 \varepsilon^{-6} \Big(\|u_{tx}(t)\|^2 + \|\theta_{tx}(t)\|^2 + \|\mathbf{b}_{tx}(t)\|^2 \Big). \tag{2.4.36}$$

Thus we infer from (2.4.34)–(2.4.36) that

$$B_0 \leq \varepsilon^2 \Big(\|u_{txxx}(t)\|^2 + \|u_{txx}(t)\|^2 + \|\mathbf{b}_{txx}(t)\|^2 + \|\theta_{txx}(t)\|^2 \Big)$$

$$+ C_3 \varepsilon^{-6} \Big(\|u_{tx}(t)\|^2 + \|\theta_{tx}(t)\|^2 + \|\mathbf{b}_{tx}(t)\|^2 + \|\theta_t(t)\|^2$$

$$+ \|u_x(t)\|_{H^2}^2 + \|\mathbf{b}_t(t)\|^2 \Big),$$

which, together with Theorems 2.1.1–2.1.2 and Lemmas 2.4.1–2.4.2, yields

$$\int_0^t B_0 ds \leq \varepsilon^2 \int_0^t \Big(\|u_{txxx}\|^2 + \|u_{txx}\|^2 + \|\mathbf{b}_{txx}\|^2 + \|\theta_{txx}\|^2 \Big)(s) ds + C_3 \varepsilon^{-6}. \tag{2.4.37}$$

Similarly, by Theorems 2.1.1–2.1.2, Lemmas 2.4.1–2.4.2 and the embedding theorem, we have

$$B_1 \leq (2C_3)^{-1}\|u_{txx}(t)\|^2 \tag{2.4.38}$$
$$+ C_3\Big(\|u_{tx}(t)\|^2 + \|\mathbf{b}_{tx}(t)\|^2 + \|\theta_t(t)\|_{H^1}^2 + \|u_x(t)\|^2 + \|u_x(t)\|_{H^1}^2\Big),$$

which, combined with (2.4.33), (2.4.37), (2.4.38), Theorems 2.1.1–2.1.2 and Lemmas 2.4.1–2.4.2, gives that for $\varepsilon \in (0,1)$ small enough,

$$\|u_{tx}(t)\|^2 + \int_0^t \|u_{txx}(s)\|^2 ds \leq C_3 \varepsilon^{-6} + C_2\varepsilon^2 \int_0^t \Big(\|\mathbf{b}_{txx}\|^2 + \|\theta_{txx}\|^2 + \|u_{txxx}\|^2\Big)(s)ds. \tag{2.4.39}$$

On the other hand, differentiating (2.1.2) with respect to x and t and using again Theorems 2.1.1–2.1.2 and Lemmas 2.4.1–2.4.2, we have

$$\|u_{txxx}(t)\| \leq C_1\|u_{ttx}(t)\| + C_2\Big(\|u_{xx}(t)\|_{H^2}^2 + \|\theta_x(t)\|_{H^1}^2 + \|\tau_x(t)\|_{H^1}^2$$
$$+ \|\mathbf{b}_x(t)\|_{H^1}^2 + \|\theta_t(t)\|_{H^2}^2 + \|\mathbf{b}_t(t)\|_{H^2}^2\Big). \tag{2.4.40}$$

Thus inserting (2.4.40) into (2.4.39) yields estimate (2.4.29).
Analogously, we can obtain estimates (2.4.30)–(2.4.32). □

Lemma 2.4.4. *Under the assumptions of Theorem 2.1.3, the following estimates hold for any $t > 0$:*

$$\|u_{tt}(t)\|^2 + \|u_{tx}(t)\|^2 + \|\mathbf{w}_{tt}(t)\|^2 + \|\mathbf{w}_{tx}(t)\|^2 + \|\mathbf{b}_{tt}(t)\|^2 + \|\mathbf{b}_{tx}(t)\|^2 + \|\theta_{tt}(t)\|^2$$
$$+ \|\theta_{tx}(t)\|^2 + \int_0^t \Big(\|u_{ttx}\|^2 + \|u_{txx}\|^2 + \|\mathbf{w}_{ttx}\|^2 + \|\mathbf{w}_{txx}\|^2 + \|\mathbf{b}_{ttx}\|^2$$
$$+ \|\mathbf{b}_{txx}\|^2 + \|\theta_{ttx}\|^2 + \|\theta_{txx}\|^2\Big)(s)ds \leq C_4, \tag{2.4.41}$$

$$\|\tau_{xxx}(t)\|_{H^1}^2 + \|\tau_{xx}(t)\|_{W^{1,\infty}}^2 + \int_0^t \Big(\|\tau_{xxx}\|_{H^1}^2 + \|\tau_{xx}\|_{W^{1,\infty}}^2\Big)(s)ds \leq C_4, \tag{2.4.42}$$

$$\|u_{xxx}(t)\|_{H^1}^2 + \|u_{xx}(t)\|_{W^{1,\infty}}^2 + \|\mathbf{w}_{xxx}(t)\|_{H^1}^2 + \|\mathbf{w}_{xx}(t)\|_{W^{1,\infty}}^2 + \|\mathbf{b}_{xxx}(t)\|_{H^1}^2$$
$$+ \|\mathbf{b}_{xx}(t)\|_{W^{1,\infty}}^2 + \|\theta_{xxx}(t)\|_{H^1}^2 + \|\theta_{xx}(t)\|_{W^{1,\infty}}^2 + \|\tau_{txxx}(t)\|^2 + \|u_{txx}(t)\|^2$$
$$+ \|\mathbf{w}_{txx}(t)\|^2 + \|\mathbf{b}_{txx}(t)\|^2 + \|\theta_{txx}(t)\|^2 + \int_0^t \Big(\|u_{tt}\|^2 + \|\mathbf{w}_{tt}\|^2 + \|\mathbf{b}_{tt}\|^2$$
$$+ \|\theta_{tt}\|^2 + \|u_{xx}\|_{W^{2,\infty}}^2 + \|\mathbf{w}_{xx}\|_{W^{2,\infty}}^2 + \|\mathbf{b}_{xx}\|_{W^{2,\infty}}^2 + \|\theta_{xx}\|_{W^{2,\infty}}^2 + \|\theta_{txx}\|_{H^1}^2$$
$$+ \|u_{txx}\|_{H^1}^2 + \|\mathbf{w}_{txx}\|_{H^1}^2 + \|\mathbf{b}_{txx}\|_{H^1}^2 + \|\theta_{tx}\|_{W^{1,\infty}}^2 + \|u_{tx}\|_{W^{1,\infty}}^2 + \|\mathbf{w}_{tx}\|_{W^{1,\infty}}^2$$
$$+ \|\mathbf{b}_{tx}\|_{W^{1,\infty}}^2 + \|\tau_{txxx}\|_{H^1}^2\Big)(s)ds \leq C_4, \tag{2.4.43}$$

$$\int_0^t \Big(\|u_{xxxx}\|_{H^1}^2 + \|\mathbf{w}_{xxxx}\|_{H^1}^2 + \|\mathbf{b}_{xxxx}\|_{H^1}^2 + \|\theta_{xxxx}\|_{H^1}^2\Big)(s)ds \leq C_4. \tag{2.4.44}$$

Proof. Adding up (2.4.29) through(2.4.32), taking $\varepsilon \in (0,1)$ enough small, and using Lemmas 2.4.1–2.4.3 and Gronwall's inequality, we get

$$\|u_{tx}(t)\|^2 + \|\mathbf{w}_{tx}(t)\|^2 + \|\mathbf{b}_{tx}(t)\|^2 + \|\theta_{tx}(t)\|^2$$

$$+ \int_0^t \left(\|u_{txx}\|^2 + \|\mathbf{w}_{txx}\|^2 + \|\mathbf{b}_{txx}\|^2 + \|\theta_{txx}\|^2 \right)(s)ds$$

$$\le C_3 \varepsilon^{-6} + C_2 \varepsilon^2 \int_0^t \Big(\|u_{ttx}\|^2 + \|\mathbf{w}_{ttx}\|^2 + \|\mathbf{b}_{ttx}\|^2$$

$$+ \|\theta_{ttx}\|^2 + \|\theta_{tx}\|\|\theta_{xxx}\| \Big)(s)ds, \qquad (2.4.45)$$

which, combined with (2.4.24)–(2.4.27), yields (2.4.41).

Differentiating (2.1.2) with respect to x, and using $\tau_{txx} = u_{xxx}$, we have

$$\lambda \frac{\partial}{\partial t} \left(\frac{\tau_{xx}}{\tau} \right) + \frac{R\theta}{\tau} \frac{\tau_{xx}}{\tau} = u_{tx} + E(x,t), \qquad (2.4.46)$$

where

$$E(x,t) = \left(\frac{R}{\tau} + \frac{4a}{3}\theta^3 \right) \theta_{xx} - \frac{2R\tau_x \theta_x}{\tau^2} + \frac{2\tau_x^2 R\theta}{\tau^3} + 4a\theta^2 \theta_x^2 + \mathbf{b} \cdot \mathbf{b}_{xx} + |\mathbf{b}_x|^2$$

$$+ \frac{2\lambda \tau_x u_{xx}}{\tau^2} - \frac{2\lambda \tau_x^2 u_x}{\tau^3}.$$

Differentiating equation (2.4.46) with respect to x, we obtain

$$\lambda \frac{\partial}{\partial t} \left(\frac{\tau_{xxx}}{\tau} \right) + \frac{R\theta}{\tau} \frac{\tau_{xxx}}{\tau} = u_{txx} + E_x(x,t) + \frac{2\theta \tau_x \tau_{xx}}{\tau^3} - \frac{\theta_x \tau_{xx}}{\tau^2} + \lambda \left(\frac{\tau_x \tau_{xx}}{\tau^2} \right)_t$$

$$\equiv E_1(x,t). \qquad (2.4.47)$$

Obviously, we can infer from Lemmas 2.4.1–2.4.3 that

$$\|E_1(t)\| \le C_2 \Big(\|u_{txx}(t)\| + \|\theta_x(t)\|_{H^2} + \|u_x(t)\|_{H^2} + \|\mathbf{b}_x(t)\|_{H^2} + \|\tau_x(t)\|_{H^1} \Big), \qquad (2.4.48)$$

whence

$$\int_0^t \|E_1(s)\|^2 ds \le C_4. \qquad (2.4.49)$$

Multiplying (2.4.47) by $\frac{\tau_{xxx}}{\tau}$, we get

$$\frac{d}{dt} \left\| \frac{\tau_{xxx}}{\tau}(t) \right\|^2 + C_1^{-1} \left\| \frac{\tau_{xxx}}{\tau}(t) \right\|^2 \le C_1 \|E_1(t)\|^2. \qquad (2.4.50)$$

Combining this with (2.4.49) and using Lemmas 2.4.1–2.4.3, we have

$$\|\tau_{xxx}(t)\|^2 + \int_0^t \|\tau_{xxx}(s)\|^2 ds \le C_4. \qquad (2.4.51)$$

Using (2.4.4), (2.4.6), (2.4.8), (2.4.10), (2.4.12), (2.4.14), (2.4.17), (2.4.19), (2.4.41), (2.4.51), Lemmas 2.4.1–2.4.3, and the embedding theorem, we have

$$\|u_{xxx}(t)\|^2 + \|u_{xx}(t)\|_{L^\infty}^2 + \|\mathbf{w}_{xxx}(t)\|^2 + \|\mathbf{w}_{xx}(t)\|_{L^\infty}^2 + \|\mathbf{b}_{xxx}(t)\|^2 + \|\mathbf{b}_{xx}(t)\|_{L^\infty}^2$$
$$+ \|\theta_{xxx}(t)\|^2 + \|\theta_{xx}(t)\|_{L^\infty}^2 + \int_0^1 \Big(\|u_{xx}\|_{W^{1,\infty}}^2 + \|\mathbf{w}_{xx}\|_{W^{1,\infty}}^2 + \|\mathbf{b}_{xx}\|_{W^{1,\infty}}^2$$
$$+ \|\theta_{xx}\|_{W^{1,\infty}}^2 + \|\theta_{xxx}\|_{H^1}^2 + \|u_{xxx}\|_{H^1}^2 + \|\mathbf{w}_{xxx}\|_{H^1}^2 + \|\mathbf{b}_{xxx}\|_{H^1}^2 \Big)(s)ds \le C_4.$$
$$(2.4.52)$$

Differentiating equations (2.1.2)–(2.1.5) with respect to t and using (2.4.41) and Lemmas 2.4.1–2.4.3, we get

$$\|u_{txx}(t)\| \le C_1 \|u_{tt}(t)\| + C_1(\|u_{tx}(t)\| + \|\mathbf{b}_{tx}(t)\| + \|\theta_{tx}(t)\|) \le C_4, \quad (2.4.53)$$
$$\|\mathbf{w}_{txx}(t)\| \le C_1 \|\mathbf{w}_{tt}(t)\| + C_1(\|\mathbf{w}_{tx}(t)\| + \|\mathbf{b}_{tx}(t)\|) \le C_4, \quad (2.4.54)$$
$$\|\mathbf{b}_{txx}(t)\| \le C_1 \|\mathbf{b}_{tt}(t)\| + C_1(\|\mathbf{w}_{tx}(t)\| + \|\mathbf{b}_{tx}(t)\|) \le C_4, \quad (2.4.55)$$
$$\|\theta_{txx}(t)\| \le C_1 \|\theta_{tt}\| + C_1(\|u_{tx}\| + \|\mathbf{w}_{tx}(t)\| + \|\mathbf{b}_{tx}(t)\| + \|\theta_{tx}(t)\|) \le C_4,$$
$$(2.4.56)$$

which, combined with (2.4.6), (2.4.10), (2.4.14) and (2.4.19), implies

$$\|u_{xxxx}(t)\| + \|\mathbf{w}_{xxxx}(t)\| + \|\mathbf{b}_{xxxx}(t)\| + \|\theta_{xxxx}(t)\|$$
$$+ \int_0^t (\|u_{txx}\|^2 + \|\mathbf{w}_{txx}\|^2 + \|\mathbf{b}_{txx}\|^2 + \|\theta_{txx}\|^2$$
$$+ \|u_{xxxx}\|^2 + \|\mathbf{w}_{xxxx}\|^2 + \|\mathbf{b}_{xxxx}\|^2 + \|\theta_{xxxx}\|^2)(s)ds \le C_4. \quad (2.4.57)$$

Next, using (2.4.52), (2.4.57) and the embedding theorem, we obtain

$$\|u_{xxx}(t)\|_{L^\infty} + \|\mathbf{w}_{xxx}(t)\|_{L^\infty} + \|\mathbf{b}_{xxx}(t)\|_{L^\infty} + \|\theta_{xxx}(t)\|_{L^\infty}$$
$$+ \int_0^t \left(\|u_{xxx}\|_{L^\infty}^2 + \|\mathbf{w}_{xxx}\|_{L^\infty}^2 + \|\mathbf{b}_{xxx}\|_{L^\infty}^2 + \|\theta_{xxx}\|_{L^\infty}^2 \right)(s)ds \le C_4. \quad (2.4.58)$$

Further, differentiating equation (2.4.47) with respect to x we obtain

$$\lambda \frac{\partial}{\partial t} \left(\frac{\tau_{xxxx}}{\tau} \right) + \frac{R\theta}{\tau} \frac{\tau_{xxxx}}{\tau} = E_2(x, t) \quad (2.4.59)$$

where

$$E_2(x, t) = E_{1x}(x, t) + \lambda \frac{\partial}{\partial t} \left(\frac{\tau_x \tau_{xxx}}{\tau^2} \right) + \frac{R\theta \tau_x \tau_{xxx}}{\tau^3} - \left(\frac{R\theta}{\tau} \right)_x \frac{\tau_{xxx}}{\tau}.$$

Using the embedding theorem and Lemmas 2.4.1–2.4.3, we conclude that

$$\|E_2(t)\| \le C_2 \left(\|u_{txxx}(t)\| + \|\theta_x(t)\|_{H^3} + \|u_x(t)\|_{H^3} + \|\mathbf{b}_x(t)\|_{H^3} + \|\tau_x(t)\|_{H^2} \right).$$
$$(2.4.60)$$

We infer from (2.4.20)–(2.4.23) that

$$\int_0^t \left(\|u_{tt}\|^2 + \|\mathbf{w}_{tt}\|^2 + \|\mathbf{b}_{tt}\|^2 + \|\theta_{tt}\|^2 \right)(s)ds \le C_4, \tag{2.4.61}$$

which, together with Lemma 2.4.3 and (2.4.41), gives

$$\int_0^t \left(\|u_{txxx}\|^2 + \|\mathbf{w}_{txxx}\|^2 + \|\mathbf{b}_{txxx}\|^2 + \|\theta_{txxx}\|^2 \right)(s)ds \le C_4. \tag{2.4.62}$$

Thus it follows from (2.4.41), (2.4.60) (2.4.62) and Lemmas 2.4.1–2.4.3 that

$$\int_0^t \|E_2(s)\|^2 ds \le C_4. \tag{2.4.63}$$

Multiplying (2.4.59) by $\frac{\tau_{xxxx}}{\tau}$, we get

$$\frac{d}{dt} \left\| \frac{\tau_{xxxx}}{\tau}(t) \right\|^2 + C_1 \left\| \frac{\tau_{xxxx}}{\tau}(t) \right\|^2 \le C_1 \|E_2(t)\|^2, \tag{2.4.64}$$

and so, by (2.4.63),

$$\|\tau_{xxxx}(t)\|^2 + \int_0^t \|\tau_{xxxx}(s)\|^2 ds \le C_4. \tag{2.4.65}$$

Differentiating (2.1.2) with respect to x three times and using Lemmas 2.4.1–2.4.3 and the Poincaré inequality, we have

$$\|u_{xxxxx}(t)\| \le C_3 \|u_{txxx}(t)\|$$
$$+ C_3 \left(\|\tau_x(t)\|_{H^3} + \|u_x(t)\|_{H^3} + \|\theta_x(t)\|_{H^3} + \|\mathbf{b}_x(t)\|_{H^3} \right). \tag{2.4.66}$$

Thus we conclude from (2.1.1), (2.4.57), (2.4.62), (2.4.65) and (2.4.66) that

$$\int_0^t \left(\|u_{xxxxx}\|^2 + \|\tau_{txxx}\|_{H^1}^2 \right)(s)ds \le C_4. \tag{2.4.67}$$

Similarly, we can deduce from (2.1.3)–(2.1.5) that

$$\int_0^t \left(\|\mathbf{b}_{xxxxx}\|^2 + \|\mathbf{w}_{xxxxx}\|^2 + \|\theta_{xxxxx}\|^2 \right)(s)ds \le C_4, \tag{2.4.68}$$

which together with (2.4.52) and (2.4.67) gives

$$\int_0^t \left(\|u_{xx}\|_{W^{2,\infty}}^2 + \|\mathbf{w}_{xx}\|_{W^{2,\infty}}^2 + \|\mathbf{b}_{xx}\|_{W^{2,\infty}}^2 + \|\theta_{xx}\|_{W^{2,\infty}}^2 \right)(s)ds \le C_4. \tag{2.4.69}$$

Finally, using (2.1.1), (2.4.51)–(2.4.57), (2.4.65), (2.4.67)–(2.4.69) and Sobolev's interpolation inequality, we get the desired estimates (2.4.42)–(2.4.44). $\qquad \Box$

Lemma 2.4.5. *Under assumptions of Theorem 2.1.3, for any* $(\tau_0, u_0, \mathbf{w}_0, \mathbf{b}_0, \theta_0) \in H_+^4$, *there exists a positive constant* $\gamma_4^{(1)} = \gamma_4^{(1)}(C_4) \leq \gamma_2(C_2)$ *such that, for any fixed* $\gamma \in (0, \gamma_4^{(1)}]$, *the following estimates hold for any* $t > 0$ *and* $\varepsilon \in (0, 1)$ *small enough:*

$$e^{\gamma t} \|u_{tt}(t)\|^2 + \int_0^t e^{\gamma s} \|u_{ttx}\|^2(s)ds$$

$$\leq C_3 + C_3 \int_0^t e^{\gamma s}(\|\mathbf{b}_{txx}\|^2 + \|\theta_{txx}\|^2)(s)ds, \tag{2.4.70}$$

$$e^{\gamma t} \|\mathbf{w}_{tt}(t)\|^2 + \int_0^t e^{\gamma s} \|\mathbf{w}_{ttx}(s)\|^2 ds$$

$$\leq C_3 + C_3 \int_0^t e^{\gamma s} \|\mathbf{b}_{txx}(s)\|^2 ds, \tag{2.4.71}$$

$$e^{\gamma t} \|\mathbf{b}_{tt}(t)\|^2 + \int_0^t e^{\gamma s} \|\mathbf{b}_{ttx}(s)\|^2 ds$$

$$\leq C_3 + C_3 \int_0^t e^{\gamma s} \|\mathbf{w}_{txx}\|^2(s)ds, \tag{2.4.72}$$

$$e^{\gamma t} \|\theta_{tt}(t)\|^2 + \int_0^t e^{\gamma s} \|\theta_{ttx}(s)\|^2 ds$$

$$\leq C_3 + C_2 \varepsilon^{-1} \int_0^t e^{\gamma s} \|\theta_{txx}\|^2(s)ds + C_1 \varepsilon \int_0^t e^{\gamma s} \Big(\|u_{txx}\|^2 + \|u_{ttx}\|^2$$

$$+ \|\mathbf{w}_{txx}\|^2 + \|\mathbf{w}_{ttx}\|^2 + \|\mathbf{b}_{txx}\|^2 + \|\mathbf{b}_{ttx}\|^2 \Big)(s)ds. \tag{2.4.73}$$

Proof. The proofs of estimates (2.4.70)–(2.4.73) are basically same as those of (2.4.24)–(2.4.27). The difference here is that one estimates (2.4.70)–(2.4.73) with the exponential weight function $e^{\gamma t}$. Multiplying (2.4.28) by $e^{\gamma t}$ and using (2.4.20) and Theorem 2.1.2, we have

$$\frac{1}{2} e^{\gamma t} \|u_{tt}(t)\|^2 \tag{2.4.74}$$

$$\leq C_4 - (C_1^{-1} - C_1 \gamma) \int_0^t e^{\gamma s} \|u_{ttx}(s)\|^2 ds + C_2 \int_0^t e^{\gamma s} \Big(\|\mathbf{b}_{tt}\|^2 + \|\theta_{tt}\|^2 \Big)(s)ds$$

$$\leq C_4 - (C_1^{-1} - C_1 \gamma) \int_0^t e^{\gamma s} \|u_{ttx}(s)\|^2 ds + C_2 \int_0^t e^{\gamma s} \Big(\|\mathbf{b}_{txx}\|^2 + \|\theta_{txx}\|^2 \Big)(s)ds,$$

which gives (2.4.70) if we take $\gamma > 0$ so small that $0 < \gamma \leq \min\left[\frac{1}{4C_1^2}, \gamma_2(C_2)\right]$.

Similarly, we can get (2.4.71)–(2.4.73). The proof is now complete. $\square$

Lemma 2.4.6. *Under assumptions of Theorem 2.1.3, for any* $(\tau_0, u_0, \mathbf{w}_0, \mathbf{b}_0, \theta_0) \in H_+^4$, *there exists a positive constant* $\gamma_4^{(2)} \leq \gamma_4^{(1)}$ *such that, for any fixed* $\gamma \in (0, \gamma_4^{(2)}]$, the following estimates hold for any $t > 0$ and $\varepsilon \in (0,1)$ small enough:

$$e^{\gamma t}\|u_{tx}(t)\|^2 + \int_0^t e^{\gamma s}\|u_{txx}(s)\|^2 ds$$

$$\leq C_3 \varepsilon^{-6} + C_2 \varepsilon^2 \int_0^t e^{\gamma s}\Big(\|\mathbf{b}_{txx}\|^2 + \|\theta_{txx}\|^2 + \|u_{ttx}\|^2\Big)(s)ds, \qquad (2.4.75)$$

$$e^{\gamma t}\|\mathbf{w}_{tx}(t)\|^2 + \int_0^t e^{\gamma s}\|\mathbf{w}_{txx}(s)\|^2 ds$$

$$\leq C_3 \varepsilon^{-6} + C_2 \varepsilon^2 \int_0^t e^{\gamma s}\Big(\|\mathbf{w}_{txx}\|^2 + \|\mathbf{b}_{ttx}\|^2\Big)(s)ds, \qquad (2.4.76)$$

$$e^{\gamma t}\|\mathbf{b}_{tx}(t)\|^2 + \int_0^t e^{\gamma s}\|\mathbf{b}_{txx}(s)\|^2 ds$$

$$\leq C_3 \varepsilon^{-6} + C_2 \varepsilon^2 \int_0^t e^{\gamma s}\Big(\|\mathbf{b}_{txx}\|^2 + \|\mathbf{w}_{ttx}\|^2\Big)(s)ds, \qquad (2.4.77)$$

$$e^{\gamma t}\|\theta_{tx}(t)\|^2 + \int_0^t e^{\gamma s}\|\theta_{txx}(s)\|^2 ds$$

$$\leq C_3 \varepsilon^{-6} + C_2 \varepsilon^2 \int_0^t e^{\gamma s}\Big(\|\mathbf{b}_{txx}\|^2 + \|u_{txx}\|^2 + \|\mathbf{w}_{txx}\|^2$$

$$+ \|\theta_{ttx}\|^2 + \|\theta_{xxx}\|\|\theta_{tx}\|\Big)(s)ds. \qquad (2.4.78)$$

Proof. Multiplying (2.4.33) by $e^{\gamma t}$ and using (2.4.34), (2.4.38) and Theorem 2.1.2, we infer that for any $\varepsilon \in (0,1)$ small enough,

$$e^{\gamma t}\|u_{tx}(t)\|^2 \leq C_3 \varepsilon^{-6} - \Big[(2C_1)^{-1} - \varepsilon^2 - C_1\gamma\Big]\int_0^t e^{\gamma s}\|u_{txx}(s)\|^2 ds$$

$$+ \varepsilon^2 \int_0^t e^{\gamma s}\Big(\|\mathbf{b}_{txx}\|^2 + \|\theta_{txx}\|^2 + \|u_{txxx}\|^2\Big)(s)ds, \qquad (2.4.79)$$

which, combined with (2.4.40), gives (2.4.75) if we take $\gamma > 0$ and $\varepsilon \in (0,1)$ so small that $0 < \varepsilon < \min[1, 1/(8C_1)]$ and $0 < \gamma \leq \min[\gamma_4^{(1)}, 1/(8C_1^2)] \equiv \gamma_4^{(2)}$. In the same manner, we easily derive (2.4.76)–(2.4.78). $\qquad\square$

Lemma 2.4.7. *Under assumptions of Theorem 2.1.3, for any* $(\tau_0, u_0, \mathbf{w}_0, \mathbf{b}_0, \theta_0) \in H_+^4$, *there exists a positive constant* $\gamma_4 \leq \gamma_4^{(2)}$ *such that, for any fixed* $\gamma \in (0, \gamma_4]$, *the following estimates hold for any* $t > 0$:

$$e^{\gamma t}\Big(\|u_{tt}(t)\|^2 + \|u_{tx}(t)\|^2 + \|\mathbf{w}_{tt}(t)\|^2 + \|\mathbf{w}_{tx}(t)\|^2 + \|\mathbf{b}_{tt}(t)\|^2 + \|\mathbf{b}_{tx}(t)\|^2$$

$$+ \|\theta_{tt}(t)\|^2 + \|\theta_{tx}(t)\|^2\Big) + \int_0^t e^{\gamma s}\Big(\|u_{ttx}\|^2 + \|u_{txx}\|^2 + \|\mathbf{w}_{ttx}\|^2 + \|\mathbf{w}_{txx}\|^2$$

$$+ \|\mathbf{b}_{ttx}\|^2 + \|\mathbf{b}_{txx}\|^2 + \|\theta_{ttx}\|^2 + \|\theta_{txx}\|^2\Big)(s)ds \leq C_4, \qquad (2.4.80)$$

$$e^{\gamma t}\left(\|\tau_{xxx}(t)\|_{H^1}^2 + \|\tau_{xx}(t)\|_{W^{1,\infty}}^2\right) + \int_0^t e^{\gamma s}\left(\|\tau_{xxx}\|_{H^1}^2 + \|\tau_{xx}\|_{W^{1,\infty}}^2\right)(s)ds \leq C_4,$$

$$(2.4.81)$$

$$e^{\gamma t}\Big(\|u_{xxx}(t)\|_{H^1}^2 + \|u_{xx}(t)\|_{W^{1,\infty}}^2 + \|\mathbf{w}_{xxx}(t)\|_{H^1}^2 + \|\mathbf{w}_{xx}(t)\|_{W^{1,\infty}}^2 + \|\mathbf{b}_{xxx}(t)\|_{H^1}^2$$

$$+ \|\mathbf{b}_{xx}(t)\|_{W^{1,\infty}}^2 + \|\theta_{xxx}(t)\|_{H^1}^2 + \|\theta_{xx}(t)\|_{W^{1,\infty}}^2 + \|\tau_{txxx}(t)\|^2 + \|u_{txx}(t)\|^2$$

$$+ \|\mathbf{w}_{txx}(t)\|^2 + \|\mathbf{b}_{txx}(t)\|^2 + \|\theta_{txx}(t)\|^2\Big) + \int_0^t e^{\gamma s}\Big(\|u_{tt}\|^2 + \|\mathbf{w}_{tt}\|^2 + \|\mathbf{b}_{tt}\|^2$$

$$+ \|\theta_{tt}\|^2 + \|u_{xx}\|_{W^{2,\infty}}^2 + \|\mathbf{w}_{xx}\|_{W^{2,\infty}}^2 + \|\mathbf{b}_{xx}\|_{W^{2,\infty}}^2 + \|\theta_{xx}\|_{W^{2,\infty}}^2 + \|\theta_{txx}\|_{H^1}^2$$

$$+ \|u_{txx}\|_{H^1}^2 + \|\mathbf{w}_{txx}\|_{H^1}^2 + \|\mathbf{b}_{txx}\|_{H^1}^2 + \|\theta_{tx}\|_{W^{1,\infty}}^2 + \|u_{tx}\|_{W^{1,\infty}}^2 + \|\mathbf{w}_{tx}\|_{W^{1,\infty}}^2$$

$$+ \|\mathbf{b}_{tx}\|_{W^{1,\infty}}^2 + \|\tau_{txxx}\|_{H^1}^2\Big)(s)ds \leq C_4,$$

$$(2.4.82)$$

$$\int_0^t e^{\gamma s}\left(\|u_{xxxx}\|_{H^1}^2 + \|\mathbf{w}_{xxxx}\|_{H^1}^2 + \|\mathbf{b}_{xxxx}\|_{H^1}^2 + \|\theta_{xxxx}\|_{H^1}^2\right)(s)ds \leq C_4.$$

$$(2.4.83)$$

Proof. Multiplying (2.4.70) through (2.4.73) by ε, ε, ε and $\varepsilon^{3/2}$, respectively, adding the resulting inequalities, and then taking $\varepsilon > 0$ small enough, we can obtain the desired estimate (2.4.80).

Next, multiplying (2.4.50) by $e^{\gamma t}$, using (2.4.48), (2.4.80) and Theorem 2.1.2, and choosing $\gamma > 0$ so small that $0 < \gamma \leq \gamma_4 \equiv \min[1/(2C_1), \gamma_4^{(2)}]$, we conclude that for any $t > 0$,

$$e^{\gamma t}\left\|\frac{\tau_{xxx}}{\tau}(t)\right\|^2 + \frac{1}{2C_1}\int_0^t e^{\gamma s}\left\|\frac{\tau_{xxx}}{\tau}(s)\right\|^2 ds \leq C_3 + C_1 \int_0^t e^{\gamma s}\|E_1(s)\|^2 ds \leq C_4,$$

whence

$$e^{\gamma t}\|\tau_{xxx}(t)\|^2 + \int_0^t e^{\gamma s}\|\tau_{xxx}(s)\|^2 ds \leq C_4.$$

$$(2.4.84)$$

Similarly to (2.4.52), (2.4.57)–(2.4.58), (2.4.61)–(2.4.62), using (2.4.80), (2.4.84) and Theorem 2.1.2, we have that for any fixed $\gamma \in (0, \gamma_4]$ and for any $t > 0$,

$$e^{\gamma t}\Big(\|u_{xxx}(t)\|_{H^1}^2 + \|u_{xx}(t)\|_{W^{1,\infty}}^2 + \|\mathbf{w}_{xxx}(t)\|_{H^1}^2 + \|\mathbf{w}_{xx}(t)\|_{W^{1,\infty}}^2 + \|\mathbf{b}_{xxx}(t)\|_{H^1}^2$$

$$+ \|\mathbf{b}_{xx}(t)\|_{W^{1,\infty}}^2 + \|\theta_{xxx}(t)\|_{H^1}^2 + \|\theta_{xx}(t)\|_{W^{1,\infty}}^2 + \|u_{txx}(t)\|^2 + \|\mathbf{w}_{txx}(t)\|^2$$

$$+ \|\mathbf{b}_{txx}(t)\|^2\Big) + \int_0^t e^{\gamma s}\Big(\|u_{xx}\|_{W^{2,\infty}}^2 + \|\mathbf{w}_{xx}\|_{W^{2,\infty}}^2 + \|\mathbf{b}_{xx}\|_{W^{2,\infty}}^2 + \|\theta_{xx}\|_{W^{2,\infty}}^2$$

$$+ \|\theta_{txx}\|_{H^1}^2 + \|u_{txx}\|_{H^1}^2 + \|\mathbf{w}_{txx}\|_{H^1}^2 + \|\mathbf{b}_{txx}\|_{H^1}^2 + \|\theta_{tx}\|_{W^{1,\infty}}^2 + \|u_{tx}\|_{W^{1,\infty}}^2$$

$$+ \|\mathbf{w}_{tx}\|_{W^{1,\infty}}^2 + \|\mathbf{b}_{tx}\|_{W^{1,\infty}}^2\Big)(s)ds \leq C_4$$

$$(2.4.85)$$

and

$$\int_0^t e^{\gamma s}\Big(\|u_{tt}\|^2 + \|\mathbf{w}_{tt}\|^2 + \|\mathbf{b}_{tt}\|^2 + \|\theta_{tt}\|^2 + \|\theta_{txx}\|^2$$

$$+ \|u_{txx}\|^2 + \|\mathbf{w}_{txx}\|^2 + \|\mathbf{b}_{txx}\|^2\Big)(s)ds \le C_4. \tag{2.4.86}$$

Multiplying (2.4.64) by $e^{\gamma t}$ and using (2.4.60), (2.4.80), (2.4.84)–(2.4.86) and Theorem 2.1.2, we get that, for any fixed $\gamma \in (0, \gamma_4]$,

$$e^{\gamma t}\left\|\frac{\tau_{xxxx}}{\tau}(t)\right\|^2 + \frac{1}{2C_1^{-1}}\int_0^t e^{\gamma s}\left\|\frac{\tau_{xxxx}}{\tau}(s)\right\|^2 ds \le C_4 + C_1 \int_0^t e^{\gamma s}\|E_2(s)\|^2 ds \le C_4,$$

that is,

$$e^{\gamma t}\|\tau_{xxxx}(t)\|^2 + \int_0^t e^{\gamma s}\|\tau_{xxxx}(s)\|^2 ds \le C_4, \quad \forall t > 0. \tag{2.4.87}$$

Similarly to (2.4.67)–(2.4.69), we easily derive that for any fixed $\gamma \in (0, \gamma_4]$,

$$\int_0^t e^{\gamma s}\Big(\|u_{xxxxx}\|^2 + \|\mathbf{w}_{xxxxx}\|^2 + \|\mathbf{b}_{xxxxx}\|^2 + \|\theta_{xxxxx}\|^2 + \|\tau_{txxx}\|_{H^1}^2 \tag{2.4.88}$$

$$+ \|u_{xx}\|_{W^{2,\infty}}^2 + \|\theta_{xx}\|_{W^{2,\infty}}^2 + \|\mathbf{w}_{xx}\|_{W^{2,\infty}}^2 + \|\mathbf{b}_{xx}\|_{W^{2,\infty}}^2\Big)(s)ds \le C_4, \quad \forall t > 0.$$

Finally, we combine estimates (2.4.80), (2.4.84)–(2.4.88) with the interpolation inequality to derive the required estimates (2.4.81)–(2.4.83). The proof is now complete. □

Thus we have completed the proof of Theorem 2.1.3. □

2.5 Bibliographic Comments

Let us mention a number of previous works in this direction. For the one-dimensional ideal gas, i.e.,

$$e = C_v\theta, \quad \sigma = -\frac{R\theta}{\tau} + \frac{\mu}{\tau}u_x, \quad Q = -\kappa\frac{\theta_x}{\tau}, \quad \mathbf{w} = \mathbf{b} \equiv \mathbf{0}, \tag{2.5.1}$$

with suitable positive constants C_v, R, Kazhikhov [60, 61], Kazhikhov and Shelukhin [63], and Kawashima and Nishida [57] established the existence of global smooth solutions. Zheng and Qin [150] proved the existence of maximal attractors in H^i ($i = 1, 2$). However, for very high temperatures and densities, the constitutive relations (2.5.1) become inadequate. Thus a more realistic model would be a linearly viscous gas (or Newtonian fluid) with

$$\sigma(\tau, \theta, u_x) = -p(\tau, \theta) + \frac{\mu(\tau, \theta)}{\tau}u_x, \tag{2.5.2}$$

satisfying Fourier's law of heat flux

$$Q(\tau, \theta, \theta_x) = -\frac{\kappa(\tau, \theta)}{\tau}\theta_x, \qquad (2.5.3)$$

whose internal energy e and pressure p are coupled by the standard thermodynamical relation

$$e_\tau(\tau, \theta) = -p(\tau, \theta) + \theta p_\theta(\tau, \theta). \qquad (2.5.4)$$

In this case, Kawohl [59] and Jiang [50] obtained the existence of global solutions to 1D viscous heat-conductive real gas under different growth assumptions on the pressure p, internal energy e and heat conductivity κ in terms of temperature. Qin [97, 99–101] established the regularity and asymptotic behavior of global solutions under more general growth assumptions on p, e, κ than those in [50, 59].

For the radiative and reactive gas, Ducomet [22] established the global existence and exponential decay in H^1 of smooth solutions, and Qin et al. [105] extended the results in [22], further establishing the global existence and exponential stability of solutions in H^i $(i = 1, 2, 4)$. Umehara and Tani [129], Qin et al. [103] and Qin, Hu and Wang [104] proved the global existence of smooth solutions for a self-gravitating radiative and reactive gas.

For non-radiative MHD flows (i.e., $a \equiv 0$ in (2.1.6)), there have been a number of studies by several authors, under various conditions (see, e.g., [10, 11], [33, 34, 47], [131, 132]). The existence and uniqueness of local smooth solutions were first obtained in [131], and the existence of global smooth solutions with small smooth initial data was shown in [125]. Under the technical condition that $\kappa(\rho, \theta)$ satisfies

$$0 < C^{-1}(1 + \theta^q) \le \kappa(\rho, \theta) \le C(1 + \theta^q)$$

for $q \ge 2$, Chen and Wang [10] proved the existence and continuous dependence of global strong solutions with large initial data satisfying

$$0 < \inf \rho_0 \le \rho_0(x) \le \sup \rho_0 < +\infty, \quad \rho_0, u_0, \mathbf{w}_0, \mathbf{b}_0, \theta_0 \in H^1(\Omega), \ \theta_0(x) > 0.$$

Chen and Wang [11] also investigated a free boundary problem with general large initial data. Wang [132] established the existence of large solutions to the initial-boundary value problem for planar magnetohydrodynamics. Under a technical condition upon $\kappa(\rho)$, namely

$$\kappa(\rho, \theta) \equiv \kappa(\rho) > \frac{C}{\rho},$$

Fan, Jiang and Nakamura [33] investigated the uniqueness of the weak solutions of MHD with initial data in Lebesgue spaces. Fan, Jiang and Nakamura [34] also considered one-dimensional plane MHD compressible flows, and proved that as the shear viscosity goes to zero, global weak solutions converge to a solution of the original equations with zero shear viscosity. The uniqueness and continuous dependence of weak solutions for the Cauchy problem have been proved by Hoff and Tsyganov [47].

For compressible and radiative MHD flows (i.e., $a > 0$ in (2.1.6)) with self-gravitation, Ducomet and Feireisl [25] proved the existence of global-in-time solutions with arbitrarily large initial data and conservative boundary conditions on a bounded spatial domain in $\mathbb{R}^3$. Under the technical condition that $\kappa(\rho, \theta)$ satisfies

$$k_1(1 + \theta^q) \leq \kappa(\rho, \theta), \quad |\kappa_\rho(\rho, \theta)| \leq k_2(1 + \theta^q),$$

for some $q > \frac{5}{2}$, Zhang and Xie [145] investigated the existence of global smooth solutions to problem (2.1.1)–(2.1.8). In this chapter, we established the global existence and exponential stability of solutions in H^i ($i = 1, 2, 4$) to problem (2.1.1)–(2.1.8). However, the large-time behavior is still open even for the non-self-gravitating case, i.e., (2.1.1)–(2.1.8).

Chapter 3

Global Smooth Solutions for 1D Thermally Radiative Magnetohydrodynamics with Self-gravitation

3.1 Introduction

In this chapter, we shall consider the one-dimensional motion of a compressible, thermally radiative fluid with magnetic diffusion. This motion is described in the Euler coordinates by the following equations, corresponding to the conservation laws of mass, momentum and energy:

$$\rho_t + (\rho u)_x = 0, \tag{3.1.1}$$

$$(\rho u)_t + \left(\rho u^2 + p + \frac{1}{2}|\mathbf{b}|^2\right)_x = (\lambda u_x)_x + \rho \psi_x, \tag{3.1.2}$$

$$(\rho \mathbf{w})_t + (\rho u \mathbf{w} - \mathbf{b})_x = (\mu \mathbf{w}_x)_x, \tag{3.1.3}$$

$$\mathbf{b}_t + (u\mathbf{b} - \mathbf{w})_x = (\nu \mathbf{b}_x)_x, \tag{3.1.4}$$

$$\mathcal{E}_t + \left(u\left(\mathcal{E} + p + \frac{1}{2}|\mathbf{b}|^2\right) - \mathbf{w} \cdot \mathbf{b}\right)_x + Q_x$$

$$= (\lambda u u_x + \mu \mathbf{w} \cdot \mathbf{w}_x + \nu \mathbf{b} \cdot \mathbf{b}_x)_x + \rho \psi_x u \tag{3.1.5}$$

where ρ is the density, $u \in \mathbb{R}$ the longitudinal velocity, $\mathbf{w} = (w_1, w_2) \in \mathbb{R}^2$ the transverse velocity, $\mathbf{b} = (b_1, b_2) \in \mathbb{R}^2$ the transverse magnetic field, and θ the temperature, $p = p(\rho, \theta)$ the total pressure and $e = e(\rho, \theta)$ the internal energy. Further, $\rho \psi_x$ represents the gravitational force, where the function ψ is determined by the boundary value problem

$$\begin{cases} -\psi_{xx} = G\rho, & (x,t) \in \Omega \times (0,T), \\ \psi|_{\partial\Omega} = 0, \end{cases} \tag{3.1.6}$$

where $\Omega \subset \mathbb{R}$ is a bounded domain. The total energy $\mathcal{E}$ is given by

$$\mathcal{E} = \rho\left(e + \frac{1}{2}(u^2 + |\mathbf{w}|^2)\right) + \frac{1}{2}|\mathbf{b}|^2, \tag{3.1.7}$$

where $\rho(u^2 + |\mathbf{w}|^2)/2$ is the kinetic energy and $|\mathbf{b}|^2/2$ is the magnetic energy. The heat flux Q takes the form

$$Q = Q_F + Q_R = -k\theta_x, \tag{3.1.8}$$

where $k = k(\rho,\theta)$ is the heat conductivity coefficient, Q_F is given by the Fourier's law, and Q_R is the radiation heat flux.

In agreement with the classical Boyle's law that applies in the non-degenerate area of high temperatures and low densities, we may assume that $p(\rho,\theta)$ and $e(\rho,\theta)$ take the forms

$$p(\rho,\theta) = R\rho\theta + \frac{a}{3}\theta^4, \qquad e(\rho,\theta) = C_V\theta + \frac{a}{\rho}\theta^4, \tag{3.1.9}$$

and the heat conductivity $\kappa = \kappa(\rho,\theta)$ satisfies the growth condition

$$\kappa_1(1+\theta^q) \le k(\rho,\theta), \qquad |\kappa_\rho(\rho,\theta)| \le \kappa_2(1+\theta^q) \qquad \text{for any } q > (2+\sqrt{211})/9 \tag{3.1.10}$$

where $R > 0$ is the perfect gas constant and $C_V > 0$ is the specific heat at constant volume, respectively.

We shall consider the initial-boundary value problem (3.1.1)–(3.1.6) in a bounded spatial domain $\Omega = (0,1)$ subject to the following initial and boundary conditions:

$$\begin{cases} (\rho, u, \mathbf{w}, \mathbf{b}, \theta)|_{t=0} = (\rho_0, u_0, \mathbf{w}_0, \mathbf{b}_0, \theta_0)(x), \\ (u, \mathbf{w}, \mathbf{b}, \theta_x)|_{x=0} = (u, \mathbf{w}, \mathbf{b}, \theta_x)|_{x=1} = 0, \end{cases} \tag{3.1.11}$$

where the initial data satisfy certain compatibility conditions as usual. Note that the boundary conditions in $(3.1.11)_2$ imply that the boundary is non-slip, impermeable, and thermally insulated.

In this chapter, we shall consider problems (3.1.1)–(3.1.11), which is similar to a model in ideal MHD considered for specially physical interests by Ojha and Singh [92], where the effects of radiation and magnetic fields were discussed in the cases of plane, cylindrical and spherical flows.

We shall prove the global existence of a unique classical solution of system (3.1.1)–(3.1.6) with initial-boundary conditions (3.1.11) for $q > (2 + \sqrt{211})/9$, improving the result in [145] for $q > \frac{5}{2}$.

For the sake of simplicity, we assume that the viscosity coefficients λ, μ and the magnetic diffusivity ν are constants, and assume the heat conductivity $\kappa = \kappa(\rho, \theta)$ is strictly positive and continuously differentiable on $\mathbb{R}_+ \times \mathbb{R}_+$, $\mathbb{R}_+ = [0, +\infty)$.

Our main result in this chapter is as follows.

Theorem 3.1.1. *Assume that the total pressure $p = p(\rho, \theta)$, the internal energy $e = e(\rho, \theta)$ and the heat conductivity $\kappa = \kappa(\rho, \theta)$ satisfy (3.1.9) and (3.1.10). Assume also that the initial data $(\rho_0, u_0, \mathbf{w}_0, \mathbf{b}_0, \theta_0)$ satisfy the compatibility conditions and, for some $\alpha \in (0, 1)$,*

$$(\rho_0, u_0, \mathbf{w}_0, \mathbf{b}_0, \theta_0) \in C^{1+\alpha}(\Omega) \times \left(C^{2+\alpha}(\Omega) \right)^6,$$

and there exists a constant $C_0 > 0$ such that, for any $x \in (0, 1)$,

$$0 < C_0^{-1} \leq \rho_0(x) \leq C_0, \qquad 0 < C_0^{-1} \leq \theta_0(x) \leq C_0.$$

Then there exists a unique classical solution $(\rho, u, \mathbf{w}, \mathbf{b}, \theta)$ of the initial-boundary value problem (3.1.1)–(3.1.11) such that, for any $T > 0$,

$$\rho(x, t) > 0 \quad and \quad \theta(x, t) > 0 \quad for \ any \quad (x, t) \in Q_T = \Omega \times (0, T), \tag{3.1.12}$$

$$(\rho, \rho_x, \rho_t) \in \left(C_{x,t}^{\alpha, \alpha/2}(Q_T) \right)^3, \qquad (u, \mathbf{w}, \mathbf{b}, \theta) \in \left(C_{x,t}^{2+\alpha, 1+\alpha/2}(Q_T) \right)^6, \tag{3.1.13}$$

and $\psi \geq 0$ satisfying that $\psi_{xx} \in C_{x,t}^{1+\alpha, 1+\alpha/2}(Q_T)$ is determined by the boundary value problem (3.1.6).

The existence of global-in-time solutions will be proved by continuing the local solutions with respect to time based on global a priori estimates. The existence and uniqueness of the local solution to the initial-boundary value problem (3.1.1)–(3.1.10) can be proved by the standard method based on the Banach fixed-point theorem. We omit the details of the proof of local existence here. Note that ψ can be solved from (3.1.6) in terms of ρ. To prove the global existence, it is thus sufficient to establish global a priori estimates for the solutions. The main difficulty is caused by the high-temperature radiation terms appearing in (3.1.2) and (3.1.5), which makes the upper bound for θ more complicated than that in the works mentioned above. This will be done by proper inequalities to reduce the higher order of θ in the equations (3.1.1)–(3.1.5).

Throughout this chapter we denote by $C^{m+\alpha}(\Omega)$ and $C^{2m+\alpha, m+\alpha/2}(Q_T)$ with $m \in \mathbb{Z}_+, 0 < \alpha < 1$, the standard Hölder spaces, and by $W^{m,p}(\Omega)$ ($W^{0,p}(\Omega) \equiv L^p(\Omega), W^{m,2}(\Omega) \equiv H^m(\Omega)$) with $1 \leq p \in \mathbb{R}$, $m \in \mathbb{Z}_+$ the usual Sobolev space. For simplicity, we also use the following abbreviations:

$$\| \cdot \|_{H^m} = \| \cdot \|_{H^m(\Omega)}, \quad \| \cdot \|_{L^p} = \| \cdot \|_{L^p(\Omega)}, \quad \| \cdot \|_{L^p(0,T;H^m)} = \| \cdot \|_{L^p(0,T;H^m(\Omega))}.$$

Moreover, the same letter C denotes various generic positive constants appearing in the estimates, which may depend on T.

3.2   A Priori Estimates

In order to establish the global existence of solutions, we need some a priori estimates for the solution and its derivatives on $(0,1) \times (0,T)$ for any fixed $T > 0$, which will be derived by detailed analysis of the equations. As was mentioned above, the main difficulty, caused by the effect of high-temperature radiation in momentum and energy equations, lies in obtaining a global uniform upper bound for θ which then plays an important role in the derivation of a priori estimates on the second derivatives of the quantities $(u, \mathbf{w}, \mathbf{b}, \theta)$.

To prove Theorem 3.1.1, we shall establish several lemmas concerning estimates of the solution and its derivatives. Our methods are mainly based on the techniques in Qin [101, 103–105], that is, we carefully estimate the solution and its higher derivatives in terms of functions A, X, Y and Z (see their definitions below) and resort to delicate interpolation techniques.

Lemma 3.2.1. *For any $t \in [0,T]$, it holds that*

$$m(t) := \int_0^1 \rho(x,t)dx = \int_0^1 \rho_0(x)dx := m_0, \tag{3.2.1}$$

$$\sup_{t \in (0,T)} \int_0^1 \psi_x^2(x,t)dx \leq C, \tag{3.2.2}$$

$$0 \leq \psi(x,t) \leq C, \qquad \forall (x,t) \in Q_T, \tag{3.2.3}$$

$$\|\psi_{xx}\|_{L^2}^2 \leq C, \qquad \|\psi_x\|_{L^\infty}^2 \leq C, \tag{3.2.4}$$

$$\int_0^1 \left(\rho(\theta + u^2 + |\mathbf{w}|^2) + \theta^4 + |\mathbf{b}|^2 \right)(x,t)dx \leq C, \tag{3.2.5}$$

$$\int_0^1 \rho(\ln \rho + |\ln \theta|)(x,t)dx$$
$$+ \int_0^t \int_0^1 \left(\frac{\kappa \theta_x^2}{\theta^2} + \frac{\lambda u_x^2 + \mu|\mathbf{w}_x|^2 + \nu|\mathbf{b}_x|^2}{\theta} \right) dxds \leq C, \tag{3.2.6}$$

$$0 < C^{-1} \leq \rho(x,t) \leq C, \quad \forall (x,t) \in [0,1] \times [0,T], \tag{3.2.7}$$

$$\int_0^T \|\theta(t)\|_{L^\infty}^{q+4}dt \leq C, \tag{3.2.8}$$

$$\int_0^T \left(\|\theta\|_{L^\infty} + \|\mathbf{b}\|_{L^\infty}^2 + \|\theta\|_{L^\infty}^4 \right) dt \leq C, \tag{3.2.9}$$

$$\int_0^T \int_0^1 (u_x^2 + |w_x|^2 + |\mathbf{b}_x|^2)(x,t)dxdt \leq C, \tag{3.2.10}$$

$$\int_0^1 \rho_x^2 dx + \int_0^t \int_0^1 \theta \rho_x^2 dxds \leq C. \tag{3.2.11}$$

$$\sup_{t\in(0,T)} \int_0^1 (|\mathbf{w}|^4 + |\mathbf{b}|^4)(x,t)dx$$

$$+ \int_0^T \int_0^1 (|\mathbf{w}|^2|\mathbf{w}_x|^2 + |\mathbf{b}|^2|\mathbf{b}_x|^2)dxdt \le C, \tag{3.2.12}$$

$$\sup_{t\in(0,T)} \int_0^1 |(\mathbf{w}_x,\mathbf{b}_x)|^2 dx + \int_0^T \int_0^1 |(\mathbf{w}_t,\mathbf{w}_{xx},\mathbf{b}_t,\mathbf{b}_{xx})|^2 dxds \le C. \tag{3.2.13}$$

Proof. See, e.g., [145]. $\square$

Now we define the following functions:

$$|u|^{(0)} := \sup_{(x,t)\in Q_T} |u(x,t)|, \qquad A = A(t) := \sup_{0\le s\le t} \|\theta(s)\|_{L^\infty(\Omega)},$$

$$X = X(t) := \int_0^t \int_0^1 (1+\theta^{q+3})\theta_t^2 dxds, \quad Y = Y(T) := \max_{t\in[0,T]} \int_0^1 (1+\theta^{2q})\theta_x^2 dx,$$

$$Z = Z(T) := \max_{t\in[0,T]} \|u_{xx}(t)\|^2.$$

From these definitions, we can obtain the following estimates.

Lemma 3.2.2. *For any $t \in [0,T]$, it holds that*

$$A \le |\theta|^{(0)} \le C + CY^{\frac{1}{2q+6}}, \tag{3.2.14}$$

$$\max_{t\in[0,T]} \|u_x\|^2 \le C + CZ^{\frac{1}{2}}, \tag{3.2.15}$$

$$|u_x|^{(0)} \le C + CZ^{\frac{3}{8}}. \tag{3.2.16}$$

Proof. Firstly, from (3.2.5) and (3.2.7), we have

$$\int_0^1 u^2(x,t)dx \le C, \qquad \int_0^1 \theta(x,t)dx \le C, \tag{3.2.17}$$

By using the mean value theorem, we find that, for each $t \in [0,T]$ there exists a point $\xi^*(t) \in [0,1]$ such that

$$\theta(\xi^*,t) = \int_0^1 \theta(\xi,t)d\xi \le C,$$

which, combined with (3.2.5), yields that for any $x \in [0,1]$, and for any $\varepsilon > 0$,

$$\theta^{2q+6}(x,t) = \theta^{2q+6}(\xi^*,t) + (2q+6)\int_{\xi^*}^x \theta^{2q+5}\theta_\eta(\eta,t)d\eta$$

$$\le C + C\int_0^1 \theta^{2q+5}|\theta_\eta|d\eta$$

$$\leq C + C \left(\int_0^1 (1+\theta^{2q})\theta_\eta^2 d\eta \right)^{\frac{1}{2}} \left(\int_0^1 (1+\theta)^{2q+10} d\eta \right)^{\frac{1}{2}}$$

$$\leq C + \varepsilon \int_0^1 (1+\theta)^{2q+10} d\eta + C(\varepsilon) \int_0^1 (1+\theta^{2q})\theta_\eta^2 d\eta$$

$$\leq C + \varepsilon \left(|\theta|^{(0)} \right)^{2q+6} \int_0^1 \theta^4 d\eta + C(\varepsilon) \int_0^1 (1+\theta^{2q})\theta_\eta^2 d\eta$$

$$\leq C + C\varepsilon \left(|\theta|^{(0)} \right)^{2q+6} + CY. \tag{3.2.18}$$

Taking $\varepsilon > 0$ sufficiently small and taking the supremum over $\mathcal{Q}_T$ of the left-hand of inequality (3.2.18), we obtain

$$\left(|\theta|^{(0)} \right)^{2q+6} \leq C + CY,$$

which implies

$$A \leq |\theta|^{(0)} \leq C + CY^{\frac{1}{2q+6}}.$$

Secondly, using (3.2.17) and the Gagliardo-Nirenberg interpolation inequality, we get

$$\|u_x\|_{L^2}^2 \leq \|u_x\|_{L^\infty}^2 \leq C\|u\|_{L^2}\|u_{xx}\|_{L^2} + C\|u\|_{L^2}^2$$
$$\leq C\|u_{xx}\|_{L^2} + C,$$

from which we can derive the inequality (3.2.14).
Finally, by (3.2.14) and the Gagliardo-Nirenberg interpolation inequality, we also get

$$\|u_x\|_{L^\infty} \leq C\|u_x\|_{L^2}^{\frac{1}{2}}\|u_{xx}\|_{L^2}^{\frac{1}{2}} + C\|u_x\|_{L^2}$$
$$\leq \left(CZ^{\frac{1}{4}} + C \right) CZ^{\frac{1}{8}} + CZ^{\frac{1}{4}} + C,$$

which implies

$$\|u_x\|_{L^\infty} \leq C + CZ^{\frac{3}{8}},$$

that is,

$$|u_x|^{(0)} \leq C + CZ^{\frac{3}{8}}.$$

This completes the proof of Lemma 3.2.2. □

Lemma 3.2.3. *For any $t \in (0,T)$, the following inequalities hold:*

$$\int_0^1 u^4 dx + \int_0^t \int_0^1 u^2 u_x^2 dx ds \leq C + CA^{\gamma_1}, \tag{3.2.19}$$

$$\int_0^1 u_x^2 dx + \int_0^t \int_0^1 (u_t^2 + u_{xx}^2) dx ds \leq C + CA^{\gamma_2}, \tag{3.2.20}$$

where $\gamma_1 = \max(0, 4-q)$, and $\gamma_2 = \max(0, 8-q)$.

Proof. Multiplying (3.1.2) by $4u^3$ and integrating by parts over $(0,1) \times (0,t)$, we get

$$\int_0^1 \rho u^4 dx + 12\lambda \int_0^t \int_0^1 u^2 u_x^2 dx ds \qquad (3.2.21)$$

$$= \int_0^1 \rho_0 u_0^4 dx + 12 \int_0^t \int_0^1 \left(p + \frac{1}{2}|\mathbf{b}|^2\right) u^2 u_x dx ds + 4 \int_0^t \int_0^1 \rho \psi_x u^3 dx ds$$

which implies

$$\int_0^1 \rho u^4 dx + 12\lambda \int_0^t \int_0^1 u^2 u_x^2 dx ds$$

$$\leq C + C \left| \int_0^t \int_0^1 (p + \frac{1}{2}|\mathbf{b}|^2) u^2 u_x dx ds \right| + C \left| \int_0^t \int_0^1 \rho \psi_x u^3 dx ds \right|$$

where the terms on the right-hand side are estimated as follows.

Firstly, by Lemmas 3.2.1 and 3.2.2 and the Gagliardo-Nirenberg interpolation inequality, we get

$$\left| \int_0^t \int_0^1 \rho \psi_x u^3 dx ds \right| \leq C \left| \int_0^t \int_0^1 u^3 dx ds \right| \leq C \int_0^t \|u\|_{L^\infty} \left(\int_0^1 u^2 dx\right) ds$$

$$\leq C \int_0^t \|u\|_{L^\infty} ds \leq C \int_0^t \|u_x\|_{L^2}^2 ds + C$$

$$\leq C.$$

On the other hand, since $\|(\mathbf{b}, \theta)\|_{L^\infty(0,T;L^4)} \leq C$, recalling the definition of the total pressure p and using Lemmas 3.2.1 and 3.2.2 again, we find that, for any $\varepsilon > 0$,

$$\left| \int_0^t \int_0^1 \left(p + \frac{1}{2}|\mathbf{b}|^2\right) u^2 dx ds \right|$$

$$\leq \varepsilon \int_0^t \int_0^1 u^2 u_x^2 dx ds + C(\varepsilon) \int_0^t \int_0^1 \left(p + \frac{1}{2}|\mathbf{b}|^2\right)^2 u^2 dx ds$$

$$\leq \varepsilon \int_0^t \int_0^1 u^2 u_x^2 dx ds + C(\varepsilon) \int_0^t \left(\|p + \frac{1}{2}|\mathbf{b}|^2\|_{L^\infty}^2\right) \left(\int_0^1 u^2 dx\right) ds$$

$$\leq \varepsilon \int_0^t \int_0^1 u^2 u_x^2 dx ds + C(\varepsilon) \int_0^t \left(\|p + \frac{1}{2}|\mathbf{b}|^2\|_{L^\infty}^2\right) ds$$

$$\leq \varepsilon \int_0^t \int_0^1 u^2 u_x^2 dx ds + C(\varepsilon) + C(\varepsilon) \int_0^t \left(\|1 + \theta\|_{L^\infty}^8 + \|\frac{1}{2}|\mathbf{b}|^2\|_{L^\infty}^2\right) ds$$

$$\leq \varepsilon \int_0^t \int_0^1 u^2 u_x^2 dx ds + C(\varepsilon) + C(\varepsilon) \int_0^t \left(\|1 + \theta\|_{L^\infty}^{q+4} \|1 + \theta\|_{L^\infty}^{4-q}\right) ds$$

$$\leq \varepsilon \int_0^t \int_0^1 u^2 u_x^2 dx ds + C(\varepsilon) + C(1 + A^{\gamma_1})$$

$$\leq \varepsilon \int_0^t \int_0^1 u^2 u_x^2 dx ds + C(\varepsilon) + C A^{\gamma_1}.$$

Thus we obtain the inequality (3.2.19) by taking $\varepsilon > 0$ sufficiently small.

Noting that the total pressure is given by $p(\rho, \theta) = R\rho\theta + \frac{a}{3}\theta^4$, we infer from (3.1.2) that

$$\rho u_t^2 + \frac{\lambda^2}{\rho} u_{xx}^2 - 2\lambda u_t u_{xx} = \rho^{-1}\left(\rho u u_x + R\rho_x\theta + R\rho\theta_x + \frac{4}{3}a\theta^3\theta_x + \mathbf{b}\cdot\mathbf{b}_x - \rho\psi_x\right)^2,$$

which, together with Lemmas 3.2.1 and 3.2.2, leads to

$$\lambda \int_0^1 u_x^2 dx + \int_0^t \int_0^1 \left(\rho u_t^2 + \frac{\lambda^2}{\rho} u_{xx}^2\right) dx ds$$

$$\leq C + C \int_0^t \int_0^1 (u^2 u_x^2 + \rho_x^2 \theta^2 + \rho^2 \theta_x^2 + \theta^6 \theta_x^2 + |\mathbf{b}|^2|\mathbf{b}_x|^2 + \rho^2\psi_x^2) dx ds$$

$$\leq C + C A^{\gamma_1} + C \int_0^t \|\theta(s)\|_{L^\infty}^2 \left(\int_0^1 \rho_x^2 dx\right) ds$$

$$+ C \max_{t\in(0,T)} \|\theta(t)\|_{L^\infty}^{\gamma_0} \int_0^t \int_0^1 \frac{\kappa\theta_x^2}{\theta^2} dx ds + C \max_{t\in(0,T)} \|\theta(t)\|_{L^\infty}^{\gamma_2} \int_0^t \int_0^1 \frac{\kappa\theta_x^2}{\theta^2} dx ds$$

$$\leq C + C A^{\gamma_1} + C A^{\gamma_2}$$

$$\leq C + C A^{\gamma_2},$$

here $\gamma_0 = \max(0, 2 - q)$.

The proof of Lemma 3.2.3 is complete. $\qquad\square$

Lemma 3.2.4. *For $q > 1$, there exists a constant $\delta \in [0, 1)$ such that*

$$X + Y \leq C + C Z^\delta. \tag{3.2.22}$$

Proof. In view of the equations of state (3.1.9), equation (3.1.5) can be rewritten as

$$\rho e_\theta \theta_t + \rho u e_\theta \theta_x + \theta p_\theta u_x = (\kappa\theta_x)_x + \lambda u_x^2 + \mu|\mathbf{w}_x|^2 + \nu|\mathbf{b}_x|^2, \tag{3.2.23}$$

which, multiplied by $\kappa\theta_t$ and integrated over $(0, 1) \times (0, t)$, yields

$$\int_0^t \int_0^1 \rho e_\theta \kappa\theta_t^2 dx ds + \int_0^t \int_0^1 \kappa\theta_x(\kappa\theta_t)_x dx ds$$

$$= \int_0^t \int_0^1 \left(\lambda u_x^2 + \mu|\mathbf{w}_x|^2 + \nu|\mathbf{b}_x|^2 - \rho u e_\theta \theta_x - \theta p_\theta u_x\right) \kappa\theta_t dx ds. \tag{3.2.24}$$

Note that (3.1.1) also implies

$$(\kappa\theta_t)_x = (\kappa\theta_x)_t + \kappa_x\theta_t - \kappa_t\theta_x$$
$$= (\kappa\theta_x)_t + \kappa_\rho\rho_x\theta_t - \kappa_\rho\rho_t\theta_x$$
$$= (\kappa\theta_x)_t + \kappa_\rho\rho_x\theta_t + \kappa_\rho\theta_x(\rho_x u + \rho u_x),$$

and so we can rewrite (3.2.24) as

$$\int_0^t\int_0^1 \rho e_\theta\kappa\theta_t^2 dx ds + \frac{1}{2}\int_0^1 (\kappa\theta_x)^2 dx - \frac{1}{2}\int_0^1 (\kappa\theta_x)^2 dx\Big|_{t=0}$$
$$= \int_0^t\int_0^1 (\lambda u_x^2 + \mu|w_x|^2 + \nu|b_x|^2 - \rho u e_\theta\theta_x - \theta p_\theta u_x)\kappa\theta_t dx ds$$
$$- \int_0^t\int_0^1 \kappa_\rho\kappa\theta_x(\rho_x\theta_t + \rho u_x\theta_x + u\rho_x\theta_x)dx ds := \sum_{i=1}^{5} L_i. \qquad (3.2.25)$$

As for the terms on the left-hand side, recalling that $C^{-1} \le e_\theta(\rho,\theta)/(1+\theta)^3 \le C$ (because of (3.1.9) and Lemma 3.2.2) and using Lemma 3.2.2 and (3.1.11), we have

$$\int_0^t\int_0^1 \rho\kappa e_\theta\theta_t^2 dx ds + \frac{1}{2}\int_0^1 (\kappa\theta_x)^2 dx - \frac{1}{2}\int_0^1 (\kappa\theta_x)^2 dx\Big|_{t=0}$$
$$\ge C^{-1}\left(\int_0^1 (1+\theta)^{2q}\theta_x^2 dx + \int_0^t\int_0^1 (1+\theta)^{q+3}\theta_t^2 dx ds\right) - C$$
$$\ge C^{-1}(X+Y) - C.$$

To bound the right-hand side of (3.2.25), we firstly use the Cauchy-Schwarz inequality and Lemma 3.2.3 to deduce that, for $1 < q \le 8$ and for any $\varepsilon > 0$,

$$|L_1| := \left|\int_0^t\int_0^1 (\lambda u_x^2 + \mu|w_x|^2 + \nu|b_x|^2)\kappa\theta_t dx ds\right|$$
$$\le C\left|\int_0^t\int_0^1 (\lambda u_x^2 + \mu|w_x|^2 + \nu|b_x|^2)(1+\theta)^q\theta_t dx ds\right|$$
$$\le \varepsilon\int_0^t\int_0^1 (1+\theta)^{q+3}\theta_t^2 dx ds + C\int_0^t\int_0^1 (1+\theta)^{q-3}(u_x^4 + |w_x|^4 + |b_x|^4)dx ds$$
$$\le \varepsilon X + C(1+A^{\gamma_3})\int_0^t\int_0^1 (u_x^4 + |w_x|^4 + |b_x|^4)dx ds$$
$$\le \varepsilon X + C(1+A^{\gamma_3})|u_x^2|^{(0)}\int_0^t\int_0^1 u_x^2 dx ds + C$$
$$\le \varepsilon X + C(1+A^{\gamma_3})(C + CZ^{\frac{3}{4}}) + C$$
$$\le \varepsilon X + C + CA^{\gamma_3} + CA^{\gamma_3}Z^{\frac{3}{4}} + CZ^{\frac{3}{4}}$$

$$\leq \varepsilon(X+Y)+C+CZ^{\delta_1}.$$

Here $\gamma_3 = \max(0, q-3)$, $\delta_1 = \frac{3}{2} \cdot \frac{q+3}{2q+6-\gamma_3}$, and we have used the following facts

$$\int_0^T \int_0^1 |(\mathbf{w}_x, \mathbf{b}_x)|^4 dxdt \leq C\left(1 + \int_0^T \|(\mathbf{w}_x, \mathbf{b}_x)(t)\|_{L^\infty}^2 \int_0^1 |(\mathbf{w}_x, \mathbf{b}_x)|^2 dxdt\right)$$

$$\leq C\left(1 + \int_0^T \|(\mathbf{w}_{xx}, \mathbf{b}_{xx})(t)\|_{L^2}^2 dt\right) \leq C$$

and, since $\gamma_3 < 2q + 6$, by Lemma 3.2.1,

$$CA^{\gamma_3} \leq \frac{1}{2}\varepsilon Y + C \quad \text{and} \quad CA^{\gamma_3} Z^{\frac{3}{4}} \leq \frac{1}{2}\varepsilon Y + CZ^{\delta_1}.$$

For $q \geq 8$, we have and for any $\varepsilon > 0$,

$$|L_1| \leq \varepsilon X + C \int_0^t \int_0^1 (1+\theta)^{q-3}(u_x^4 + |\mathbf{w}_x|^4 + |\mathbf{b}_x|^4)dxds$$

$$\leq \varepsilon X + C(1 + A^{q-3}) + C$$

$$\leq \varepsilon(X+Y)+C,$$

where we used the facts that $q - 3 < 2q + 6$, and

$$CA^{q-3} \leq \varepsilon Y + C.$$

Secondly, using the Cauchy-Schwarz inequality and the Gagliardo-Nirenberg interpolation inequality, we can easily deduce from Lemmas 3.2.1–3.2.3 that, for $q > 1$, and for any $\varepsilon > 0$,

$$|L_2| := \left|\int_0^t \int_0^1 (\rho u e_\theta \theta_x + \theta p_\theta u_x)\kappa\theta_t dxds\right|$$

$$\leq C \int_0^t \int_0^1 |(\rho u\theta_x + \theta u_x)(1+\theta)^{q+3}\theta_t|dxds$$

$$\leq \varepsilon \int_0^t \int_0^1 (1+\theta)^{q+3}\theta_t^2 dxds + C \int_0^t \int_0^1 (1+\theta)^{q+3}(\rho^2 u^2\theta_x^2 + \theta^2 u_x^2)dxds$$

$$\leq \varepsilon X + C \int_0^t \int_0^1 (1+\theta)^{q+3}u^2\theta_x^2 dxds + C \int_0^t \int_0^1 (1+\theta)^{q+5}u_x^2 dxds$$

$$\leq \varepsilon X + C(1+A^5)\int_0^t \|u\|_{L^2}\|u_x\|_{L^2}\left(\int_0^1 \frac{\kappa\theta_x^2}{\theta^2}dx\right)ds + C(1+A^{q+5})$$

$$\leq \varepsilon X + C(1+A^5)|u_x|^{(0)}\int_0^t \int_0^1 \frac{\kappa\theta_x^2}{\theta^2}dxds + C(1+A^{q+5})$$

$$\leq \varepsilon X + C(1+A^5)(C + CZ^{\frac{3}{8}}) + C + CA^{q+5}$$

$$\leq \varepsilon(X + Y) + CZ^{\delta_2},$$

where $\delta_2 = \frac{3}{4} \cdot \frac{q+3}{2q+1} < 1$ and here we know that, since $5 < 2q + 6$,

$$CA^5 \leq \frac{1}{2}\varepsilon Y + C, \quad \text{and} \quad CA^5 Z^{\frac{3}{8}} \leq \frac{1}{2}\varepsilon Y + CZ^{\delta_2}.$$

Thirdly, in view of Lemma 3.2.2, by the Hölder and the Cauchy-Schwarz inequalities we have that, for any $\varepsilon > 0$,

$$|L_3| := \left| \int_0^t \int_0^1 \kappa\theta_x \kappa_\rho \theta_t \rho_x dx ds \right|$$

$$\leq C \int_0^t \int_0^1 |\kappa\theta_x (1+\theta)^q \theta_t \rho_x| \, dx ds$$

$$\leq C \int_0^t \max_{x \in (0,1)} \left((1+\theta)^{\frac{q-3}{2}} |\kappa\theta_x| \right) \int_0^1 \left| (1+\theta)^{\frac{q+3}{2}} \theta_t \rho_x \right| dx ds$$

$$\leq C \int_0^t \max_{x \in (0,1)} \left((1+\theta)^{\frac{q-3}{2}} |\kappa\theta_x| \right) \left(\int_0^1 (1+\theta)^{q+3} \theta_t^2 dx \right)^{\frac{1}{2}} \left(\int_0^1 \rho_x^2 dx \right)^{\frac{1}{2}} ds$$

$$\leq \varepsilon \int_0^t \int_0^1 (1+\theta)^{q+3} \theta_t^2 dx ds + C \int_0^t \max_{x \in (0,1)} \left((1+\theta)^{q-3} |\kappa\theta_x|^2 \right) ds$$

$$\leq \varepsilon X + C \int_0^t \max_{x \in (0,1)} \left((1+\theta)^{q-3} |\kappa\theta_x|^2 \right) ds.$$

In a similar manner, by Lemmas 3.2.1–3.2.3, we find that, for $1 < q \leq 8$, and for any $\varepsilon > 0$,

$$|L_4| := \left| \int_0^t \int_0^1 \rho\kappa_\rho \kappa\theta_x^2 u_x dx ds \right|$$

$$\leq C \int_0^t \int_0^1 |\kappa\theta_x^2 (1+\theta)^q u_x| \, dx ds$$

$$\leq C \int_0^t \max_{x \in (0,1)} \left((1+\theta)^{\frac{q-3}{2}} |\kappa\theta_x| \right) \left[\int_0^1 \left| \theta_x (1+\theta)^{\frac{q+3}{2}} u_x \right| dx \right] ds$$

$$\leq C \int_0^t \max_{x \in (0,1)} \left((1+\theta)^{\frac{q-3}{2}} |\kappa\theta_x| \right) \left(\int_0^1 (1+\theta)^{q+3} \theta_x^2 dx \right)^{\frac{1}{2}} \left(\int_0^1 u_x^2 dx \right)^{\frac{1}{2}} ds$$

$$\leq C \max_{t \in (0,T)} \|u_x\|_{L^2} \int_0^t \max_{x \in (0,1)} \left((1+\theta)^{\frac{q-3}{2}} |\kappa\theta_x| \right) \left(\int_0^1 (1+\theta)^{q+3} \theta_x^2 dx \right)^{\frac{1}{2}} ds$$

$$\leq (C + CZ^{\frac{1}{4}}) \left[\int_0^t \max_{x \in (0,1)} \left((1+\theta)^{q-3} |\kappa\theta_x|^2 \right) ds + \int_0^t \int_0^1 (1+\theta)^{q+3} \theta_x^2 dx ds \right]$$

$$\leq (C + CZ^{\frac{1}{4}})(C + CA^5) + (C + CZ^{\frac{1}{4}}) \int_0^t \max_{x \in (0,1)} \left((1+\theta)^{q-3} |\kappa\theta_x|^2 \right) ds$$

$$\leq C + CZ^{\frac{1}{4}} + CA^5 Z^{\frac{1}{4}} + CA^5 + (C + CZ^{\frac{1}{4}}) \int_0^t \max_{x \in (0,1)} \left((1+\theta)^{q-3}|\kappa\theta_x|^2\right) ds$$

$$\leq \varepsilon Y + CZ^{\frac{1}{2} \cdot \frac{q+3}{2q+1}} + (C + CZ^{\frac{1}{4}}) \int_0^t \max_{x \in (0,1)} \left((1+\theta)^{q-3}|\kappa\theta_x|^2\right) ds$$

$$\leq \varepsilon Y + CZ^{2\delta_2/3} + (C + CZ^{\frac{1}{4}}) \int_0^t \max_{x \in (0,1)} \left((1+\theta)^{q-3}|\kappa\theta_x|^2\right) ds,$$

while for $q \geq 8$, and for any $\varepsilon > 0$, it holds that

$$|L_4| \leq C \max_{t \in (0,T)} \|u_x\|_{L^2} \int_0^t \max_{x \in (0,1)} \left((1+\theta)^{\frac{q-3}{2}}|\kappa\theta_x|\right) \left(\int_0^1 (1+\theta)^{q+3}\theta_x^2 dx\right)^{\frac{1}{2}} ds$$

$$\leq C \int_0^t \max_{x \in (0,1)} \left((1+\theta)^{\frac{q-3}{2}}|\kappa\theta_x|\right) \left(\int_0^1 (1+\theta)^{q+3}\theta_x^2 dx\right)^{\frac{1}{2}} ds$$

$$\leq \varepsilon Y + C + C \int_0^t \max_{x \in (0,1)} \left((1+\theta)^{q-3}|\kappa\theta_x|^2\right) ds.$$

Finally, by virtue of Lemmas 3.2.1–3.2.3 and the Gagliardo-Nirenberg interpolation inequality, we derive, for any $\varepsilon > 0$, the estimates

$$|L_5| := \left|\int_0^t \int_0^1 \kappa_\rho u \rho_x \kappa \theta_x^2 dx ds\right| \leq C \int_0^t \int_0^1 \left|(1+\theta)^q u \rho_x \kappa \theta_x^2\right| dx ds$$

$$\leq C \max_{t \in (0,T)} \|u\|_{L^\infty} \int_0^t \max_{x \in (0,1)} \left((1+\theta)^{\frac{q-3}{2}}|\kappa\theta_x|\right) \left(\int_0^1 (1+\theta)^{\frac{q+3}{2}} \rho_x \theta_x dx\right) ds$$

$$\leq C \max_{t \in (0,T)} \|u\|_{L^\infty} \int_0^t \max_{x \in (0,1)} \left((1+\theta)^{q-3}|\kappa\theta_x|\right)$$

$$\times \left(\int_0^1 (1+\theta)^{q+3}\theta_x^2 dx\right)^{\frac{1}{2}} \left(\int_0^1 \rho_x^2 dx\right)^{\frac{1}{2}} ds$$

$$\leq C \max_{t \in (0,T)} \|u\|_{L^\infty} \left(\int_0^t \max_{x \in (0,1)} \left((1+\theta)^{q-3}|\kappa\theta_x|^2\right) ds\right)^{\frac{1}{2}}$$

$$\times \left(\int_0^t \int_0^1 (1+\theta)^{q+3}\theta_x^2 dx ds\right)^{\frac{1}{2}}$$

$$\leq C \max_{t \in (0,T)} \|u\|_{L^\infty}^2 \int_0^t \int_0^1 (1+\theta)^{q+3}\theta_x^2 dx ds$$

$$+ C \int_0^t \max_{x \in (0,1)} \left((1+\theta)^{q-3}|\kappa\theta_x|^2\right) ds$$

$$\leq C \max_{t \in (0,T)} \|u_x\|_{L^2} \|u\|_{L^2} (C + CA^5) + C \int_0^t \max_{x \in (0,1)} \left((1+\theta)^{q-3}|\kappa\theta_x|^2\right) ds$$

$$\leq (C + CZ^{\frac{1}{4}})(C + CA^5) + C \int_0^t \max_{x \in (0,1)} \left((1+\theta)^{q-3}|\kappa\theta_x|^2\right) ds$$

$$\leq C + CZ^{\frac{1}{4}} + CA^5 Z^{\frac{1}{4}} + CA^5 + C \int_0^t \max_{x \in (0,1)} \left((1+\theta)^{q-3} |\kappa \theta_x|^2 \right) ds$$

$$\leq \varepsilon Y + CZ^{\frac{1}{2} \cdot \frac{q+3}{2q+1}} + C + C \int_0^t \max_{x \in (0,1)} \left((1+\theta)^{q-3} |\kappa \theta_x|^2 \right) ds$$

$$\leq \varepsilon Y + CZ^{2\delta_2/3} + C + C \int_0^t \max_{x \in (0,1)} \left((1+\theta)^{q-3} |\kappa \theta_x|^2 \right) ds.$$

By the estimates of L_i $(i = 1, 2, 3, 4, 5)$, and choosing ε appropriately small, we conclude that for any $t \in (0, T)$, for $1 < q \leq 8$,

$$X + Y \leq C + CZ^{\delta_1} + CZ^{\delta_2} + (C + CZ^{\frac{1}{4}}) \int_0^t \max_{x \in (0,1)} \left((1+\theta)^{q-3} |\kappa \theta_x|^2 \right) ds, \quad (3.2.26)$$

and for $q \geq 8$,

$$X + Y \leq C + CZ^{\delta_2} + C \int_0^t \max_{x \in (0,1)} \left((1+\theta)^{q-3} |\kappa \theta_x|^2 \right) ds. \quad (3.2.27)$$

It is clear from (3.2.26) and (3.2.27) that we still need to deal with the term $(1+\theta)^{q-3} |\kappa \theta_x|^2$. To this end, we observe from (3.2.23) that

$$|(\kappa \theta_x)_x|^2 = \left(\rho e_\theta \theta_t + \rho u e_\theta \theta_x + \theta p_\theta u_x - \lambda u_x^2 - \mu |\mathbf{w}_x|^2 - \nu |\mathbf{b}_x|^2 \right)^2. \quad (3.2.28)$$

Multiplying (3.2.28) by $(1+\theta)^{q-3}$ and integrating the resulting equation over $(0,1) \times (0,t)$, we deduce that

$$\int_0^t \int_0^1 (1+\theta)^{q-3} |(\kappa \theta_x)_x|^2 \, dx ds$$

$$= \int_0^t \int_0^1 (1+\theta)^{q-3} \left(\rho e_\theta \theta_t + \rho u e_\theta \theta_x + \theta p_\theta u_x - \lambda u_x^2 - \mu |\mathbf{w}_x|^2 - \nu |\mathbf{b}_x|^2 \right)^2 dx ds$$

$$\leq C \int_0^t \int_0^1 \left[(1+\theta)^{q-3} (\theta_t^2 + \theta_x^2 u^2) + (1+\theta)^{q+5} u_x^2 \right]$$

$$+ C \int_0^t \int_0^1 (1+\theta)^{q-3} (u_x^4 + |\mathbf{w}_x|^4 + |\mathbf{b}_x|^4) dx ds. \quad (3.2.29)$$

By the Sobolev inequality and Lemmas 3.2.1–3.2.3, we can easily see that, for any $\varepsilon > 0$,

$$\int_0^t \int_0^1 (1+\theta)^{q+3} \theta_x^2 u^2 dx ds \leq \int_0^t \|u\|_{L^\infty}^2 \int_0^1 (1+\theta)^{q+3} \frac{\theta^2}{k} \cdot \frac{\kappa \theta_x^2}{\theta^2} dx ds$$

$$\leq C \int_0^t \|u\|_{L^2} \|u_x\|_{L^2} \int_0^1 (1+\theta^5) \frac{\kappa \theta_x^2}{\theta^2} dx ds \leq C \int_0^t \|u_x\|_{L^2} \int_0^1 (1+\theta^5) \frac{\kappa \theta_x^2}{\theta^2} dx ds$$

$$\leq C \sup_{t \in (0,T)} \|u_x\|_{L^2} (1+A^5) \int_0^t \int_0^1 \frac{\kappa \theta_x^2}{\theta^2} dx ds \leq (C + CZ^{\frac{1}{4}})(1+A^5)$$

$$\leq \varepsilon Y + C + CZ^{\frac{1}{2} \cdot \frac{q+3}{2q+1}} \leq \varepsilon Y + C + CZ^{\delta_2},$$

$$\int_0^t \int_0^1 (1+\theta)^{q+5} u_x^2 dx ds \le (C+CA^{q+5}) \int_0^t \int_0^1 u_x^2 dx ds$$

$$\le C + CA^{q+5} \le \varepsilon Y + C,$$

and from the derivation of the estimate for L_1, we also find that for $1 < q \le 8$,

$$\int_0^t \int_0^1 (1+\theta)^{q-3} (u_x^4 + |\mathbf{w}_x|^4 + |\mathbf{b}_x|^4) dx ds \le \varepsilon(X+Y) + CZ^{\delta_1} + C,$$

and for $q \ge 8$,

$$\int_0^t \int_0^1 (1+\theta)^{q-3} (u_x^4 + |\mathbf{w}_x|^4 + |\mathbf{b}_x|^4) dx ds \le \varepsilon(X+Y) + C.$$

Hence, inserting these estimates into (3.2.29), we get for $1 < q \le 8$,

$$\int_0^t \int_0^1 (1+\theta)^{q-3} |(\kappa\theta_x)_x|^2 \, dx ds$$

$$\le C \int_0^t \int_0^1 (1+\theta)^{q+3} \theta_t^2 dx ds + C + \varepsilon(X+Y) + CZ^{\delta_1} + CZ^{\delta_2}, \quad (3.2.30)$$

and for $q \ge 8$,

$$\int_0^t \int_0^1 (1+\theta)^{q-3} |(\kappa\theta_x)_x|^2 \, dx ds$$

$$\le C \int_0^t \int_0^1 (1+\theta)^{q+3} \theta_t^2 dx ds + C + \varepsilon(X+Y) + CZ^{\delta_2}. \quad (3.2.31)$$

Multiplying (3.2.26) and (3.2.27) by $C+1$ and adding to (3.2.30) and (3.2.31), we conclude that for $1 < q \le 8$,

$$X + Y + \int_0^t \int_0^1 (1+\theta)^{q-3} |(\kappa\theta_x)_x|^2 \, dx ds \quad (3.2.32)$$

$$\le C + CZ^{\delta_1} + CZ^{\delta_2} + (C + CZ^{\frac{1}{4}}) \int_0^t \max_{x\in(0,1)} \left((1+\theta)^{q-3}|\kappa\theta_x|^2\right) ds,$$

and for $q \ge 8$,

$$X + Y + \int_0^t \int_0^1 (1+\theta)^{q-3} |(\kappa\theta_x)_x|^2 \, dx ds$$

$$\le C + +CZ^{\delta_2} + C \int_0^t \max_{x\in(0,1)} \left((1+\theta)^{q-3}|\kappa\theta_x|^2\right) ds. \quad (3.2.33)$$

On the other hand, since

$$\max_{x\in(0,1)} |\kappa\theta_x|^2 \le C \int_0^1 |\kappa\theta_x||(\kappa\theta_x)_x| dx$$

the last term on the right-hand side of (3.2.32) and (3.2.33) can be bounded as
follows, using the Cauchy-Schwarz inequality and Lemmas 3.2.1–3.2.3, for $1 < q \leq 8$, and for any $\varepsilon > 0$:

$$\int_0^t \max_{x \in (0,1)} \left((1+\theta)^{q-3} |\kappa \theta_x|^2 \right) ds \leq C(1+A^{\gamma_3}) \int_0^t \int_0^1 |\kappa \theta_x| \, |(\kappa \theta_x)_x| \, dx ds$$

$$\leq \varepsilon \int_0^t \int_0^1 (1+\theta)^{q-3} |(\kappa \theta_x)_x|^2 \, dx ds + C\varepsilon^{-1}(1+A^{2\gamma_3}) \int_0^t \int_0^1 (1+\theta)^{q+3} \theta_x^2 dx ds$$

$$\leq \varepsilon \int_0^t \int_0^1 (1+\theta)^{q-3} |(\kappa \theta_x)_x|^2 \, dx ds + C\varepsilon^{-1}(1+A^{2\gamma_3})(1+A^5) \int_0^t \int_0^1 \frac{\kappa \theta_x^2}{\theta^2} dx ds$$

$$\leq \varepsilon \int_0^t \int_0^1 (1+\theta)^{q-3} |(\kappa \theta_x)_x|^2 \, dx ds + C\varepsilon^{-1}(1+A^{2\gamma_3+5}).$$

If we set here $\varepsilon = \frac{1}{2}(C + CZ^{\frac{1}{4}})^{-1}$, we get

$$(C+CZ^{\frac{1}{4}}) \int_0^t \max_{x \in (0,1)} \left((1+\theta)^{q-3} |\kappa \theta_x|^2 \right) ds$$

$$\leq \frac{1}{2} \int_0^t \int_0^1 (1+\theta)^{q-3} |(\kappa \theta_x)_x|^2 \, dx ds + C(1+Z^{\frac{1}{4}})(1+A^{2\gamma_3+5}), \quad (3.2.34)$$

and for $q \geq 8$, and for any $\varepsilon > 0$,

$$C \int_0^t \max_{x \in (0,1)} \left((1+\theta)^{q-3} |\kappa \theta_x|^2 \right) ds$$

$$\leq C(1+A)^{q-3} \int_0^t \int_0^1 |\kappa \theta_x| \, |(\kappa \theta_x)_x| \, dx ds$$

$$\leq \frac{1}{2} \int_0^t \int_0^1 (1+\theta)^{q-3} |(\kappa \theta_x)_x|^2 \, dx ds + C(1+A^{2q-6}) \int_0^t \int_0^1 (1+\theta)^{q+3} \theta_x^2 dx ds$$

$$\leq \frac{1}{2} \int_0^t \int_0^1 (1+\theta)^{q-3} |(\kappa \theta_x)_x|^2 \, dx ds + C + CA^{2q-1}$$

$$\leq \frac{1}{2} \int_0^t \int_0^1 (1+\theta)^{q-3} |(\kappa \theta_x)_x|^2 \, dx ds + \varepsilon Y + C. \quad (3.2.35)$$

Inserting the inequalities (3.2.34) and (3.2.35) into (3.2.32) and (3.2.33), we can
deduce that, for $1 < q \leq 8$,

$$X + Y + \int_0^t \int_0^1 (1+\theta)^{q-3} |(\kappa \theta_x)_x|^2 \, dx ds$$

$$\leq C + CZ^{\delta_1} + CZ^{\delta_2} + (C + CZ^{\frac{1}{4}})(1 + A^{2\gamma_3+5})$$

$$\leq C + CZ^{\delta_1} + CZ^{\delta_2} + CZ^{\frac{1}{4}} A^{2\gamma_3+5} + CZ^{2\gamma_3+5} + CZ^{\frac{1}{4}}$$

$$\leq \varepsilon Y + C + CZ^{\frac{1}{4}} + CZ^{\delta_1} + CZ^{\delta_2} + CZ^{\frac{1}{2} \cdot \frac{q+3}{2q+1-2\gamma_3}},$$

and for $q \geq 8$,

$$X + Y + \int_0^t \int_0^1 (1+\theta)^{q-3} |(\kappa\theta_x)_x|^2 \, dx ds \leq C + CZ^{\delta_2}.$$

Hence, we can easily get that, for $1 < q \leq 8$,

$$X + Y \leq C + CZ^{\delta}, \tag{3.2.36}$$

where $\delta = \max(\delta_1, \delta_2, \frac{1}{4}, \frac{1}{2} \cdot \frac{q+3}{2q+1-2\gamma_3})$, and for $q \geq 8$,

$$X + Y \leq C + CZ^{\delta}, \tag{3.2.37}$$

where $\delta = \delta_2$.

Thus the proof of Lemma 3.2.4 is complete.   □

Applying the previous lemmas, we can get the boundedness of Z.

Lemma 3.2.5. *For $q > (2 + \sqrt{211})/9$, we have*

$$Z \leq C. \tag{3.2.38}$$

Proof. Since

$$\rho u_t^2 + \frac{\lambda^2}{\rho} u_{xx}^2 - 2\lambda u_t u_{xx} = \rho^{-1} \left(\rho u u_x + R\rho_x \theta + R\rho\theta_x + \frac{4}{3}a\theta^3\theta_x + \mathbf{b} \cdot \mathbf{b}_x - \rho\psi_x \right)^2,$$

we derive

$$\frac{\lambda^2}{\rho} u_{xx}^2 = 2\lambda u_t u_{xx} - \rho u_t^2 + \rho^{-1} \left(\rho u u_x + R\rho_x \theta + R\rho\theta_x + \frac{4}{3}a\theta^3\theta_x + \mathbf{b} \cdot \mathbf{b}_x - \rho\psi_x \right)^2. \tag{3.2.39}$$

Integrating the equality (3.2.39) on $(0,1)$ with respect to x, we get

$$\int_0^1 u_{xx}^2 dx \leq C \int_0^1 (u^2 u_x^2 + \rho_x^2 \theta^2 + \theta_x^2 + \theta^6\theta_x^2 + |\mathbf{b}|^2|\mathbf{b}_x|^2 + \psi_x^2) dx + C \int_0^1 u_t^2 dx$$

$$\leq C + CZ^{\frac{3}{4}} + CA^2 + CZ^{\delta} + CA^{\gamma_4} + CA^{\gamma_4}Z^{\delta} + C\|u_t\|_{L^2}^2. \tag{3.2.40}$$

Here we have used the inequalities

$$\int_0^1 u^2 u_x^2 dx \leq |u_x^2|^{(0)} \int_0^1 u^2 dx \leq C|u_x^2|^{(0)} \leq C + CZ^{\frac{3}{4}},$$

$$\int_0^1 \rho_x^2 \theta^2 dx \leq CA^2 \int_0^1 \rho_x^2 dx \leq CA^2,$$

$$\int_0^1 \theta_x^2 dx \leq C + CZ^{\delta},$$

$$\int_0^1 \theta^6 \theta_x^2 dx \le A^{\gamma_4} \int_0^1 \theta^{2q} \theta_x^2 dx \le (C + CZ^\delta) A^{\gamma_4} \le CA^{\gamma_4} + CA^{\gamma_4} Z^\delta,$$

where $\gamma_4 = \max(6 - 2q, 0)$, and

$$\int_0^1 \psi_x^2 dx \le C,$$

$$\int_0^1 |\mathbf{b}|^2 |\mathbf{b}_x|^2 dx \le \|\mathbf{b}\|_{L^\infty}^2 \int_0^1 |\mathbf{b}_x|^2 dx \le C \int_0^1 |\mathbf{b}_x|^2 dx \le C. \tag{3.2.41}$$

By (3.1.2), we can derive

$$\rho u_t + \rho u_x u + \left(p + \frac{1}{2}|\mathbf{b}|^2 \right)_x = \lambda u_{xx} + \rho \psi_x. \tag{3.2.42}$$

Differentiating (3.2.42) with respect to t, we deduce

$$\rho u_{tt} - (\rho u)_x u u_x + \rho u u_{xt} + \left(p + \frac{1}{2}|\mathbf{b}|^2 \right)_{xt} = \lambda u_{xxt} + (\rho \psi_x)_t. \tag{3.2.43}$$

Multiplying (3.2.43) by $2u_t$ and integrating with respect to x, t on $(0,1) \times (0,t)$, we arrive at

$$\int_0^1 \rho u_t^2 dx + 2\lambda \int_0^t \int_0^1 u_{xt}^2 dx ds$$

$$= -2 \int_0^t \int_0^1 \rho u(u u_x u_t)_x dx ds + 2 \int_0^t \int_0^1 \left(p + \frac{1}{2}|\mathbf{b}|^2 \right)_t u_{xt} dx ds$$

$$+ 2 \int_0^t \int_0^1 \rho u(\psi_x u_t)_x dx ds + 2 \int_0^t \int_0^1 \rho \psi_{xt} u_t dx ds. \tag{3.2.44}$$

Using the Young inequality and the Gagliardo-Nirenberg interpolation inequality, we can obtain the following estimates for any $\varepsilon > 0$:

$$\int_0^t \int_0^1 \rho u(u u_x u_t)_x dx ds = \int_0^t \int_0^1 \left(\rho u u_x^2 u_t + \rho u^2 u_{xx} u_t + \rho u^2 u_x u_{xt} \right) dx ds, \tag{3.2.45}$$

$$\int_0^t \int_0^1 \rho u u_x^2 dx ds \le \varepsilon \int_0^t \int_0^1 \rho u_t^2 dx ds + C \int_0^t \int_0^1 u^2 u_x^4 dx ds$$

$$\le \varepsilon \int_0^t \int_0^1 \rho u_t^2 dx ds + C|u_x^2|^{(0)} \int_0^t \int_0^1 u^2 u_x^2 dx ds$$

$$\le \varepsilon \int_0^t \int_0^1 \rho u_t^2 dx ds + (C + CZ^{\frac{3}{4}})(C + CA^{\gamma_1})$$

$$\le \varepsilon \int_0^t \int_0^1 \rho u_t^2 dx ds + C + CA^{\gamma_1} + CA^{\gamma_1} Z^{\frac{3}{4}} + CZ^{\frac{3}{4}}, \tag{3.2.46}$$

$$\int_0^t \int_0^1 \rho u^2 u_{xx} u_t \, dx \, ds \le \frac{\varepsilon}{T} \int_0^t \int_0^1 u_{xx}^2 \, dx \, ds + C \int_0^t \int_0^1 u^4 u_t^2 \, dx \, ds$$

$$\le \varepsilon Z + C \sup_{t \in (0,T)} \|u\|_{L^\infty}^4 \int_0^t \int_0^1 \rho u_t^2 \, dx \, ds$$

$$\le \varepsilon Z + C \sup_{t \in (0,T)} \|u\|_{L^2}^2 \|u_x\|_{L^2}^2 \int_0^t \int_0^1 \rho u_t^2 \, dx \, ds$$

$$\le \varepsilon Z + C \sup_{t \in (0,T)} \|u_x\|_{L^2}^2 \int_0^t \int_0^1 \rho u_t^2 \, dx \, ds$$

$$\le \varepsilon Z + (C + CA^{\gamma_2}) \int_0^t \int_0^1 \rho u_t^2 \, dx \, ds, \qquad (3.2.47)$$

$$\int_0^t \int_0^1 \rho u^2 u_x u_{xt} \, dx \, ds \le \varepsilon \int_0^t \int_0^1 u_{xt}^2 \, dx \, ds + C \int_0^t \int_0^1 u^4 u_x^2 \, dx \, ds$$

$$\le \varepsilon \int_0^t \int_0^1 u_{xt}^2 \, dx \, ds + C \sup_{t \in (0,T)} \|u\|_{L^\infty}^4 \int_0^t \int_0^1 u_x^2 \, dx \, ds$$

$$\le \varepsilon \int_0^t \int_0^1 u_{xt}^2 \, dx \, ds + C \sup_{t \in (0,T)} \|u\|_{L^2}^2 \|u_x\|_{L^2}^2$$

$$\le \varepsilon \int_0^t \int_0^1 u_{xt}^2 \, dx \, ds + C \sup_{t \in (0,T)} \|u_x\|_{L^2}^2$$

$$\le \varepsilon \int_0^t \int_0^1 u_{xt}^2 \, dx \, ds + C + CA^{\gamma_2}. \qquad (3.2.48)$$

Inserting the inequalities (3.2.46)–(3.2.48) into (3.2.45), we deduce that for any $\varepsilon > 0$,

$$\int_0^t \int_0^1 \rho u (u u_x u_t)_x \, dx \, ds \le \varepsilon \int_0^t \int_0^1 u_{xt}^2 \, dx \, ds + (C + CA^{\gamma_2}) \int_0^t \int_0^1 \rho u_t^2 \, dx \, ds$$

$$+ CA^{\gamma_2} + CA^{\gamma_1} + CA^{\gamma_1} Z^{\frac{3}{4}} + CZ^{\frac{3}{4}} + \varepsilon Z + C. \qquad (3.2.49)$$

Similarly, we can also show that, for any $\varepsilon > 0$,

$$\int_0^t \int_0^1 \rho u (\psi_x u_t)_x \, dx \, ds = \int_0^t \int_0^1 \rho u (\psi_{xx} u_t + \psi_x u_{xt}) \, dx \, ds$$

$$\le \varepsilon \int_0^t \int_0^1 \rho u_t^2 \, dx \, ds + \varepsilon \int_0^t \int_0^1 u_{xt}^2 \, dx \, ds$$

$$+ C \int_0^t u^2 \psi_{xx}^2 \, dx \, ds + C \int_0^t \int_0^1 u^2 \psi_x^2 \, dx \, ds$$

$$\leq \varepsilon \int_0^t \int_0^1 \rho u_t^2 dx ds + \varepsilon \int_0^t \int_0^1 u_{xt}^2 dx ds + C \int_0^t \|u\|_{L^\infty}^2 \int_0^1 \psi_{xx}^2 dx ds + C$$

$$\leq \varepsilon \int_0^t \int_0^1 \rho u_t^2 dx ds + \varepsilon \int_0^t \int_0^1 u_{xt}^2 dx ds + C \int_0^t \|u\|_{L^\infty}^2 ds + C$$

$$\leq \varepsilon \int_0^t \int_0^1 \rho u_t^2 dx ds + \varepsilon \int_0^t \int_0^1 u_{xt}^2 dx ds + C \int_0^t \|u_x\|_{L^2}^2 ds + C$$

$$\leq \varepsilon \int_0^t \int_0^1 \rho u_t^2 dx ds + \varepsilon \int_0^t \int_0^1 u_{xt}^2 dx ds + C, \tag{3.2.50}$$

and

$$\int_0^t \int_0^1 \rho \psi_{xt} u_t dx ds \leq \varepsilon \int_0^t \int_0^1 \rho u_t^2 dx ds + C \int_0^t \int_0^1 \psi_{xt}^2 dx ds. \tag{3.2.51}$$

By (3.1.6), we know that

$$-\psi_x = \int_0^x G\rho d\xi - \psi_x(x,t)\Big|_{x=0}, \quad \text{and} \quad -\psi_{xt} = -G\rho u - \psi_{xt}(x,t)\Big|_{x=0}.$$

By the Sobolev inequality, we have that, for any $\varepsilon > 0$,

$$\int_0^1 \psi_{xt}^2 dx \leq C \int_0^1 (G\rho u)^2 dx + C \int_0^1 \psi_{xt}^2 \big|_{x=0} dx$$

$$\leq C \int_0^1 \rho^2 u^2 dx + C\|\psi_{xt}\|_{L^\infty}^2$$

$$\leq C \int_0^1 \rho^2 u^2 dx + C\|\psi_{xt}\|_{L^2}\|\psi_{xxt}\|_{L^2}$$

$$\leq \varepsilon\|\psi_{xt}\|_{L^2}^2 + C\|\psi_{xxt}\|_{L^2}^2 + C \int_0^1 \rho^2 u^2 dx$$

$$\leq \varepsilon\|\psi_{xt}\|_{L^2}^2 + C\|\psi_{xxt}\|_{L^2}^2 + C, \tag{3.2.52}$$

and

$$\int_0^1 \psi_{xxt}^2 dx = \int_0^1 (-G\rho)_t^2 dx = \int_0^1 G^2(\rho u)_x^2 dx$$

$$\leq C \int_0^1 (\rho_x^2 u^2 + \rho^2 u_x^2) dx \leq C\|u\|_{L^2}^2 \int_0^1 \rho_x^2 dx + C \int_0^1 u_x^2 dx$$

$$\leq C \int_0^1 u_x^2 dx + C \leq C Z^{\frac{1}{2}} + C. \tag{3.2.53}$$

Inserting (3.2.53) into (3.2.52), we get

$$\int_0^1 \psi_{xt}^2 dx \leq C Z^{\frac{1}{2}} + C. \tag{3.2.54}$$

Next, combining (3.2.51) and (3.2.54), the inequality (3.2.51) can be estimated, for any $\varepsilon > 0$, by

$$\int_0^t \int_0^1 \rho \psi_{xt} u_t \, dx \, ds \leq \varepsilon \int_0^t \int_0^1 \rho u_t^2 \, dx \, ds + C Z^{\frac{1}{2}} + C. \tag{3.2.55}$$

By the Young inequality, we have for any $\varepsilon > 0$,

$$\left| \int_0^t \int_0^1 \left(p + \frac{1}{2}|\mathbf{b}|^2 \right)_t u_{xt} \, dx \, ds \right| \leq \varepsilon \int_0^t \int_0^1 u_{xt}^2 \, dx \, ds + C \int_0^t \int_0^1 \left(p + \frac{1}{2}|\mathbf{b}|^2 \right)_t^2 \, dx \, ds$$

$$\leq \varepsilon \int_0^t \int_0^1 u_{xt}^2 \, dx \, ds + C \int_0^t \int_0^1 (\rho_t \theta + \rho \theta_t + \theta^3 \theta_t + \mathbf{b} \cdot \mathbf{b}_t)^2 \, dx \, ds$$

$$\leq \varepsilon \int_0^t \int_0^1 u_{xt}^2 \, dx \, ds + C + C A^2 + C Z^\delta + C A^{\gamma_5} + C A^{\gamma_5} Z^\delta$$

$$+ C Z^{\delta \cdot \frac{2q+6}{2q+6-\gamma_5}} + C Z^{\delta \cdot \frac{2q+6}{2q+6-\gamma_5}}, \tag{3.2.56}$$

where $\gamma_5 = \max(3 - q, 0)$, and we used the following facts:

$$\int_0^t \int_0^1 (\rho_t \theta + \rho \theta_t + \theta^3 \theta_t + |\mathbf{b} \cdot \mathbf{b}_t|)^2 \, dx \, ds$$

$$\leq C \int_0^t \int_0^1 (\rho_t^2 \theta^2 + \rho^2 \theta_t^2 + \theta^6 \theta_t^2 + |\mathbf{b}|^2 |\mathbf{b}_t|^2) \, dx \, ds,$$

$$\int_0^t \int_0^1 \rho_t^2 \theta^2 \, dx \, ds = \int_0^t \int_0^1 (\rho u)_x^2 \theta^2 \, dx \, ds$$

$$\leq C \int_0^t \int_0^1 \left(\rho_x^2 u^2 \theta^2 + \rho^2 u_x^2 \theta^2 \right) dx \, ds \leq C A^2 \int_0^t \int_0^1 \left(\rho_x^2 u^2 + \rho^2 u_x^2 \right) dx \, ds$$

$$\leq C A^2 \int_0^t \|u\|_{L^\infty}^2 \int_0^1 \rho_x^2 \, dx \, ds + C A^2 \int_0^t \int_0^1 \rho u_x^2 \, dx \, ds$$

$$\leq C A^2 + C A^2 \int_0^t \|u\|_{L^\infty}^2 \, ds + C \leq C A^2 + C,$$

$$\int_0^t \int_0^1 \rho^2 \theta_t^2 \, dx \, ds \leq C \int_0^t \int_0^1 \theta_t^2 \, dx \, ds \leq C + C Z^\delta,$$

$$\int_0^t \int_0^1 \theta^6 \theta_t^2 \, dx \, ds = \int_0^t \int_0^1 \theta^6 \theta^{q+3} \theta_t^2 \theta^{-3-q} \, dx \, ds$$

$$\leq (C + C A^{\gamma_5}) \int_0^t \int_0^1 \theta^{q+3} \theta_t^2 \, dx \, ds$$

$$\leq (C + C A^{\gamma_5})(C + C Z^\delta)$$

$$\leq (C + C A^{\gamma_5})(C + C Z^\delta)$$

$$\leq C + C A^{\gamma_5} + C A^{\gamma_5} Z^\delta + C Z^\delta,$$

and

$$\int_0^t \int_0^1 |\mathbf{b}|^2 |\mathbf{b}_t|^2 \, dx ds \leq \sup_{t \in (0,T)} \|\mathbf{b}\|_{L^\infty}^2 \int_0^t \int_0^1 |\mathbf{b}_t|^2 \, dx ds$$

$$\leq C \sup_{t \in (0,T)} \|\mathbf{b}\|_{L^\infty}^2 \leq C \sup_{t \in (0,T)} \|\mathbf{b}_x\|_{L^\infty}^2 \leq C.$$

Inserting the estimates (3.2.49), (3.2.50), (3.2.55) and (3.2.56) into (3.2.44), we get, for any $\varepsilon > 0$,

$$\int_0^1 \rho u_t^2 \, dx + 2\lambda \int_0^t \int_0^1 u_{xt}^2 \, dx ds \leq (C + CA^{\gamma_2}) \int_0^t \int_0^1 \rho u_t^2 \, dx ds + CA^{\gamma_2} + CA^{\gamma_1}$$

$$+ CA^{\gamma_1} Z^{\frac{3}{4}} + CZ^{\frac{3}{4}} + \varepsilon Z + CZ^{\delta \cdot \frac{2q+6}{2q+6-\gamma_5}} + C,$$

$$(3.2.57)$$

where $\gamma_2 < 2q + 6, \gamma_1 < 2q + 6$. Further,

$$CA^{\gamma_2} \leq CZ^\delta + C, \quad CA^{\gamma_1} \leq CZ^\delta + C. \tag{3.2.58}$$

Inserting (3.2.58) into (3.2.57), we get

$$\int_0^1 \rho u_t^2 \, dx + \int_0^t \int_0^1 u_{xt}^2 \, dx ds$$

$$\leq (C + CZ^\delta) \int_0^t \int_0^1 \rho u_t^2 \, dx ds + C + CZ^\delta + CZ^{\frac{3}{4} \cdot \frac{2q+6}{2q+6-\gamma_1}}$$

$$+ CZ^{\frac{3}{4}} + \varepsilon Z + CZ^{\delta \cdot \frac{2q+6}{2q+6-\gamma_5}}$$

$$\leq (C + CZ^\delta) \int_0^t \int_0^1 \rho u_t^2 \, dx ds + C + CZ^{\frac{3}{4} \cdot \frac{2q+6}{2q+6-\gamma_1}} + \varepsilon Z + CZ^{\delta \cdot \frac{2q+6}{2q+6-\gamma_5}}.$$

By the Gronwall inequality, we conclude from the above inequality that, for any $t \in (0, T)$,

$$\int_0^1 \rho u_t^2 \, dx + \int_0^t \int_0^1 u_{xt}^2 \, dx ds \leq C + \varepsilon Z + CZ^{\frac{3}{4} \cdot \frac{2q+6}{2q+6-\gamma_1}} + CZ^{\delta \cdot \frac{2q+6}{2q+6-\gamma_5}}. \quad (3.2.59)$$

Combining (3.2.59) and (3.2.40), we obtain

$$\|u_{xx}\|_{L^2}^2 \leq C + CZ^{\frac{3}{4}} + CZ^\delta + CZ^{\delta \cdot \frac{2q+6}{2q+6-\gamma_4}} + CZ^{\frac{3}{4} \cdot \frac{2q+6}{2q+6-\gamma_1}} + CZ^{\delta \cdot \frac{2q+6}{2q+6-\gamma_5}} + \varepsilon Z,$$

whence

$$Z \leq C + CZ^{\delta \cdot \frac{2q+6}{2q+6-\gamma_4}} + CZ^{\delta \cdot \frac{2q+6}{2q+6-\gamma_5}} + CZ^{\delta \cdot \frac{2q+6}{2q+6-\gamma_1}}. \tag{3.2.60}$$

In order to get the estimate $Z \leq C$, we must require that

$$\begin{cases} 0 \leq \delta \cdot \dfrac{2q+6}{2q+6-\gamma_1} < 1, \\[2mm] 0 \leq \delta \cdot \dfrac{2q+6}{2q+6-\gamma_4} < 1, \\[2mm] 0 \leq \delta \cdot \dfrac{2q+6}{2q+6-\gamma_5} < 1, \end{cases} \tag{3.2.61}$$

that is,

$$\begin{cases} q > \frac{2+\sqrt{211}}{9}, \\ q > \frac{9}{5}, \\ q > \frac{\sqrt{21}}{3} \end{cases}$$

which implies

$$q > (2 + \sqrt{211})/9.$$

This completes the proof of Lemma 3.2.5. □

Applying the previous lemmas, we can further obtain estimates on $(u, \mathbf{w}, \mathbf{b})$.

Lemma 3.2.6. *For any $t \in (0, T)$, we have*

$$\theta(x, t) \geq C, \quad \text{for any } (x, t) \in (0, 1) \times (0, T), \tag{3.2.62}$$

$$\int_0^1 \left(|(u_t, \mathbf{w}_t, \mathbf{b}_t)|^2 + |(u_{xx}, \mathbf{w}_{xx}, \mathbf{b}_{xx})|^2 \right) dx$$
$$+ \int_0^t \int_0^1 |(u_{xt}, \mathbf{w}_{xt}, \mathbf{b}_{xt})|^2 \, dx dt \leq C, \tag{3.2.63}$$

$$\sup_{t \in (0,T)} \int_0^1 \left(\theta_{xx}^2 + \theta_t^2 \right) dx + \int_0^T \int_0^1 \theta_{xt}^2 dx ds \leq C. \tag{3.2.64}$$

Proof. See, e.g., [130]. □

By using the global a priori estimates established in Lemmas 3.2.1–3.2.6, we can prove the main theorem in a standard way. Indeed, by Lemmas 3.2.5–3.2.6,

$$(u, \mathbf{w}, \mathbf{b}, \theta) \in \left(C^{1,0}(Q_T) \right)^6.$$

To show that $(u, \mathbf{w}, \mathbf{b}, \theta) \in \left(C^{1,1/2}(Q_T) \right)^6$, that $(u_x, \mathbf{w}_x, \mathbf{b}_x, \theta_x) \in \left(C^{1/2,1/4}(Q_T) \right)^6$ and that $(\rho, \psi_x) \in \left(C^{1/2,1/4}(Q_T) \right)^2$, we can refer to [130]. By applying the classical theory on parabolic equations in [69] and the results in [68], we obtain the desired Hölder continuity of the solutions asserted in Theorem 3.1.1, and thus, the existence of classical solutions to problem (3.1.1)–(3.1.5), (3.1.12) and (3.1.13) is proved. The uniqueness and stability of the solution in the class in Theorem 3.1.1 can be established in a quite standard way. This completes the proof of Theorem 3.1.1.

3.3 Bibliographic Comments

One-dimensional problems for compressible fluids have been extensively studied under various conditions by a number of authors(see, e.g., [8, 16, 50, 59, 63, 87, 103–105, 119, 124, 129, 130, 145]). Qin [103, 104] and Umehara and Tani [129, 130]

proved the global existence of a unique classical solution to the following free-
boundary problem for governing the flows of the one-dimensional system for self-
gravitating viscous radiative and reactive gases (see also Chapter 4):

$$v_t = u_x, \tag{3.3.1}$$

$$u_t = \left(-p + \mu \frac{u_x}{v} \right)_x - G \left(x - \frac{1}{2} \right), \tag{3.3.2}$$

$$e_t = \left(-p + \mu \frac{u_x}{v} \right) u_x + \left(\kappa \frac{\theta_x}{v} \right)_x + \lambda \phi z, \tag{3.3.3}$$

$$z_t = d \left(\frac{z_x}{v^2} \right)_x - \phi z \tag{3.3.4}$$

in $\Omega \times (0, +\infty)$, with $\Omega = (0, 1)$. Here the specific volume $v = v(x, t)$, the velocity
$u = u(x, t)$, the absolute temperature $\theta = \theta(x, t)$ and mass fraction of the reactant
$z = z(x, t)$ are the unknown quantities, and the positive constants μ, G, d and λ
are the bulk viscosity, the Newtonian gravitational constant, the species diffusion
coefficient and the difference in heat between the reactant and the product, respectively. The pressure p and the internal energy per unit mass e are defined by
(3.1.9), and the thermal conductivity $\kappa = \kappa(v, \theta)$ takes the form

$$\kappa(v, \theta) = \kappa_1 + \kappa_2 v \theta^q, \tag{3.3.5}$$

with positive constants κ_1, κ_2 and q. Furthermore, the reaction rate function $\phi = \phi(\theta)$ is defined, from the Arrhenius law, by

$$\phi(\theta) = K \theta^\beta e^{-A/\theta}, \tag{3.3.6}$$

where the positive constants K and A are the coefficient of rate of the reactant
and the activation energy, respectively, and β is a non-negative real number.

In [129], Umehara and Tani proved the global existence of a unique classical
solutions to (3.3.1)–(3.1.6) with the same state equations (3.1.9) when $4 \leq q \leq
16, 0 \leq \beta \leq 13/2$. In [130], in the case when $q \geq 3$ and $0 \leq \beta < q + 9$, they
also proved the existence of global smooth solutions, which improved the results
in [129]. Qin et al. [103, 104] improved the results of Umehara and Tani [129, 130],
that is, when $\frac{9}{4} < q < 3$, $0 \leq \beta < q + 9$, Qin et al. [103] proved the existence of
global solutions, and later on, Qin [104] also proved the existence of global smooth
solutions when $q \geq 2$, $0 \leq \beta < 2q + 6$ or $q \geq 3$, $0 \leq \beta < q + 9$. Chen and Wang [10]
and Wang [132] considered a 1D model problem for plane magnetohydrodynamics
flows, where the effect of self-gravitation and the influence of high temperature
radiation were not taken into account (i.e., $G = a = 0$ in (3.1.1)–(3.1.9)). In [10],
they established the existence, uniqueness, and regularity of global solutions with
large initial data in H^1, and also established an existence theorem of global solutions with large discontinuous initial data under certain assumptions on e, p and
κ, in the case where $r \in [0, 1]$, $q \geq 2 + 2r$. Recently, Zhang and Xie [145] considered
an initial-boundary value problem for nonlinear planar magnetohydrodynamics in

the case where the effect of self-gravitation and the influence of radiation on the dynamics at high temperature regimes are taken into account, i.e., when $q > \frac{5}{2}$, and proved the global existence of a unique classical solution with large initial data to the problem under quite general assumptions on the heat and conductivity. Ducomet and Feireisl [25] considered a three-dimensional mathematical model derived from the classical principles of continuum mechanics and electrodynamics, and proved the existence of global-in-time solutions of considered the corresponding problem.

Chapter 4

Global Smooth Solutions to a 1D Self-gravitating Viscous Radiative and Reactive Gas

4.1 Introduction

In this chapter we are concerned with the free-boundary problem describing the motion of a compressible, viscous and heat-conducting gas which is self-gravitating, radiative and chemically reactive. The motion of such a gas, especially in the case of unimolecular reactions with first-order kinetics, is described in Lagrangian mass coordinates by the following equations:

$$v_t = u_x, \tag{4.1.1}$$

$$u_t = \left(-p + \mu \frac{u_x}{v}\right)_x - G\left(x - \frac{1}{2}\right), \tag{4.1.2}$$

$$e_t = \left(-p + \mu \frac{u_x}{v}\right) u_x + \left(\kappa \frac{\theta_x}{v}\right)_x + \lambda \phi z, \tag{4.1.3}$$

$$z_t = d\left(\frac{z_x}{v^2}\right)_x - \phi z, \tag{4.1.4}$$

in $\Omega \times (0, +\infty)$, with $\Omega = (0, 1)$. Here the specific volume $v = v(x, t)$, the velocity $u = u(x, t)$, the absolute temperature $\theta = \theta(x, t)$ and mass fraction of the reactant $z = z(x, t)$ are the unknown quantities, and the positive constants μ, G, d and λ are the bulk viscosity, the Newtonian gravitational constant, the species diffusion coefficient and the difference in heat between the reactant and the product, respectively. The pressure p and the internal energy per unit mass of e are defined by

$$p = p(v, \theta) = \frac{R\theta}{v} + \frac{a}{3}\theta^4, \quad e = e(v, \theta) = c_V \theta + av\theta^4 \tag{4.1.5}$$

where the positive constants R, c_V and a are the perfect gas constant, the specific heat capacity at constant volume and the radiation-density constant, respectively. The second terms on the right-hand side of both relations in (4.1.5) stand for the effect of radiation phenomena, whose forms are given by the famous Stefan-Boltzmann law. In the radiating regime, we naturally take into account the heat flux from the radiative contribution, and not only the one from the heat-conductive contribution. To simplify (see, e.g., [22, 129, 130]), we assume here that the thermal conductivity $\kappa = \kappa(v, \theta)$ takes the form

$$\kappa(v, \theta) = \kappa_1 + \kappa_2 v \theta^q \qquad (4.1.6)$$

with positive constants κ_1, κ_2 and q. Furthermore, we assume that the reaction rate function $\phi = \phi(\theta)$ is defined, from the Arrhenius law, by

$$\phi(\theta) = K \theta^\beta e^{-A/\theta}, \qquad (4.1.7)$$

where the positive constants K and A are the coefficient of rate of the reactant and the activation energy, respectively, and β is a non-negative real number.

We consider (4.1.1)–(4.1.4) subject to the following boundary condition

$$(\sigma, \theta_x, z_x) \mid_{x=0,1} = (-Pe, 0, 0), \quad t > 0, \qquad (4.1.8)$$

with the stress $\sigma = -p + \mu \frac{u_x}{v}$ and the external pressure Pe (a positive constant), and the initial condition

$$(v, u, \theta, z) \mid_{t=0} = (v_0(x), u_0(x), \theta_0(x), z_0(x)), \quad x \in [0, 1]. \qquad (4.1.9)$$

Without loss of generality, we may assume that (see, e.g., [129, 130])

$$\int_0^1 u_0(x) dx = 0. \qquad (4.1.10)$$

The notation in this chapter will be as follows:

Let m be a non-negative integer and $0 < \alpha, \alpha' < 1$. By $C^{m+\alpha}(\Omega)$ we denote the spaces of functions $u = u(x)$ which have bounded derivatives up to order m and $d^m u/dx^m$ is uniformly Hölder continuous with exponent α. Let T be a positive constant and $Q_T = \Omega \times (0, T)$. For a function u defined on Q_T, we say that $u \in C_{x,t}^{\alpha, \alpha'}(Q_T)$ if

$$|u|^{(0)} = \sup_{(x,t) \in Q_T} |u(x,t)| < +\infty \qquad (4.1.11)$$

and u is uniformly Hölder continuous in x and t with exponents α and α', respectively. Its norm is denoted by $|\cdot|_{\alpha,\alpha'}$. We also say that $u \in C_{x,t}^{2+\alpha, 1+\alpha/2}(Q_T)$ if u is bounded, has bounded derivative u_x, and $(u_{xx}, u_t) \in \left(C_{x,t}^{\alpha, \alpha/2}(Q_T) \right)^2$. Its norm is denoted by $|\cdot|_{2+\alpha,1+\alpha/2}$. $L^p, 1 \le p \le +\infty, W^{m,p}, m \in \mathbb{N}, H^1 = W^{1,2}, H_0^1 = W_0^{1,2}$ denote the usual (Sobolev) spaces on $[0,1]$. In addition, $\|\cdot\|_B$ denotes the norm in the space B; we also put $\|\cdot\| = \|\cdot\|_{L^2(0,1)}$. We use C_0 to denote the generic positive constant depending on the initial data, but independent of t. Constant $C > 0$ stands for the generic positive constant depending only on the initial data and $T > 0$. Without danger of confusion, we shall use the same symbol to denote the state functions as well as their values along a dynamic process, e. g., $p(v,\theta), p(v(x,t), \theta(x,t))$ and $p(x,t)$.

Let $E = E_1 \cup E_2$ be a set in the (q,β)-plane in $\mathbb{R}^2$, where

$$E_1 = \left\{ (q,\beta) \in \mathbb{R}^2 : \frac{9}{4} < q < 3, \ 0 \le \beta < 2q+6 \right\},$$

$$E_2 = \left\{ (q,\beta) \in \mathbb{R}^2 : 3 \le q, \ 0 \le \beta < q+9 \right\}.$$

Our results read as follows:

Theorem 4.1.1. *Let* $(q,\beta) \in E$ *and* $\alpha \in (0,1)$. *Assume that the initial data*

$$(v_0, u_0, \theta_0, z_0) \in C^{1+\alpha}(\Omega) \times (C^{2+\alpha}(\Omega))^3 \tag{4.1.12}$$

satisfies the compatibility conditions, (4.1.10) *and* $v_0(x) > 0, \theta_0(x) > 0, 0 \le z_0(x) \le 1$ *for any* $x \in [0,1]$. *Then there exists a unique solution* (v, u, θ, z) *of the initial boundary value problem* (4.1.1)–(4.1.4), (4.1.8)–(4.1.9) *such that for any* $T > 0$,

$$(v(t), v_x(t), v_t(t)) \in \left(C_{x,t}^{\alpha,\alpha/2}(Q_T) \right)^3, \quad (v(t), \theta(t), z(t)) \in \left(C_{x,t}^{2+\alpha,1+\alpha/2}(Q_T) \right)^3. \tag{4.1.13}$$

Moreover, for any $(x,t) \in \overline{Q_T}$,

$$v(x,t) > 0, \quad \theta(x,t) > 0, \ 0 \le z(x,t) \le 1. \tag{4.1.14}$$

The aim of this chapter is to further improve upon those results in [129, 130] by providing a larger range of $(q,\beta) \in E = E_1 \cup E_2$ than those in [129, 130]. This means that our results have improved those in [105, 127].

The main mathematical difficulty arises from the higher-order nonlinear dependence on the temperature θ of $p(v,\theta), e(v,\theta)$ and $\kappa(v,\theta)$ in (4.1.5) and (4.1.6). To overcome this one, we shall first use delicate interpolation techniques to reduce the order in θ, then bound the norms of $(v(t), u(t), \theta(t), z(t))$ and their derivatives in terms of expression of the from

$$A^\Lambda \equiv \left(\sup_{0 \le s \le t} \|\theta(s)\|_{L^\infty} \right)^\Lambda \tag{4.1.15}$$

with Λ being a positive constant depending only on q and β.

The local existence of solutions can be established by the standard method (see, e.g., [126, 127]). Therefore to prove our results, it suffices to continue the local solutions based on the following a priori estimates.

Theorem 4.1.2. *Let $T > 0$ be an arbitrarily given constant. Under assumptions of Theorem 4.1.1, assume that problem* (4.1.1)–(4.1.4), (4.1.8)–(4.1.9) *possesses a global smooth solution* $(v(t), u(t), \theta(t), z(t))$ *such that*

$$(v(t), v_x(t), v_t(t)) \in \left(C^{\alpha,\alpha/2}_{x,t}(Q_T)\right)^3, \quad (u(t), \theta(t), z(t)) \in \left(C^{2+\alpha,1+\alpha/2}_{x,t}(Q_T)\right)^3.$$
(4.1.16)

Then there exists a positive constant C depending on the initial data and $T > 0$, such that

$$|(v, v_x, v_t)|_{\alpha,\alpha/2} + |(u, \theta, z)|_{2+\alpha,1+\alpha/2} \le C$$
(4.1.17)

and for any $(x,t) \in \overline{Q_T}$,

$$v(x,t), \theta(x,t) \ge C^{-1} > 0, \ 0 \le z(x,t) \le 1.$$
(4.1.18)

4.2 Proof of the Main Results

In this section, we shall finish the proof of Theorem 4.1.1. We begin with the following lemma.

Lemma 4.2.1. *Under the assumptions of Theorem 4.1.1, the following estimates hold for any* $t \in [0,T]$:

$$\int_0^1 \left[\frac{1}{2}u^2 + e + \lambda z + f(x)v\right] dx \equiv E_0,$$
(4.2.1)

$$U(t) + \int_0^t V(s)\,ds \le C_0,$$
(4.2.2)

$$\int_0^1 z(x,t)\,dx + \int_0^t \int_0^1 \phi z\,dxds = \int_0^1 z_0(x)\,dx,$$
(4.2.3)

$$\frac{1}{2}\int_0^1 z^2(x,t)\,dx + \int_0^t \int_0^1 \left(\frac{d}{v^2}z_x^2 + \phi z^2\right) dxds = \frac{1}{2}\int_0^1 z_0^2\,dx,$$
(4.2.4)

$$\int_0^t \left(\max_{x \in \overline{\Omega}} \theta(x,s)\right)^\gamma ds \le C, \ \text{for} \ 0 \le \gamma \le q + 4 \ \ q \ge 0,$$
(4.2.5)

$$\int_0^t \|u_x(s)\|^2\,ds \le C,$$
(4.2.6)

$$\|v_x(t)\|^2 + \int_0^t \int_0^1 (\theta v_x^2 + v_x^2)(x,s)dxds \le C, \ \text{for} \ q \ge 2,$$
(4.2.7)

$$\int_0^t \|u_x(s)\|_{L^3(0,1)}^3 ds \le C, \ \text{for} \ q \ge 4,$$
(4.2.8)

and, for any $(x,t) \in \overline{Q_T}$,

$$0 \le z(x,t) \le 1, \ 0 < C^{-1} \le v(x,t) \le C,$$
(4.2.9)

where

$$E_0 = \int_0^1 \left[\frac{1}{2} u_0^2 + e_0 + \lambda z_0 + f(x) v_0 \right] dx, \tag{4.2.10}$$

$$U(t) = \int_0^1 \left[R(v - \log v - 1) + c_V(\theta - \log \theta - 1) \right] dx, \tag{4.2.11}$$

$$V(t) = \int_0^1 \left(\frac{\mu u_x^2}{v\theta} + \frac{\kappa \theta_x^2}{v\theta^2} + \lambda \frac{\phi z}{\theta} \right) dx \tag{4.2.12}$$

and

$$e_0 = c_V \theta + a v_0 \theta_0^4, \quad f(x) = Pe + \frac{1}{2} Gx(1 - x). \tag{4.2.13}$$

Proof. See, e.g., [129, 130]. □

Now we begin to derive higher-order estimates. To this end, we define

$$X = \int_0^t \int_0^1 (1 + \theta^{q+3}) \theta_x^2 dx ds,$$

$$Y = \max_{t \in [0,T]} \int_0^1 (1 + \theta^{2q}) \theta_x^2 dx,$$

$$Z = \max_{t \in [0,T]} \| u_{xx} \|^2.$$

Then by Lemma 4.2.1 and standard interpolation techniques, we can derive the following result.

Lemma 4.2.2. *For any $t \in [0, T]$, we have*

$$A \leq |\theta|^{(0)} \leq C + CY^{1/(2q+6)}, \tag{4.2.14}$$

$$\max_{t \in [0,T]} \| u_x \|^2 \leq C + CZ^{1/2}, \tag{4.2.15}$$

$$|u_x|^{(0)} \leq C + CZ^{3/8}, \tag{4.2.16}$$

where $A = \sup_{0 \leq s \leq t} \| \theta(s) \|_{L^\infty}$.

Proof. See, e.g., [130]. □

Lemma 4.2.3. *For $(q, \beta) \in E$, we have*

$$X + Y \leq C + CZ^{7/8} + CZ^{\delta_0}, \tag{4.2.17}$$

where $\delta_0 = \delta \equiv \frac{3}{4} p' \in (0, 1), (q - 3)p/(2q + 6) < 1$ with $1 < p, p' < +\infty$ for $3 < q < 4$, and $\delta_0 = 3/4$ for $2 \leq q \leq 3$ or $4 \leq q$.

Proof. Let

$$H(x,t) = H(v,\theta) = \int_0^\theta \frac{\kappa(v,\xi)}{v} \, d\xi = \frac{\kappa_1 \theta}{v} + \frac{\kappa_2 \theta^{q+1}}{q+1}.$$

Then it is easy to verify that

$$H_t = H_v u_x + \frac{\kappa(v,\theta)\theta_t}{v},$$

$$H_{xt} = \left[\frac{\kappa\theta_x}{v}\right]_t + H_v u_{xx} + H_{vv} v_x u_x + \left(\frac{\kappa}{v}\right)_v v_x \theta_t,$$

$$|H_v| + |H_{vv}| \le C\theta.$$

Multiplying (4.1.3) by H_t and integrating the resulting equation over $Q_t = (0,1) \times (0,t)$, we obtain

$$\int_0^t \int_0^1 \left(e_\theta \theta_t + \theta p_\theta u_x - \mu \frac{u_x^2}{v}\right) H_t \, dxds + \int_0^t \int_0^1 \frac{\kappa(v,\theta)}{v} \theta_x H_{tx} \, dxds$$

$$= \int_0^t \int_0^1 \lambda \phi z H_t \, dxds. \tag{4.2.18}$$

Now let us estimate each term in (4.2.19) by using Lemmas 4.2.1 and 4.2.2.

We have first

$$\int_0^t \int_0^1 \frac{e_\theta \theta_t^2 \kappa(v,\theta)}{v} \, dxds \ge CX, \tag{4.2.19}$$

$$\int_0^t \int_0^1 \frac{\kappa\theta_x}{v} \left(\frac{\kappa\theta_x}{v}\right)_t \, dxds \ge CY - C, \tag{4.2.20}$$

with $\kappa_0 = \kappa_1 + \kappa_2 v_0 \theta_0^q$, and for any $\varepsilon > 0$,

$$\left|\int_0^t \int_0^1 e_\theta \theta_t H_u u_x \, dxds\right| \le \varepsilon X + C(\varepsilon) \int_0^t \int_0^1 (1+\theta)^{5-q} u_x^2 \, dxds$$

$$\le \varepsilon X + C(\varepsilon)|u_x^2|^{(0)} \int_0^t (1 + \|\theta\|_{L^\infty}^{\max(1-q,0)}) \int_0^1 (1+\theta)^4 dxds$$

$$\le \varepsilon X + C(\varepsilon)(1 + Z^{3/4}), \tag{4.2.21}$$

$$\left|\int_0^t \int_0^1 \frac{\kappa}{v} \theta_x H_v u_{xx} \, dxds\right| \le \varepsilon X + C(\varepsilon)(1 + Z^{3/4}), \tag{4.2.22}$$

$$\left|\int_0^t \int_0^1 \frac{\kappa}{v} \theta_x H_{vv} v_x u_x \, dxds\right| \le \varepsilon Y + C(\varepsilon)(1 + Z^{3/4}), \tag{4.2.23}$$

$$\left|\int_0^t \int_0^1 \theta p_\theta u_x \frac{\kappa}{v} \theta_t \, dxds\right| \le \varepsilon X + C(\varepsilon)|1 + \theta^{q+5}|^{(0)} \int_0^t \|u_x\|^2 \, ds$$

$$\leq \varepsilon(X+Y)+C(\varepsilon), \tag{4.2.24}$$

$$\left| \int_0^t \int_0^1 \theta p_\theta u_x H_v u_x \, dxds \right| \leq C|1+\theta^5|^{(0)} \int_0^t \|u_x\|^2 \, ds$$

$$\leq \varepsilon Y + C(\varepsilon), \tag{4.2.25}$$

$$\left| \int_0^t \int_0^1 \frac{\mu u_x^2}{v} H_v u_x \, dxds \right| \leq C|u_x|^{(0)} \max_{t\in[0,T]} \|u_x\|^2 \int_0^t \max_{x\in[0,1]} \theta \, ds$$

$$\leq C(1+Z^{7/8}), \tag{4.2.26}$$

$$\left| \int_0^t \int_0^1 \lambda \phi z H_v u_x \, dxds \right| \leq C|\theta u_x|^{(0)} \int_0^t \int_0^1 \phi z \, dxds$$

$$\leq \varepsilon Y + C(\varepsilon)(1+Z^{3/4}). \tag{4.2.27}$$

The next estimate improves the one in [130]: for any $\varepsilon > 0$,

$$\left| \int_0^t \int_0^1 \frac{\kappa\theta_x}{v} \left(\frac{\kappa}{v}\right)_v v_x \theta_t \, dxds \right| \leq \varepsilon X + C(\varepsilon) \int_0^t \|\frac{\kappa\theta_x}{v}\|_{L^\infty}^2 \|v_x\| \, ds + C(\varepsilon)$$

$$\leq \varepsilon X + C(\varepsilon) + C(\varepsilon) \left(\int_0^t \|\frac{\kappa\theta_x}{v}\|^2 \, ds \right)^{1/2} \left(\int_0^t \|(\frac{\kappa\theta_x}{v})_x\|^2 \, ds \right)^{1/2}. \tag{4.2.28}$$

Further, we can estimate

$$\int_0^t \|\frac{\kappa\theta_x}{v}\|^2 \, ds \leq C(1+A)^{q+2} \int_0^t V_1(s) \, ds$$

$$\leq C(1+Y^{(q+2)/(2q+6)}) \tag{4.2.29}$$

with $V_1(t) = \int_0^1 \frac{\kappa\theta_x^2}{v\theta^2} \, dx$, and

$$\int_0^t \left\| \left(\frac{\kappa\theta_x}{v}\right)_x \right\|^2 \, ds \leq C \int_0^t \left(\|e_\theta\theta_t\|^2 + \|\theta p_\theta v_x\|^2 + \|u_x^2\|^2 + \|\phi\|^2 \right) ds$$

$$\leq C \int_0^t \int_0^1 \left[(1+\theta^6)\theta_t^2 + (1+\theta^8)u_x^2 + u_x^4 + z\theta^\beta\phi \right] dxds$$

$$\leq C(1+A)^{q_1} X + C \int_0^t |u_x^2|^{(0)}(1+\|\theta\|_{L^\infty})^4 \int_0^1 (1+\theta)^4 \, dxds$$

$$+ C \int_0^t |u_x|^{(0)} \|u_x\|^2 \, ds + CA^\beta \int_0^t \int_0^1 z\phi \, dxds$$

$$\leq CX + CY^{q_1/(2q+6)} X + C + CZ^{3/4} + CY^{\beta/(2q+6)},$$

which, together with (4.2.30) and (4.2.31), implies that, for any $\varepsilon > 0$,

$$\left| \int_0^t \int_0^1 \frac{\kappa\theta_x}{v} \left(\frac{\kappa}{v}\right)_v v_x \theta_t \, dxds \right|$$

$$\leq \varepsilon X + C(\varepsilon)(1 + Y^{(q/2+1)/(2q+6)})\Big\{ X^{1/2} + Y^{q_1/[2(2q+6)]} X^{1/2} + 1$$

$$+ Z^{3/8} + Y^{\beta/(2q+6)}\Big\}$$

$$\leq 2\varepsilon X + C(\varepsilon)\Big\{ 1 + Y^{(q/2+1+q_1/2)/(2q+6)} X^{1/2} + Y^{(q/2+1)/(2q+6)} Z^{3/8}$$

$$+ Z^{3/8} + Y^{(\beta/2+q/2+1)/(2q+6)}\Big\}$$

$$\leq 3\varepsilon X + \varepsilon Y + C(\varepsilon) + C(\varepsilon) Z^{3(q+3)/[2(4q+11)]}, \tag{4.2.30}$$

if $2 \leq q$ and $0 \leq \beta < 3q + 10$.

Moreover, by Lemmas 4.2.1 and 4.2.2, we easily deduce that, for any $\varepsilon > 0$,

$$\left| \int_0^t \int_0^1 \lambda \phi z \frac{\kappa \theta_t}{v} \, dx ds \right| \leq \varepsilon X + C(\varepsilon)(1 + Y)^{(q_2+\beta)/(2q+6)} \int_0^t \int_0^1 \phi z^2 \, dx ds$$

$$\leq \varepsilon(X + Y) + C(\varepsilon) \tag{4.2.31}$$

for $2 \leq q \leq 3$ and $0 \leq \beta < 2q+6$ or $3 \leq q$ and $0 \leq \beta < q+9$ with $q_2 = \max(q-3, 0)$, which is an improved estimate;

$$\left| \int_0^t \int_0^1 \frac{\mu u_x^2}{v} \frac{\kappa \theta_t}{v} \, dx ds \right| \leq \varepsilon X + C(\varepsilon) \int_0^t \int_0^1 (1 + \theta)^{q-3} u_x^4 \, dx ds$$

$$\leq \varepsilon X + C(\varepsilon) |u_x^2|^{(0)} \int_0^t \|u_x\|^2 \, ds$$

$$\leq \varepsilon X + C(\varepsilon)(1 + Z^{3/4}) \tag{4.2.32}$$

for $2 \leq q \leq 3$,

$$\left| \int_0^t \int_0^1 \frac{\mu u_x^2}{v} \frac{\kappa \theta_t}{v} \, dx ds \right| \leq \varepsilon X + C(\varepsilon) |(1 + \theta)^{q-3}|^{(0)} |u_x^2|^{(0)} \int_0^t \|u_x\|^2 \, ds$$

$$\leq \varepsilon X + C(\varepsilon) \left(1 + Y^{(q-3)/(2q+6)} + Y^{(q-3)/(2q+6)} Z^{3/4} + Z^{3/4} \right)$$

$$\leq \varepsilon(X + Y) + C(\varepsilon)(1 + Z^{\delta}) \tag{4.2.33}$$

for $3 < q < 4$ with $\delta = \frac{3}{4}p' \in (0,1)$, $(q-3)p/(2q+6) < 1, p^{-1} + p'^{-1} = 1, 1 < p, p' < +\infty$;

$$\left| \int_0^t \int_0^1 \frac{\mu u_x^2}{v} \frac{\kappa \theta_t}{v} \, dx ds \right| \leq \varepsilon X + C(\varepsilon) |(1 + \theta)^{q-3}|^{(0)} |u_x|^{(0)} \int_0^t \|u_x\|_{L^3(\Omega)}^3 \, ds$$

$$\leq \varepsilon X + C(\varepsilon) \left(1 + Y^{(q-3)/(2q+6)} + Y^{(q-3)/(2q+6)} Z^{3/8} + Z^{3/8} \right)$$

$$\leq \varepsilon(X + Y) + C(\varepsilon)(1 + Z^{3/4})$$

for $4 \leq q$, which, together with (4.2.34) and (4.2.35), gives, for any $\varepsilon > 0$,

$$\left| \int_0^t \int_0^1 \frac{\mu u_x^2}{v} \frac{\kappa \theta_t}{v} \, dx ds \right| \leq \varepsilon(X + Y) + C(\varepsilon)(1 + Z^{\delta_0}). \tag{4.2.34}$$

Therefore it follows from (4.2.19)–(4.2.36) that

$$X + Y \leq C\varepsilon(X + Y) + C(\varepsilon)(Z^{7/8} + Z^{\delta_0}) \tag{4.2.35}$$

which gives (4.2.18) by choosing $\varepsilon > 0$ small enough. $\square$

Lemma 4.2.4. *If $(q, \beta) \in E$, then for any $t \in [0, T]$, there holds that*

$$\|(u_x, u_{xx}, u_t, \theta_x, \theta_{xx}, \theta_t, z_x, z_{xx}, z_t)\|^2 + \int_0^t \|(u_{xt}, \theta_{xt}, z_{xt})\|^2 \, ds \leq C, \tag{4.2.36}$$

$$|u_x|^{(0)} + |u|^{(0)} + |\theta|^{(0)} \leq C, \tag{4.2.37}$$

$$\theta(x, t) \geq C^{-1} > 0 \quad \text{for any } (x, t) \in \overline{Q_T}. \tag{4.2.38}$$

Proof. Differentiating (4.1.2) with respect to t, multiplying the result by u_t, and then integrating with respect to x, we arrive at

$$\frac{1}{2}\frac{d}{dt}\|u_t\|^2 + \int_0^1 \frac{\mu u_{xt}^2}{v} \, dx = \int_0^1 \left(p_t u_{xt} + \frac{\mu u_x^2 u_{xt}}{v^2} \right) dx. \tag{4.2.39}$$

Noting that $p_t = \left(\frac{R}{v} + \frac{4}{3}a\theta^3\right)\theta_t - \frac{R\theta u_x}{v^2}$ and using Lemmas 4.2.1–4.2.3, we deduce from (4.2.41) that

$$\|u_t\|^2 + \int_0^t \|u_{xt}\|^2 \, ds \leq C\left\{1 + \int_0^t \int_0^1 [p_t^2 + u_x^4] \, dx ds\right\}$$

$$\leq C\left\{1 + (1 + A)^{q_1} X + Z^{3/4}\right\} \leq C\left\{1 + X + XY^{q_1/(2q+6)} + Z^{3/4}\right\}, \tag{4.2.40}$$

with $q_1 = \max(3 - q, 0)$.

Noting that $p_x = \left(\frac{R}{v} + \frac{4}{3}a\theta^3\right)\theta_x - \frac{R\theta v_x}{v^2}$ and using (4.1.2) and Lemma 4.2.3, we conclude from (4.2.42) that, for any $t \in [0, T]$ and for $(q, \beta) \in E$,

$$\|u_{xx}\|^2 \leq C\left\{1 + \|u_t\|^2 + \int_0^1 (1 + \theta)^6 \theta_x^2 \, dx + \int_0^1 (\theta^2 + u_x^2)v_x^2 \, dx\right\}$$

$$\leq C\left\{1 + \|u_t\|^2 + (1 + A)^{2q_1}Y + (A^2 + |u_x^2|^{(0)})\|v_x\|\right\}$$

$$\leq C\left\{1 + X + XY^{2q_1/(2q+6)} + Z^{3/4} + Y + Y^{q_1/(q+3)+1}\right\}$$

$$\leq C\left\{1 + Z^{7/8} + Z^{\delta_0} + (1 + Z^{7/8} + Z^{\delta_0})^{q_1/(2q+6)+1}\right.$$

$$\left. + (1 + Z^{7/8} + Z^{\delta_0})^{q_1/(q+3)+1}\right\}$$

$$\leq C\left\{1 + Z^{7(q_1+q+3)/[8(q+3)]} + Z^{\delta_0[q_1/(q+3)+1]}\right\}$$

which, by the Young inequality, gives

$$Z \leq C \tag{4.2.41}$$

if $q > 9/4$ (i.e., $7(q_1 + q + 3)/[8(q + 3)] < 1$, and $\delta_0[q_1/(q + 3) + 1] < 1$ is satisfied automatically). Hence from (4.2.37), we have

$$X + Y \leq C. \tag{4.2.42}$$

The other quantities in (4.2.38)–(4.2.40) can be estimated in the same way as in [129, 130]. The Hölder estimates can be also derived from the classical Schauder estimates by using Lemmas 4.2.1–4.2.4. We omit here the details (see, e.g., [129] for details). The proofs of Theorem 4.1.2 and Theorem 4.1.1 are complete.   □

4.3 Bibliographic Comments

Radiation hydrodynamics [78, 95, 134] describes the propagation of thermal radiation through a fluid or gas. Similarly to ordinary fluid mechanics, the equations of motion are derived from conservation laws for macroscopic quantities. However, when radiation is present, the classical "material" flow has to be coupled with the radiation, which is an assembly of photons and needs a priori a relativistic treatment (the photons are massless particles travelling at the speed of light). The whole problem under consideration when the matter is in local thermodynamical equilibrium is thus a coupling between standard hydrodynamics for the matter and a radiative transfer equation for the photon distribution. Through a suitable description, like in plasma when the radiation is in local thermodynamical equilibrium with matter and velocities are not too large, a non-relativistic one LTE temperature description is possible [78, 134]. Moreover, if the matter is extremely radiatively opaque, so that the matter free-path of photons is much smaller than the typical length of the flow, we obtain a simplified description (radiation hydrodynamics in the diffusion limit), which amounts to solving a standard hydrodynamical (compressible Navier-Stokes) system with additional correction terms in the pressure, the internal energy and the thermal conduction. To describe richer physical processes, for simplicity we may consider the fluid as reactive and couple the dynamics with the first-order chemical kinetics of combustion type, namely the first-order Arrhenius kinetics. Although it is simplified, this model can be proved to describe correctly some astrophysical situations of interest, such as stellar evolution or interstellar medium dynamics (see, e.g., [21]). In recent years, the heat-conducting radiative viscous gas system has drawn attention of mathematicians (see, e.g., [19, 21–30, 115, 118, 129, 130]). For the compressible viscous and heat-conducting model in one space dimension, the global existence and large-time behavior of smooth (strong, weak) solutions have been established by many authors. Among them, Antontsev, Kazhikhov and Monakhov [1], Chen [8], Kazhikhov and Shelukhin [63], Ducomet [21], and Ducomet and Zlotnik [29, 30] studied one-dimensional gaseous models similar to ours, i.e., radiative and reactive models with free boundary in an external force field. However, in a series of papers [21, 29, 30], they adopted a special form of self-gravitation that does not depend on the time variable explicitly in Lagrangian mass coordinate system, not the exact form (see (1.8) in [130]) these are called "pancakes model" and are relevant to some large-scale structure of the universe (see, e.g., [118]). Qin [101] established the global existence, exponential stability and the existence of global attractors for a 1D viscous heat-conducting real gas. Moreover, we note that the global existence of solutions to some initial-boundary value problem (4.1.1)–(4.1.4) without pure free-boundary case (4.1.8), but with partially Dirichlet boundary conditions. For $q \geq 4, \beta > 0$, Ducomet [22] proved the global existence and exponential decay in H^1 of solutions to (4.1.1)–(4.1.4) with the boundary conditions

$$(u, \theta_x, z_x)\,|_{x=0,1} = 0. \tag{4.3.1}$$

However, there exists some defects in the proofs of the main results in [22]. Recently, Qin et al. [105] corrected these defects and established the global existence and exponential stability of solutions in H^i ($i = 1, 2, 4$) to (4.1.1)–(4.1.4) with boundary condition (4.1.15), which has improved the range of (q, β) considered in [22]. For our problem, Umehara and Tani [129] proved the global solvability of smooth solutions for $4 \leq q \leq 16$ and $0 \leq \beta \leq 13/2$. Later on, they further improved their results in [130] with the larger range of $(q, \beta) \in E_2$ compared with that in [129].

Chapter 5

The Cauchy Problem for a 1D Compressible Viscous Micropolar Fluid Model

5.1 Main Results

In this chapter, we shall study the global existence and large-time behavior of H^i-global solutions ($i = 1, 2, 4$) to a kind of Navier-Stokes equations for a one-dimensional compressible viscous heat-conducting micropolar fluid, which is assumed to be thermodynamically perfect and polytropic. In Lagrangian coordinates, the system studied her can be written as follows:

$$\eta_t = v_x, \tag{5.1.1}$$

$$v_t = \sigma_x, \tag{5.1.2}$$

$$w_t = A\left(\frac{w_x}{\eta}\right)_x - A\eta w, \tag{5.1.3}$$

$$e_t = \pi_x + \sigma v_x + \frac{w_x^2}{\eta} + \eta w^2, \tag{5.1.4}$$

where $(x, t) \in \mathbb{R} \times \mathbb{R}_+$ are the Lagrangian mass and time coordinates and

$$\sigma = \frac{\nu v_x}{\eta} - P, \quad \pi = \frac{\hat{\kappa}\theta_x}{\eta}, \quad P(\eta, \theta) = \frac{R\theta}{\eta}, \quad e(\eta, \theta) = \theta \tag{5.1.5}$$

denote stress, heat flux, pressure and internal energy, respectively.

We consider (5.1.1)–(5.1.5) subject to the initial condition

$$(\eta(x, 0), v(x, 0), w(x, 0), \theta(x, 0)) = (\eta_0(x), v_0(x), w_0(x), \theta_0(x)), \quad \text{for all} \quad x \in \mathbb{R}. \tag{5.1.6}$$

The unknown quantities η, v, w, θ denote the specific volume, the velocity, the microrotation velocity and the absolute temperature, respectively. Finally,

$\hat{\kappa}(x,t)$ is the heat conductivity; however, in this chapter, the heat conductivity is simply assumed to be a positive constant κ. Moreover, the total mass of the fluid is taken equal to 1 in our article. The residual ones like ν, R, etc., are also physical constants, representing the viscosity and the Boltzmann constants, etc. and we take $\nu = 1$ in this chapter.

In this chapter, we shall adopt the notation in Chapter 1.

Based on the method in [54] and [112], we can establish the global existence in H^i ($i = 1, 2, 4$) and large-time behavior in H^i ($i = 1, 2, 4$). Correspondingly, C_i ($i = 1, 2, 3, 4$) denote universal constants depending on $\min_{x \in \mathbb{R}} \eta_0(x)$, $\min_{x \in \mathbb{R}} \theta_0(x)$, the $H^i(\mathbb{R})$ ($i = 1, 2, 3, 4$) norm of $(\eta_0 - 1, v_0, w_0, \theta_0 - 1)$ and the number ϵ_0, which is defined in Theorem 5.1.1.

We now state our main results in this chapter.

Theorem 5.1.1. *Assume that $\eta_0 - 1$, v_0, w_0, $\theta_0 - 1 \in H^1(\mathbb{R})$ with $\eta_0(x)$, $\theta_0(x) > 0$ on $\mathbb{R}$. Define*

$$e_0^2 := \|\eta_0 - 1\|_{L^\infty}^2 + \int_{\mathbb{R}} (1 + x^2)^\alpha \Big[(\eta_0(x) - 1)^2 + v_0^2(x)$$
$$+ w_0^2(x) + (\theta_0(x) - 1)^2 + v_0^4(x) \Big] dx$$

with $\alpha > \frac{1}{2}$ being an arbitrary but fixed parameter. Then if $e_0 \le \epsilon_0$ where $\epsilon_0 \in (0, 1]$, the problem (5.1.1)–(5.1.6) has a unique H^1-global solution $(\eta(t), v(t), w(t), \theta(t))$ and the following inequalities hold:

$$0 < C_1^{-1} \le \eta(t, x) \le C_1^{-1} \quad \text{on } \mathbb{R} \times \mathbb{R}_+, \tag{5.1.7}$$

$$0 < C_1^{-1} \le \theta(t, x) \le C_1^{-1} \quad \text{on } \mathbb{R} \times \mathbb{R}_+, \tag{5.1.8}$$

$$\|\eta(t) - 1\|_{H^1}^2 + \|v(t)\|_{H^1}^2 + \|\theta(t) - 1\|_{H^1}^2 + \|w(t)\|_{H^1}^2$$
$$+ \int_0^t \Big[\|\eta_x\|^2 + \|v_x\|^2 + \|\theta_x\|^2 + \|w_x\|^2 + \|v_{xx}\|^2 + \|\theta_{xx}\|^2 \tag{5.1.9}$$
$$+ \|w_{xx}\|^2 + \|v_t\|^2 + \|\theta_t\|^2 + \|w_t\|^2 \Big](s)ds \le C_1.$$

Moreover, as $t \to +\infty$,

$$\|(\eta - 1, v, w, \theta - 1)(t)\|_{L^\infty} + \|(\eta_x, v_x, w_x, \theta_x)(t)\| \to 0. \tag{5.1.10}$$

Theorem 5.1.2. *Assume that $\eta_0 - 1$, v_0, w_0, $\theta_0 - 1 \in H^2(\mathbb{R})$ and $\eta_0(x) > 0$, $\theta_0(x) > 0$ on $\mathbb{R}$, and that the other conditions of Theorem 5.1.1 hold. Then for any $t > 0$, the Cauchy problem (5.1.1)–(5.1.6) has a unique H^2-global solution $(\eta(t), v(t), w(t), \theta(t))$ and the following estimate holds:*

$$\|\eta(t) - 1\|_{H^2}^2 + \|\eta(t) - 1\|_{W^{1,\infty}}^2 + \|\eta_t(t)\|_{H^1}^2 + \|v(t)\|_{H^2}^2 + \|v(t)\|_{W^{1,\infty}}^2 + \|v_t(t)\|^2$$
$$+ \|w(t)\|_{H^2}^2 + \|w(t)\|_{W^{1,\infty}}^2 + \|w_t(t)\|^2 + \|\theta(t) - 1\|_{H^2}^2 + \|\theta(t) - 1\|_{W^{1,\infty}}^2$$

$$+ \|\theta_t(t)\|^2 + \int_0^t \Big[\|\eta_x\|_{H^1}^2 + \|\eta_x\|_{L^\infty}^2 + \|\eta_t\|_{H^2}^2 + \|v_x\|_{H^2}^2 + \|v_x\|_{W^{1,\infty}}^2$$

$$+ \|v_t\|_{H^1}^2 + \|w_x\|_{H^2}^2 + \|w_x\|_{W^{1,\infty}}^2 + \|w_t\|_{H^1}^2 + \|\theta_x\|_{H^2}^2$$

$$+ \|\theta_x\|_{W^{1,\infty}}^2 + \|\theta_t\|_{H^1}^2 \Big](s)ds \leq C_2. \tag{5.1.11}$$

Moreover, as $t \to +\infty$,

$$\|\eta_t(t)\|_{H^1} + \|\eta_t(t)\|_{L^\infty} + \|v_t(t)\| + \|w_t(t)\| + \|\theta_t(t)\| \to 0, \tag{5.1.12}$$

$$\|(\eta - 1, v, w, \theta - 1)(t)\|_{W^{1,\infty}} + \|(\eta_x, v_x, w_x, \theta_x)(t)\|_{H^1} \to 0. \tag{5.1.13}$$

Theorem 5.1.3. *Assume that $\eta_0 - 1$, v_0, w_0, $\theta_0 - 1 \in H^4(\mathbb{R})$ and $\eta_0(x) > 0$, $\theta_0(x) > 0$ on $\mathbb{R}$, and that the other conditions of Theorem 5.1.2 hold. Then for any $t > 0$, the Cauchy problem (5.1.1)–(5.1.6) admits a unique H^4-global solution $(\eta(t), v(t), w(t), \theta(t))$ and the following estimates hold:*

$$\|\eta(t) - 1\|_{H^4}^2 + \|\eta(t) - 1\|_{W^{3,\infty}}^2 + \|\eta_t(t)\|_{H^3}^2 + \|\eta_{tt}(t)\|_{H^1}^2 + \|v(t)\|_{H^4}^2 + \|v(t)\|_{W^{3,\infty}}^2$$

$$+ \|v_t(t)\|_{H^2}^2 + \|v_{tt}(t)\|^2 + \|w(t)\|_{H^4}^2 + \|w(t)\|_{W^{3,\infty}}^2 + \|w_t(t)\|_{H^2}^2 + \|w_{tt}(t)\|^2$$

$$+ \|\theta(t) - 1\|_{H^4}^2 + \|\theta(t) - 1\|_{W^{3,\infty}}^2 + \|\theta_t(t)\|_{H^2}^2 + \|\theta_{tt}(t)\|^2 \leq C_4, \tag{5.1.14}$$

$$\int_0^t \Big[\|\eta_x\|_{H^3}^2 + \|\eta_t\|_{H^4}^2 + \|\eta_{tt}\|_{H^2}^2 + \|\eta_{ttt}\|^2 + \|\eta_x\|_{W^{2,\infty}}^2 + \|v_x\|_{H^4}^2 + \|v_t\|_{H^3}^2$$

$$+ \|v_{tt}\|_{H^1}^2 + \|v_x\|_{W^{3,\infty}}^2 + \|w_x\|_{H^4}^2 + \|w_t\|_{H^3}^2 + \|w_{tt}\|_{H^1}^2 + \|w_x\|_{W^{3,\infty}}^2$$

$$+ \|\theta_x\|_{H^4}^2 + \|\theta_t\|_{H^3}^2 + \|\theta_{tt}\|_{H^1}^2 + \|\theta_x\|_{W^{3,\infty}}^2 \Big](s)ds \leq C_4. \tag{5.1.15}$$

Moreover, as $t \to +\infty$,

$$\|(\eta_x, v_x, w_x, \theta_x)(t)\|_{H^3} + \|\eta_t(t)\|_{H^3} + \|\eta_t(t)\|_{W^{2,\infty}}$$

$$+ \|v_t(t)\|_{H^2} + \|v_t(t)\|_{W^{1,\infty}} + \|w_t(t)\|_{H^2} + \|w_t(t)\|_{W^{1,\infty}} + \|\theta_t(t)\|_{H^2}$$

$$+ \|\theta_t(t)\|_{W^{1,\infty}} \to 0, \tag{5.1.16}$$

$$\|\eta_{tt}\|_{H^1} + \|v_{tt}(t)\| + \|w_{tt}(t)\| + \|\theta_{tt}(t)\|$$

$$+ \|(\eta_x, v_x, w_x, \theta_x)(t)\|_{W^{2,\infty}} \to 0. \tag{5.1.17}$$

Corollary 5.1.1. *The H^4-global solution $(\eta(t), v(t), w(t), \theta(t))$ obtained in Theorem 5.1.3 is actually classical one. Precisely, $(\eta(t), v(t), w(t), \theta(t)) \in C^{3,\frac{1}{2}}(\mathbb{R})$ and as $t \to +\infty$:*

$$\|(\eta_x(t), v_x(t), w_x(t), \theta_x(t))\|_{C^{2,\frac{1}{2}}} + \|\eta_t(t)\|_{C^{2,\frac{1}{2}}} + \|(v_t(t), w_t(t), \theta_t(t))\|_{C^{1,\frac{1}{2}}}$$

$$+ \|\eta_{tt}(t)\|_{C^{\frac{1}{2}}} \to 0. \tag{5.1.18}$$

Finally, we combine a result of Mujaković [82] together with our present results to obtain a more general statement.

We assume that there exist constants m, $M \in \mathbb{R}_+$ such that

$$m \leq \eta_0(x) \leq M, \ m \leq \theta_0(x) \leq M, \ x \in \mathbb{R}, \tag{5.1.19}$$

and also constants I_1, I_2, I_3, I_4, $M_1 \in \mathbb{R}_+$, $M_1 > 1$, such that

$$\frac{1}{2}\|v_0\|^2 + \frac{1}{2A}\|w_0\|^2 + R\int_{\mathbb{R}}(\eta_0 - \log\eta_0 - 1)dx$$
$$+ \int_{\mathbb{R}}(\theta_0 - \log\theta_0 - 1)dx := I_1, \tag{5.1.20}$$

$$\frac{1}{2}\left(\|v_{0x}\|^2 + \frac{1}{A}\|w_{0x}\|^2 + \|\theta_{0x}\|^2\right) := I_2, \tag{5.1.21}$$

$$\frac{1}{2}\left\|\frac{\eta_{0x}}{\eta_0}\right\|^2 + \int_{\mathbb{R}}v_{0x}\log\eta_0 dx \leq I_3, \tag{5.1.22}$$

$$\sup_{|x|<+\infty}\theta_0(x) < M_1, \tag{5.1.23}$$

$$I_4 = 2\max\left\{\frac{R}{2\hat{\kappa}},1\right\}I_1\left(1 + M_1 + \frac{I_3}{I_1}\right), \tag{5.1.24}$$

$$I_5 = 2\left(\frac{I_1 I_4}{R}\right)^{1/2}. \tag{5.1.25}$$

According to [82], we can find real numbers $\underline{\xi}$ and $\overline{\xi}$, $\underline{\xi} < 0 < \overline{\xi}$ such that

$$\int_{\underline{\xi}}^{0}\sqrt{e^{\xi} - 1 - \xi}d\xi = \int_{0}^{\overline{\xi}}\sqrt{e^{\xi} - 1 - \xi}d\xi = I_5, \tag{5.1.26}$$

and $\underline{u} = \exp\underline{\xi}$, $\overline{u} = \exp\overline{\xi}$. We combine our results with [82] as follows.

Theorem 5.1.4. *Assume that* $\eta_0 - 1$, v_0, w_0, $\theta_0 - 1 \in H^i(\mathbb{R})$ $(i = 1, 2, 4)$.

Case I. *Assume that* $\eta_0 > 0$, $\theta_0 > 0$ *and* $e_0 \leq \epsilon_0$, *where* e_0, ϵ_0 *are as in Theorem 5.1.1.*

Case II. *Assume that* (5.1.22) *is valid and the following conditions are satisfied:*

$$I_1\|v_{0x}\|^2 < \left(\frac{\hat{\kappa}\underline{u}}{24\overline{u}}\right)^2, \tag{5.1.27}$$

$$I_1 A\|w_{0x}\|^2 < \left(\frac{\hat{\kappa}\underline{u}}{12\overline{u}}\right)^2, \tag{5.1.28}$$

$$2I_1\left\{48I_4^2 M_1 I_1\left(\frac{\overline{u}}{\underline{u}}\right)^2(8 + 9M_1) + 2RM_1 I_4 + \frac{I_1 R^2 M_1^2}{\hat{\kappa}}\left(\frac{\overline{u}}{\underline{u}} + \frac{3M_1}{2}\right)\right.$$
$$\left. + I_1 M_1\overline{u}^2\left(1 + \frac{3AI_1}{R}\left(\frac{\overline{u}}{\underline{u}} + \overline{u}^2\right)\right) + I_2\right\} < \min\left\{\left(\frac{\hat{\kappa}\underline{u}}{12A\overline{u}}\right)^2, \left(\frac{\hat{\kappa}\underline{u}}{24\overline{u}}\right)^2,\right.$$

$$\left(\int_1^{M_1} \sqrt{s-1-\log s}\,ds\right)^2, \left(\int_0^1 \sqrt{s-1-\log s}\,ds\right)^2 \right\}. \tag{5.1.29}$$

Then in Case I and Case II, for $i = 1$, the estimates (5.1.7)–(5.1.10) hold; for $i = 2$, the estimates (5.1.11)–(5.1.13) hold and for $i = 4$, the estimates (5.1.14)–(5.1.17) hold.

In fact, from $\eta_0 - 1$, v_0, w_0, $\theta_0 - 1 \in H^i(\mathbb{R})$ ($i = 1, 2, 4$) and $\eta_0(x)$, $\theta_0(x) > 0$, we can derive (5.1.19) using the embedding of $H^i(\mathbb{R})$ into $C(\mathbb{R})$.

5.2 Global Existence and Asymptotic Behavior in $H^1(\mathbb{R})$

In this section, we shall follow Jiang's subtle idea in [54] to establish global H^1 estimates for solutions (η, v, w, θ) to our system. In what follows C and $\widetilde{C}$ stand for some generic constants (≥ 1) which might depend on physical constants such as R, etc.

At first, we suppose that

$$|\eta(x,t) - 1| + \phi(t)|\theta(x,t) - 1| \leq \frac{1}{2}, \tag{5.2.1}$$

for all $(x,t) \in \mathbb{R} \times \mathbb{R}_+$ where $\phi(t) = \min\{t, 1\}$.

Lemma 5.2.1. *For all $t > 0$, there holds*

$$\frac{1}{2}\int_{\mathbb{R}} v^2 dx + \frac{1}{2A}\int_{\mathbb{R}} w^2 dx + \int_{\mathbb{R}}(\theta - \log\theta - 1)dx + R\int_{\mathbb{R}}(\eta - \log\eta - 1)dx$$
$$+ \int_0^t \int_{\mathbb{R}}\left(\frac{v_x^2}{\theta\eta} + \frac{w_x^2}{\theta\eta} + \frac{\eta w^2}{\theta} + \hat{\kappa}\frac{\theta_x^2}{\eta\theta^2}\right)dx\,ds = E_0, \tag{5.2.2}$$

where

$$E_0 := \frac{1}{2}\int_{\mathbb{R}} v_0^2 dx + \frac{1}{2A}\int_{\mathbb{R}} w_0^2 dx + \int_{\mathbb{R}}(\theta_0 - \log\theta_0 - 1)dx + R\int_{\mathbb{R}}(\eta_0 - \log\eta_0 - 1)dx. \tag{5.2.3}$$

Also, as $|x| \to +\infty$ and $t > 0$,

$$\theta(x,t) \to 1, \quad \eta(x,t) \to 1. \tag{5.2.4}$$

Moreover,

$$\frac{1}{2}\int_{\mathbb{R}} v^2 dx + \frac{1}{2A}\int_{\mathbb{R}} w^2 dx + \int_{\mathbb{R}}(\theta - \log\theta - 1)dx + R\int_{\mathbb{R}}(\eta - \log\eta - 1)dx$$
$$+ \int_1^t \int_{\mathbb{R}}\left(\frac{v_x^2}{\theta\eta} + \frac{w_x^2}{\theta\eta} + \frac{\eta w^2}{\theta} + \hat{\kappa}\frac{\theta_x^2}{\eta\theta^2}\right)dx\,ds = E_1$$

$$\leq C \int_{\mathbb{R}} \left\{ v^2 + w^2 + (\eta - 1)^2 + (\theta - 1)^2 \right\}(x,1)dx, \quad \text{for } t \geq 1, \qquad (5.2.5)$$

where,

$$E_1 := \frac{1}{2} \int_{\mathbb{R}} v^2(x,1)dx + \frac{1}{2A} \int_{\mathbb{R}} w^2(x,1)dx + \int_{\mathbb{R}} (\theta - \log \theta - 1)(x,1)dx$$

$$+ R \int_{\mathbb{R}} (\eta - \log \eta - 1)(x,1)dx. \qquad (5.2.6)$$

Proof. Estimate (5.2.2) can be borrowed directly from Lemma 3.1 of [82] and (5.2.4) can be derived from (5.6.2)–(5.6.3). Similarly as for (5.2.2) and applying the mean value theorem to the functions $\theta - \log \theta$, $\eta - \log \eta$, we can immediately get (5.2.5). The proof is complete. $\qquad \square$

Now let us estimate the L^2 norms of v, w, $\eta - 1$, $\theta - 1$ by using a weighted L^2-norm given by $\| \cdot \|_w = \left(\int_{\mathbb{R}} (1 + x^2)^\alpha | \cdot |^2 dx \right)^{1/2}$, which is basic for H^1-global estimates. Here we use the weight function $\psi(x) = (1 + x^2)^\alpha$ $(\alpha > \frac{1}{2})$ known as a weighted function.

Lemma 5.2.2. *Under the assumptions of Theorem 5.1.1 except, for the moment, the condition on e_0, the following estimate holds, for any $t > 0$:*

$$\|v(t)\|^2 + \|\eta(t) - 1\|^2 + \|w\|^2 + \|\theta(t) - 1\|^2$$

$$+ \int_0^t \left[\|v_x(s)\|^2 + \|w_x(s)\|^2 + \|w(s)\|^2 + \|\theta_x(s)\|^2 \right] ds \leq C e_0^2 \qquad (5.2.7)$$

where $e_0 \leq 1/(2C)$.

Proof. Multiplying (5.1.2) by $\psi(x)v$, integrating over $\mathbb{R}$, and using integration by parts we get

$$\frac{1}{2} \frac{\partial}{\partial t} \int_{\mathbb{R}} \psi v^2 dx = - \int_{\mathbb{R}} \frac{v_x}{\eta} (\psi v)_x dx + \int_{\mathbb{R}} \left(\frac{R\theta}{\eta} - 1 \right) (\psi v)_x dx$$

$$= - \int_{\mathbb{R}} \frac{v_x}{\eta} (\psi_x v + \psi v_x) dx + \int_{\mathbb{R}} \left(\frac{R\theta}{\eta} - 1 \right) (\psi_x v + \psi v_x) dx.$$

Then, using $|\psi_x| \leq C|\psi|$ with (5.2.1) and the mean value theorem for the function $f(\eta, \theta) = R\theta/\eta - R$, we arrive, after integrating over $(0,t)$, $t \in [0,1]$ and employing Young's inequality, at the inequality

$$\int_{\mathbb{R}} \psi v^2(x,t)dx + \int_0^t \int_{\mathbb{R}} \psi v_x^2 dx ds$$

$$\leq C e_0^2 + C \int_0^t \int_{\mathbb{R}} \left[\psi((\eta - 1)^2 + v^2 + (\theta - 1)^2) \right] dx ds, \quad \forall t \in [0,1]. \qquad (5.2.8)$$

Multiplying (5.1.1) by $\psi(x)(\eta - 1)$ and integrating in the same way, we can see that

$$\int_{\mathbb{R}} \psi(\eta-1)^2(x,t)dx \le Ce_0^2 + C\int_0^t \int_{\mathbb{R}} \psi\left((\eta-1)^2 + v^2 + (\theta-1)^2\right)dxds, \quad \forall t \in [0,1].$$
(5.2.9)

Adding (5.2.9) to (5.2.8) gives

$$\int_{\mathbb{R}} \psi\left((\eta-1)^2 + v^2\right)(x,t)dx + \int_0^t \int_{\mathbb{R}} \psi v_x^2 dxds$$

$$\le Ce_0^2 + C\int_0^t \int_{\mathbb{R}} \psi\left((\eta-1)^2 + v^2 + (\theta-1)^2\right)dxds, \quad (5.2.10)$$

for all $t \in [0,1]$.

In the same way, we multiply (5.1.3) by $\psi(x)w$ to get

$$\frac{1}{2}\frac{d}{dx}\int_{\mathbb{R}} \psi w^2 dx + A\int_{\mathbb{R}} \psi\eta w^2 dx = -A\int_{\mathbb{R}} \frac{\phi w_x^2}{\eta}dx - A\int_{\mathbb{R}} \frac{w_x w \psi_x}{\eta}dx,$$

whence

$$\int_{\mathbb{R}} \psi w^2(x,t)dx + \int_0^t \int_{\mathbb{R}} \psi\left(w^2 + w_x^2\right)dxds \le Ce_0^2, \quad \forall t \in [0,1]. \quad (5.2.11)$$

Let us introduce $h(t) = \sup_{s \in [0,t]} \int_{\mathbb{R}} \psi(x)[v^2 + (\theta-1)^2](x,s)dx$. We derive from (5.1.2) and (5.1.4) immediately that

$$\frac{1}{2}\frac{d}{dt}\int_{\mathbb{R}} \psi(\theta-1)^2(x,t)dx + \int_{\mathbb{R}} \psi\hat{\kappa}\frac{\theta_x^2}{\eta}dxds$$

$$= -\int_{\mathbb{R}} \hat{\kappa}\frac{\theta_x}{\eta}\psi_x(\theta-1)dx + \int_{\mathbb{R}}\left(\frac{v_x}{\eta} - \frac{R\theta}{\eta}\right)v_x\psi(\theta-1)dx$$

$$+ \int_0^t \int_{\mathbb{R}}\left(\frac{w_x^2}{\eta} + \eta w^2\right)\psi(\theta-1)dxds, \quad (5.2.12)$$

$$\frac{1}{4}\frac{d}{dt}\int_{\mathbb{R}} \psi v^4(x,t)dx + \int_{\mathbb{R}} \frac{3\psi v_x^2 v_2}{\eta}dx = \int_{\mathbb{R}}\left(\frac{3R\theta v^2 \psi v_x}{\eta} - \frac{v_x v^3 \psi_x}{\eta} + \frac{R\theta v^3 \psi_x}{\eta}\right)dx, \quad (5.2.13)$$

for all $t \in [0,1]$. Adding (5.2.12) to (5.2.13), integrating with respect to t over $(0,t)$, and employing (5.2.1) and (5.2.11), we obtain

$$\int_{\mathbb{R}} \psi\left((\theta-1)^2 + v^4\right)(x,t)dx + \int_0^t \int_{\mathbb{R}} \psi\left(\theta_x^2 + v^2 v_x^2\right)dxds$$

$$\le Ce_0^2 + C\int_0^t \int_{\mathbb{R}} \psi\left[(\theta-1)^4 + (\theta-1)^2 v^2\right]dxds$$

$$+ C \int_0^t \int_{\mathbb{R}} \psi v^4 dx ds + C \int_0^t \int_{\mathbb{R}} v_x^2 dx ds + C \int_0^t \int_{\mathbb{R}} \psi \left((\theta - 1)^2 + v^2 \right) dx ds$$

$$\leq C e_0^2 + C h(t) \int_0^t \max_{\mathbb{R}} (\theta - 1)^2 (\cdot, s) ds + C \int_0^t \int_{\mathbb{R}} \psi \left((\theta - 1)^2 + v^4 + v^2 \right) dx ds$$

$$\leq C e_0^2 + C h^3(t) + \frac{1}{2} \int_0^t \int_{\mathbb{R}} \psi \theta_x^2 dx ds$$

$$+ C \int_0^t \int_{\mathbb{R}} \psi \left((\theta - 1)^2 + v^4 + v^2 \right) dx ds, \quad \forall t \in [0, 1]. \tag{5.2.14}$$

To get (5.2.14) we have used the estimate

$$\max_{\mathbb{R}} (\theta - 1)^2 \leq \max_{x \in \mathbb{R}} \int_{-\infty}^x 2|\theta - 1||\theta_x| dx$$

$$\leq C \int_{\mathbb{R}} |\theta - 1||\theta_x| dx,$$

and then employed Young's inequality to the term $C h(t) \int_0^t \max_{\mathbb{R}} (\theta - 1)^2 ds$ and the definition of $h(t)$.

From (5.2.10)–(5.2.14), we find that for all $t \in [0, 1]$

$$\int_{\mathbb{R}} \psi \left((\eta - 1)^2 + (\theta - 1)^2 + v^2 + v^4 + w^2 \right) (x, t) dx$$

$$+ \int_0^t \int_{\mathbb{R}} \psi \left(\theta_x^2 + v_x^2 + v^2 v_x^2 + w_x^2 + w^2 \right) dx ds \leq C(e_0^2 + h^3(t)). \tag{5.2.15}$$

With the definition of $h(t)$, we derive that $h(t) \leq C(e_0^2 + h^3(t))$ for $t \in [0, 1]$. Actually, we can assume $1 - C h^2(t) \geq 1/2$ and get at once $h(t) \leq 2 C e_0^2$, then $h(t) \leq e_0$ if $e_0 \leq 1/(2C)$. As a result, (5.2.15) can be improved under the condition $e_0 \leq 1/(2C)$ to

$$\int_{\mathbb{R}} \psi \left((\eta - 1)^2 + (\theta - 1)^2 + v^2 + v^4 + w^2 \right) (x, t) dx$$

$$+ \int_0^t \int_{\mathbb{R}} \psi \left(\theta_x^2 + v_x^2 + v^2 v_x^2 + w_x^2 + w^2 \right) dx ds \leq C e_0^2, \quad \forall t \in [0, 1]. \tag{5.2.16}$$

Repeating what we have done when $t \in [0, 1]$ using (5.2.1), we can derive from (5.2.16) and (5.2.5) that

$$\int_{\mathbb{R}} \left((\eta - 1)^2 + (\theta - 1)^2 + v^2 + w^2 \right) (x, t) dx + \int_0^t \int_{\mathbb{R}} \left(\theta_x^2 + v_x^2 + w_x^2 + w^2 \right) dx ds$$

$$\leq C e_0^2, \quad \forall t \in [0, +\infty), \tag{5.2.17}$$

provided that $e_0 \leq 1/(2C)$, i.e., (5.2.7). This completes the proof. $\square$

Next let us obtain a global H^1 estimate for (η, v, w, θ). We define here

$$H(t) := \sup_{0 \le s \le t} \left\{ \|\eta - 1\|_{L^\infty}^2 + \phi^2 \|v_x\|^2 + \phi^2 \|w_x\|^2 + \phi^2 \|w\|^2 + \phi^4 \|\theta_x\|^2 \right\}$$

$$+ \int_0^t \left[\phi^2 \|v_t\|^2 + \phi^4 \|\theta_t\|^2 + \|v_x\|^2 + \phi^2 \|w_t\|^2 + \|w_x\|^2 \right](s)ds. \quad (5.2.18)$$

Lemma 5.2.3. *Under the assumptions of Lemma 5.2.2, we have the estimate*

$$H(t) \le e_0, \quad (5.2.19)$$

where $e_0 < \min\{1/(2C), 1/(2\tilde{c})\}$, and $\tilde{c} \ge 1$ depends on some physical constants.

Proof. Multiplying (5.1.2) by $\phi^2 v_t$, and integrating over $\mathbb{R} \times \mathbb{R}_+$, we arrive at

$$\int_0^t \int_{\mathbb{R}} \phi^2 v_t^2 dx ds + \int_0^t \int_{\mathbb{R}} \left(\frac{\phi^2 v_x^2}{\eta} \right)_t dx ds$$

$$= - \int_0^t \int_{\mathbb{R}} \left(\frac{R\theta}{\eta} \right)_x \phi^2 v_t dx ds + \int_0^t \int_{\mathbb{R}} \frac{2\phi\phi_t \cdot v_x^2}{\eta} dx ds - \int_0^t \int_{\mathbb{R}} \frac{\phi^2 v_x^3}{\eta} dx ds.$$

From the definition of $\phi(t)$, we derive that $|\phi| \le 1$, $|\phi_t| \le 1$, and then utilizing (5.2.1) and (5.2.15) and applying Young's inequality, we obtain

$$\phi^2(t) \int_{\mathbb{R}} v_x^2 dx + \int_0^t \int_{\mathbb{R}} \phi^2 v_t^2 dx ds$$

$$\le C e_0^2 + C \int_0^t \int_{\mathbb{R}} \phi^2 |v_x|^3 dx ds + C \left| \int_0^t \int_{\mathbb{R}} \phi^2 \left[\left(\frac{1}{\eta - 1} \right) \theta \right]_x v_t dx ds \right|$$

$$\le C e_0^2 + C \int_0^t \int_{\mathbb{R}} \phi^4 v_x^4 dx ds$$

$$+ \left| - \int_{\mathbb{R}} \frac{\eta + 1}{\eta} \theta \phi^2 v_x dx + \int_{\mathbb{R}} (1 - \eta)\theta_0 \phi^2 v_{0x} dx + \int_0^t \int_{\mathbb{R}} 2\phi\phi_t \left(\frac{1}{\eta} - 1 \right) \theta v_x \right.$$

$$\left. + \phi^2 \left[\left(\frac{1}{\eta} - 1 \right) \theta \right]_t v_x dx ds \right|$$

$$\le C e_0^2 + C \int_0^t \int_{\mathbb{R}} \phi^4 v_x^4 dx ds + C \int_0^t |\eta - 1||\phi||v_x| dx$$

$$+ C \int_0^t \int_{\mathbb{R}} \phi |\eta - 1||\theta||v_x| dx ds + C \int_0^t \int_{\mathbb{R}} \phi^2 \left| \left[\left(\frac{1}{\eta} - 1 \right) \theta \right]_t v_x \right| dx ds$$

$$\le C e_0^2 + C \int_0^t \int_{\mathbb{R}} \phi^2 v_t^2 dx ds + C \int_0^t \int_{\mathbb{R}} |\eta - 1|\phi^4 \theta_t^2 dx ds + \frac{1}{2} \phi^2 \int_{\mathbb{R}} v_x^2 dx,$$

i.e.,

$$\phi^2 \int_{\mathbb{R}} v_x^2 dx + \int_0^t \int_{\mathbb{R}} \phi^2 v_t^2 dx ds \le C e_0^2 + C \int_0^t \int_{\mathbb{R}} \phi^4 v_x^4 dx ds + H^2(t), \quad \forall t \ge 0.$$

$$(5.2.20)$$

Actually the term $\int_0^t \int_{\mathbb{R}} \phi^4 v_x^2 dx ds$ in (5.2.20) can be estimated with (5.2.1), (5.2.17), and by applying the interpolation inequality $\|\cdot\|_{L^\infty} \leq C\|\cdot\|_{H^1}$ to the function $v_x/\eta - R\theta/\eta + R$ and the mean value theorem to the function $R\theta/\eta - R$:

$$\int_0^t \int_{\mathbb{R}} \phi^4 v_x^2 dx ds \leq C \int_0^t \phi^4 \max_{\mathbb{R}} v_x^2 \int_{\mathbb{R}} v_x^2 dx ds$$

$$\leq C e_0^2 + C \int_0^t \phi^4 \max_{\mathbb{R}} \left(\frac{v_x}{\eta} - R\frac{\theta}{\eta} + R \right)^2 \left(\int_{\mathbb{R}} v_x^2 dx \right) ds$$

$$\leq C e_0^2 + C \int_0^t \phi^4 (\|\sigma + R\|^2 + \|\sigma_x\|^2) \|v_x\|^2 ds$$

$$\leq C e_0^2 + C \int_0^t \phi^4 (\|v_x\|^2 + \|\eta - 1\|^2 + \|\theta - 1\|^2 + \|v_t\|^2) \|v_x\|^2 ds$$

$$\leq C(e_0^2 + H^2(t)). \tag{5.2.21}$$

Combining (5.2.20) and (5.2.21), we infer that

$$\phi^2(t) \int_{\mathbb{R}} v_x^2 dx + \int_0^t \int_{\mathbb{R}} \phi^2 v_t^2 dx ds \leq C(e_0^2 + H^2(t)), \quad \forall t \geq 0. \tag{5.2.22}$$

Similarly, we can get from (5.1.3), upon multiplying (5.2.17) with $\phi^2 w_t$, that

$$\phi^2(t)\|w_x(t)\|^2 + \phi^2(t)\|w(t)\|^2 + \int_0^t \phi^2(s)\|w_t(s)\|^2 ds \leq C(e_0^2 + H^2(t)), \quad \forall t \geq 0. \tag{5.2.23}$$

Also analogously, we infer from (5.1.4), upon multiplying with $\phi^4 \theta_t$, that

$$\phi^4(t)\|\theta_x(t)\|^2 + \int_0^t \|\theta_t(s)\|^2 \phi^4 ds$$

$$\leq C e_0^2 + C \int_0^t \int_{\mathbb{R}} (\phi^4 v_x^4 + \phi^4 |v_x|\theta_x^2) dx ds + \frac{1}{2} \int_0^t \int_{\mathbb{R}} \phi^4 \theta_t^2 dx ds + CH^2(t)$$

$$\leq C e_0^2 + CH^2(t) + C \int_0^t \phi^8 \max_{\mathbb{R}} v_x^2 \int_{\mathbb{R}} \theta_x^2 dx ds + \frac{1}{2} \int_0^t \int_{\mathbb{R}} \phi^4 \theta_t^2 dx ds$$

$$\leq C e_0^2 + CH^2(t) + \frac{1}{2} \int_0^t \int_{\mathbb{R}} \phi^4 \theta_t^2 dx ds, \quad \forall t \geq 0,$$

i.e.,

$$\phi^4(t)\|\theta_x(t)\|^2 + \int_0^t \|\theta_t(s)\|^2 \phi^4 ds \leq C e_0^2 + CH^2(t), \tag{5.2.24}$$

for all $t \geq 0$.

Now let us estimate $\eta - 1$. Rewrite (5.1.2) using (5.1.1) as

$$(\log \eta)_{xt} = v_t + R \left(\frac{\theta}{\eta} - 1 \right)_x. \tag{5.2.25}$$

Integrating (5.2.25) over $(-\infty, x) \times (0, t)$ ($\forall t \in [0, 1]$), taking the absolute value and using (5.2.1), (5.2.16)–(5.2.17) and the weight $\alpha > 1/2$, we obtain

$$|\eta - 1| \leq C|\log \eta| \leq C|\eta - 1| + C \int_{-\infty}^{x} (|v| + |v_0|) dy + C \int_0^t (|\eta - 1| + |\theta - 1|) ds$$

$$\leq Ce_0 + C\|\psi^{\frac{1}{2}} v\| \|\psi^{-\frac{1}{2}}\| + C \int_0^t |\eta - 1| ds + C \int_0^t \|\theta - 1\|_{H^1} ds$$

$$\leq Ce_0 + C \int_0^t |\eta - 1| ds + C \left(\int_0^t \|\theta - 1\|_{H^1}^2 ds \right)^{\frac{1}{2}}$$

$$\leq Ce_0 + C \int_0^t |\eta - 1| ds, \quad \forall t \in [0, 1]. \tag{5.2.26}$$

Applying the Gronwall inequality to (5.2.26), we get

$$|\eta(x, t) - 1| \leq Ce_0, \quad x \in \mathbb{R}, \ t \in [0, 1]. \tag{5.2.27}$$

For $t \geq 1$, we denote $F := (v_x/\eta) - R(\eta/\theta) + R$. Obviously, $[1/\eta - 1]_t + (R\theta/\eta) \cdot [1/\eta - 1] = -F/\eta - R(\theta - 1)/\eta$. Multiplying this equality by $1/\eta - 1$, using (5.2.1), (5.1.2), (5.2.15) and the interpolation inequality, we obtain

$$\left[\frac{1}{\eta} - 1 \right]_t^2 + C^{-1} \left[\frac{1}{\eta} - 1 \right]^2 \leq C(\|F\|_{L^\infty}^2 + \|\theta - 1\|_{L^\infty}^2)$$

$$\leq C(\|F\|_{H^1}^2 + \|\theta - 1\|_{H^1}^2)$$

$$\leq Ce_0^2 + C\|(v_x, v_t, \theta_x)\|^2, \quad \forall t \geq 1,$$

which, combined with (5.2.17) and (5.2.22), yields for $x \in \mathbb{R}$, $t \geq 1$,

$$|\eta(x, t) - 1|^2 \leq Ce_0^2 + C|\eta(x, t) - 1|^2 + C \int_1^t \|(v_x, v_t, \theta_x)(s)\|^2 ds \leq C(e_0^2 + H^2(t)).$$

$$\tag{5.2.28}$$

Combining the definition of $H(t)$, (5.2.22)–(5.2.24), (5.2.27) and (5.2.28), we obtain $H(t) \leq \tilde{c}[e_0^2 + H^2(t)]$ for $t \in [0, +\infty)$, where $\tilde{c} \geq 1$ depends on C and other systematic constants. Hence similar to the estimate for $h(t)$, we still assume $H(t)$ so small that $1 - \tilde{c}H(t) \geq 1/2$, and then $H(t) \leq 2\tilde{c}e_0^2 \leq e_0$ provided that $e_0 \leq \min\{1/2C, 1/2\tilde{c}\}$. The proof is complete. □

Proof of Theorem 5.1.1. From Lemmas 5.2.2 and 5.2.3, we immediately derive that

$$|\eta(x, t) - 1| + \phi(t)|\theta(x, t) - 1| \leq A^{1/2}(t) + C\|\theta - 1\|^{1/2}\|\theta_x\|^{1/2}$$

$$\leq A^{1/2}(t) + Ce_0 A^{1/4}(t) \tag{5.2.29}$$

$$\leq \sqrt{2}\tilde{c}e_0 \left(1 + \tilde{C}\sqrt{e_0}\right) < \frac{1}{3}, \quad x \in \mathbb{R}, \ t \geq 0.$$

Actually, we assume that e_0 is small enough to assure that $1 + \tilde{C}\sqrt{e_0} < 4/3$, and then $e_0 < 1/(4\sqrt{2})$. So we set $e_0 < \min\{1/(6\sqrt{\tilde{c}}), 1/(3\tilde{C})^2, 1/(2\tilde{c}), 1/(2\tilde{C})\} := \epsilon_0$, with the hypothesis discussed in Lemmas 5.2.2 and 5.2.3.

We can see thus that $|\eta-1|+\phi(t)|\theta-1|$ can be smaller than in the hypothesis (5.2.1). As a result, we find

$$\|\eta - 1\|_{L^\infty}^2 + \|v_x\|^2 + \|w_x\|^2 + \|\theta_x\|^2 + \int_0^t \left(\|v_t\|^2 + \|\theta_t\|^2 + \|w_t\|^2\right)ds \le \epsilon_0^2 \le C_1. \tag{5.2.30}$$

Also from (5.2.29) with arguments similar to those used in [112], we can get (5.1.10) and (5.1.11). Under the assumption $(\eta - 1, v, w, \theta - 1) \in H^1(\mathbb{R})$, we rewrite (5.2.25) again as

$$\left(\frac{\eta_x}{\eta}\right)_t = v_t + \frac{R\theta}{\eta}, \tag{5.2.31}$$

multiply it by η_x/η, and integrate over $\mathbb{R} \times [0,t]$ to get, using (5.1.11), the estimate

$$\|\eta_x\|^2 + \int_0^t \int_\mathbb{R} \theta\eta_x^2 dxds \le C_1 + \frac{1}{2}\int_0^t \int_\mathbb{R} \theta\eta_x^2 dxds + C\int_0^t \left(\|v_t\|^2 + \|\theta_x\|^2\right)ds \le C_1. \tag{5.2.32}$$

From (5.1.11), it follows that

$$\|\eta_x(t)\|^2 + \int_0^t \|\eta_x(s)\|^2 ds \le C_1. \tag{5.2.33}$$

Upon rewriting (5.1.2) as

$$v_t = \frac{v_{xx}}{\eta} - \frac{v_x\eta_x}{\eta^2} - \frac{R\theta_x}{\eta} + \frac{R\theta\eta_x}{\eta^2}$$

we infer that

$$\|v_{xx}\|^2 \le C_1\left(\|v_t\|^2 + \|v_x\|_{L^\infty}^2 + \|\theta_x\|^2 + \|\eta_x\|^2\right),$$

and conclude immediately from (5.1.11), Lemmas 5.2.2–5.2.3 and the interpolation inequalities that

$$\int_0^t \|v_{xx}(s)\|^2 ds \le C_1, \quad \|v_{xx}\| \le C_1\|v_t\| + C_1.$$

Similarly, we get

$$\int_0^t \left(\|w_{xx}\|^2 + \|\theta_{xx}\|^2\right)(s)ds \le C_1,$$

$$\|w_{xx}\| \le C_1\|w_t\| + C_1,$$

$$\|\theta_{xx}\| \le C_1\|\theta_t\| + C_1. \tag{5.2.34}$$

Combining with (5.2.30), (5.1.10) and (5.1.11), we complete the proof of (5.1.12) of Theorem 5.1.1.

By integrating by parts, we can conclude further that

$$\frac{1}{2}\frac{d}{dt}\int_{\mathbb{R}} v_x^2 dx = \int_{\mathbb{R}} v_x v_{xt} dx, \tag{5.2.35}$$

$$\frac{1}{2}\frac{d}{dt}\int_{\mathbb{R}} w_x^2 dx = \int_{\mathbb{R}} w_x w_{xt} dx, \tag{5.2.36}$$

$$\frac{1}{2}\frac{d}{dt}\int_{\mathbb{R}} \theta_x^2 dx = \int_{\mathbb{R}} \theta_x \theta_{xt} dx, \tag{5.2.37}$$

and then using (5.1.1) and (5.1.12) we get

$$\int_0^{+\infty} \left| \frac{d}{dt}\left(\|(\eta_x, v_x, w_x, \theta_x)\| \right)(s) \right| ds \leq C_1. \tag{5.2.38}$$

Combining this with (5.2.35)–(5.2.38), we can immediately derive that

$$\|(\eta_x, v_x, w_x, \theta_x)(t)\| \to 0, \quad \text{as } t \to +\infty. \tag{5.2.39}$$

Finally, employing the interpolation inequality, we can also see that

$$\|(\eta - 1, v, w, \theta - 1)(t)\|_{L^\infty} \to 0, \quad \text{as } t \to +\infty. \tag{5.2.40}$$

This ends the proof of Theorem 5.1.1. □

5.3 Global Existence and Asymptotic Behavior in $H^2(\mathbb{R})$

In this section, we shall complete the proof of Theorem 5.1.2. We begin with the following lemma on the estimates in $H^1(\mathbb{R})$ obtained just now.

Lemma 5.3.1. *Under the assumptions of Theorem 5.1.1, the H^1-generalized global solution $(\eta(t), v(t), w(t), \theta(t))$ to the Cauchy problem (5.1.1)–(5.1.6) verifying (5.1.7)–(5.1.9) obeys for any $t > 0$ the bounds*

$$\|\eta(t) - 1\|_{H^1}^2 + \|v(t)\|_{H^1}^2 + \|w(t)\|_{H^1}^2 + \|\theta(t) - 1\|_{H^1}^2 + \|\eta_t(t)\|^2$$

$$+ \int_0^t \left(\|v_x\|_{H^1}^2 + \|w_x\|_{H^1}^2 + \|\theta_x\|_{H^1}^2 + \|\eta_x\|^2 + \|v_t\|^2 + \|w_t\|^2 + \|\theta_t\|^2 \right)(s)ds$$

$$\leq C_1, \tag{5.3.1}$$

$$\|\eta(t) - 1\|_{L^\infty}^2 + \|v(t)\|_{L^\infty}^2 + \|w(t)\|_{L^\infty}^2 + \|\theta(t) - \bar{\theta}\|_{L^\infty}^2$$

$$+ \int_0^t \left(\|\eta_t\|_{H^1}^2 + \|v_x\|_{L^\infty}^2 + \|w_x\|_{L^\infty}^2 + \|\theta_x\|_{L^\infty}^2 \right)(s)ds \leq C_1. \tag{5.3.2}$$

Proof. Estimate (5.3.1) is just (5.1.9). Next, by Theorem 5.1.1, we get

$$\|v(t)\|_{L^\infty} \le C\|v(t)\|_{H^1}, \qquad \|w(t)\|_{L^\infty} \le C\|w(t)\|_{H^1},$$

$$\|\theta(t) - 1\|_{L^\infty} \le C\|\theta(t) - 1\|_{H^1}, \tag{5.3.3}$$

$$\|v_x(t)\|_{L^\infty} \le C\|v_x(t)\|_{H^1}, \qquad \|\theta_x(t)\|_{L^\infty} \le C\|\theta_x(t)\|_{H^1}. \tag{5.3.4}$$

Moreover, (5.1.1) yields

$$\|\eta_t(t)\|_{H^1} = \|v_x(t)\|_{H^1}. \tag{5.3.5}$$

Thus estimate (5.3.2) follows from (5.3.1) and (5.3.3)–(5.3.5). The proof is complete. □

Lemma 5.3.2. *Under the assumptions of Theorem 5.1.2, the following estimates hold for any $t > 0$:*

$$\|\theta_t(t)\|^2 + \|v_t(t)\|^2 + \|w_t(t)\|^2 + \int_0^t (\|v_{xt}\|^2 + \|w_{xt}\|^2 + \|\theta_{xt}\|^2)(s)ds \le C_2,$$
$$\tag{5.3.6}$$

$$\|v_x(t)\|_{L^\infty}^2 + \|v_{xx}(t)\|^2 + \|w_x(t)\|_{L^\infty}^2 + \|w_{xx}(t)\|^2 + \|\theta_x(t)\|_{L^\infty}^2 + \|\theta_{xx}(t)\|^2 \le C_2,$$
$$\tag{5.3.7}$$

$$\|v(t)\|_{H^2}^2 + \|w(t)\|_{H^2}^2 + \|\theta(t) - 1\|_{H^2}^2 + \|\eta_t(t)\|_{H^1}^2 \le C_2. \tag{5.3.8}$$

Proof. Differentiating (5.1.2) with respect to t, then multiplying the resulting equation by v_t in $L^2(\mathbb{R})$ and using Lemma 5.3.1, we get

$$\frac{d}{dt}\|v_t(t)\|^2 + C_1^{-1}\|v_{xt}(t)\|^2$$

$$\le \frac{1}{2C_1}\|v_{xt}(t)\|^2 + C_2(\|v_x(t)\|^3\|v_{xx}(t)\| + \|\theta_t(t)\|^2 + \|v_x(t)\|^2)$$

$$\le \frac{1}{2C_1}\|v_{xt}(t)\|^2 + C_2(\|v_x(t)\|^2 + \|\theta_t(t)\|^2 + \|v_{xx}(t)\|^2) \tag{5.3.9}$$

whence

$$\|v_t(t)\|^2 + \int_0^t \|v_{xt}(s)\|^2 ds \le C_2 + C_1 \int_0^t (\|v_x\|^2 + \|\theta_t\|^2 + \|v_{xx}\|^2)(s)ds \le C_2. \tag{5.3.10}$$

Hence, by (5.1.2), Lemma 5.3.1, the interpolation inequalities and Young's inequality, we have

$$\|v_{xx}(t)\| \le C_1(\|v_t(t)\| + \|v_x(t)\| + \|u_x(t)\| + \|v_x(t)\|^{1/2}\|v_{xx}(t)\|)$$

$$\le \frac{1}{2}\|v_{xx}(t)\| + C_1(\|v_t(t)\| + \|v_x(t)\| + \|u_x(t)\|)$$

which, combined with (5.3.10), (5.3.1) and (5.3.2), leads to

$$\|v_{xx}(t)\| \le C_1(\|v_t(t)\| + \|v_x(t)\| + \|u_x(t)\|) \le C_2, \quad \forall t > 0, \tag{5.3.11}$$

$$\|v_x(t)\|_{L^\infty}^2 \leq C_1 \|v_x(t)\| \|v_{xx}(t)\| \leq C_2, \quad \forall t > 0. \tag{5.3.12}$$

Similarly, we derive from (5.1.3) that

$$\frac{d}{dt}\|w_t(t)\|^2 + C_1^{-1}\|w_{xt}(t)\|^2 + C_1^{-1}\|w_t(t)\|^2$$

$$\leq \frac{1}{2C_1}(\|w_{xt}(t)\|^2 + \|w_t(t)\|^2) + C_2(\|v_x(t)\|^2 + \|v_{xx}(t)\|^2) \tag{5.3.13}$$

which, combined with Lemma 5.3.1, gives

$$\|w_t(t)\|^2 + \int_0^t (\|w_{xt}\|^2 + \|w_t\|^2)(s)ds \leq C_2, \quad \forall t > 0. \tag{5.3.14}$$

Next, from (5.1.3), the interpolation inequalities and Young's inequality it follows that

$$\|w_{xx}(t)\| \leq C_1\Big(\|w_t(t)\| + \|w(t)\| + \|w_x(t)\|\|\eta_x(t)\|^2\Big) + \frac{1}{2}\|w_{xx}(t)\|,$$

whence

$$\|w_{xx}(t)\| \leq C_2, \quad \forall t > 0, \tag{5.3.15}$$

$$\|w_x(t)\|_{L^\infty}^2 \leq C_1\|w_x(t)\|\|w_{xx}(t)\| \leq C_2, \quad \forall t > 0. \tag{5.3.16}$$

Similarly, from (5.1.4) and (5.3.12) it follows that

$$\frac{d}{dt}\|\theta_t(t)\|^2 + C_1^{-1}\|\theta_{xt}(t)\|^2$$

$$\leq \frac{1}{2C_1}\|\theta_{xt}(t)\|^2 + C_2(\|\theta_x(t)\|^2 + \|v_x(t)\|^2 + \|\theta_t(t)\|^2 + \|v_{tx}(t)\|^2$$

$$+ \|w_x(t)\|_{L^\infty}^2\|\theta_t(t)\|^2 + \|w_x(t)\|_{L^\infty}^2\|w_x(t)\|^2 + \|w(t)\|_{L^\infty}^2\|w(t)\|^2$$

$$+ \|w_{xt}(t)\|^2), \tag{5.3.17}$$

which, combined with Lemma 5.3.1, gives

$$\|\theta_t(t)\|^2 + \int_0^t \|\theta_{xt}(s)\|^2 ds \leq C_2, \quad \forall t > 0. \tag{5.3.18}$$

Similarly to (5.3.11), equation (5.1.4), Lemma 5.3.1, (5.3.16) and the interpolation inequalities yield

$$\|\theta_{xx}(t)\| \leq C_1\Big(\|\theta_t(t)\| + \|\theta_x(t)\|^{1/2}\|\theta_{xx}(t)\|^{1/2}\|u_x(t)\| + \|v_x(t)\|^{3/2}\|v_{xx}(t)\|^{1/2}$$

$$+ \|v_x(t)\| + \|w_x(t)\|^{\frac{1}{2}}\|w_{xx}(t)\|^{\frac{1}{2}}\|w_x(t)\| + \|w_{xt}(t)\| + \|w(t)\|$$

$$+ \|\theta(t)\|\|w_x(t)\|^{\frac{1}{2}}\|w_{xx}(t)\|^{\frac{1}{2}}\Big)$$

$$\leq C_1 \Big(\|\theta_t(t)\| + \|\theta_x(t)\| + \|v_x(t)\| + \|v_{xx}(t)\|$$
$$+ \|w(t)\| + \|w_x(t)\| + \|w_{xt}(t)\| \Big) + \frac{1}{2} \|\theta_{xx}(t)\|$$

whence

$$\|\theta_{xx}(t)\| \leq C_1(\|\theta_t(t)\| + \|\theta_x(t)\| + \|v_x(t)\| + \|v_{xx}(t)\| + \|w(t)\|$$
$$+ \|w_x(t)\| + \|w_{tx}(t)\|) \leq C_2, \tag{5.3.19}$$

$$\|\theta_x(t)\|_{L^\infty}^2 \leq C_1 \|\theta_x(t)\| \|\theta_{xx}(t)\| \leq C_2. \tag{5.3.20}$$

Thus estimates (5.3.6)–(5.3.8) follow from (5.1.1), (5.3.10)–(5.3.12), (5.3.14)–(5.3.16), (5.3.18)–(5.3.20) and Lemma 3.5.1. The proof is complete.   □

Lemma 5.3.3. *Under the assumptions of Theorem 5.1.2, the following estimates hold for any $t > 0$:*

$$\|\eta_{xx}(t)\|^2 + \|\eta_x(t)\|_{L^\infty}^2 + \int_0^t (\|\eta_{xx}\|^2 + \|\eta_x\|_{L^\infty}^2)(s)ds \leq C_2, \tag{5.3.21}$$

$$\int_0^t (\|v_{xxx}\|^2 + \|w_{xxx}\|^2 + \|\theta_{xxx}\|^2)(s)ds \leq C_2. \tag{5.3.22}$$

Proof. Differentiating (5.1.2) with respect to x and using equation (5.1.1), we get

$$\frac{\partial}{\partial t} \left(\frac{\eta_{xx}}{\eta} \right) + \frac{R\theta \eta_{xx}}{\eta^2} = v_{tx} + \frac{R\theta_{xx}}{\eta} + \frac{2v_{xx}\eta_x - 2R\theta_x\eta_x}{\eta^2} + \frac{(2R\theta - 2\mu v_x)\eta_x^2}{\eta^3}. \tag{5.3.23}$$

Next, multiplying (5.3.23) by η_{xx}/η in $L^2(\mathbb{R})$, and using Lemmas 5.3.1–5.3.2, we deduce that

$$\frac{d}{dt} \Big\| \frac{\eta_{xx}}{\eta}(t) \Big\|^2 + C_1^{-1} \|\eta_{xx}(t)\|^2 \tag{5.3.24}$$

$$\leq \frac{1}{2C_1} \|\eta_{xx}(t)\|^2 + C_2(\|\theta_x(t)\|^2 + \|\eta_x(t)\|^2 + \|v_{xx}(t)\|^2 + \|\theta_{xx}(t)\|^2 + \|v_{tx}(t)\|^2)$$

which, together with Lemma 5.3.2, implies that for any $t > 0$,

$$\|\eta_{xx}(t)\|^2 + \int_0^t \|\eta_{xx}(s)\|^2 ds \leq C_2, \tag{5.3.25}$$

$$\|\eta_x(t)\|_{L^\infty}^2 \leq C \|\eta_x(t)\| \|\eta_{xx}(t)\| \leq C_2, \tag{5.3.26}$$

$$\int_0^t \|\eta_x(s)\|_{L^\infty}^2 ds \leq C \int_0^t (\|\eta_x\|^2 + \|\eta_{xx}\|^2)(s)ds \leq C_2. \tag{5.3.27}$$

Differentiating (5.1.2), (5.1.3) and (5.1.4) with respect to x and using Lemmas 5.3.1–5.3.2 and (5.3.26), we deduce that for any $t > 0$,

$$\|v_{xxx}(t)\| \leq C_2 \big(\|v_t(t)\| + \|v_{tx}(t)\| + \|v_{xx}(t)\| + \|\eta_{xx}(t)\| + \|v_x(t)\| + \|\theta_{xx}(t)\|$$

$$+ \|\theta_x(t)\| + \|\eta_x(t)\|), \tag{5.3.28}$$

$$\|w_{xxx}(t)\| \leq C_2(\|w_{tx}\| + \|w_{xx}\| + \|\eta_{xx}\| + \|w_x\|),$$

$$\|\theta_{xxx}(t)\| \leq C_2(\|\theta_t(t)\| + \|\theta_{tx}(t)\| + \|\theta_{xx}(t)\| + \|u_{xx}(t)\| + \|v_{xx}(t)\| + \|\theta_x(t)\|$$

$$+ \|w_{xx}(t)\| + \|w_x(t)\|). \tag{5.3.29}$$

Thus estimates (5.3.21)–(5.3.22) follow from (5.3.25)–(5.3.29) and Lemmas 5.3.1–5.3.2. $\qquad\square$

Lemma 5.3.4. *Under the assumptions of Theorem* 5.1.2, *the* H^2-*generalized global solution* $(\eta(t),\ v(t),\ w(t),\ \theta(t))$ *obtained in Lemmas* 5.3.1–5.3.3 *to the Cauchy problem* (5.1.1)–(5.1.6) *satisfies the estimates* (5.1.12)–(5.1.13).

Proof. We start from Lemma 5.3.1 and repeat the same reasoning as in the derivation of (5.3.9), (5.3.11)–(5.3.12), (5.3.13), (5.3.15)–(5.3.16), (5.3.17), (5.3.19)–(5.3.20), (5.3.24) and (5.3.26)–(5.3.27) in Lemmas 5.3.2–5.3.3 to obtain

$$\frac{d}{dt}\|v_t(t)\|^2 + (2C_1)^{-1}\|v_{tx}(t)\|^2 \leq C_2(\|v_x(t)\|^2 + \|v_{xx}(t)\|^2 + \|\theta_t(t)\|^2), \tag{5.3.30}$$

$$\frac{d}{dt}\|w_t(t)\|^2 + (2C_1)^{-1}\|w_{tx}(t)\|^2 + (2C_1)^{-1}\|w_t(t)\|^2 \leq C_2(\|w_x(t)\|^2 + \|w(t)\|^2), \tag{5.3.31}$$

$$\frac{d}{dt}\|\theta_t(t)\|^2 + (2C_1)^{-1}\|\theta_{tx}(t)\|^2 \leq C_2(\|v_x(t)\|^2 + \|\theta_x(t)\|^2 + \|\theta_t(t)\|^2 + \|v_{tx}(t)\|^2$$

$$+ \|w_t(t)\|^2 + \|w_{tx}(t)\|^2 + \|w_x(t)\|^2), \tag{5.3.32}$$

$$\frac{d}{dt}\left\|\frac{\eta_{xx}}{\eta}(t)\right\|^2 + (2C_1)^{-1}\|\eta_{xx}(t)\|^2 \leq C_2(\|\theta_x(t)\|^2 + \|\eta_x(t)\|^2 + \|v_{xx}(t)\|^2$$

$$+ \|\theta_{xx}(t)\|^2 + \|v_{tx}(t)\|^2), \tag{5.3.33}$$

$$\|v_{xx}(t)\| \leq C_1(\|v_t(t)\| + \|v_x(t)\| + \|u_x(t)\|) \leq C_2, \tag{5.3.34}$$

$$\|w_{xx}(t)\| \leq C_1(\|w_t(t)\| + \|w(t)\| + \|w_x(t)\|) \leq C_2, \tag{5.3.35}$$

$$\|\theta_{xx}(t)\| \leq C_1(\|\theta_t(t)\| + \|\theta_x(t)\| + \|v_x(t)\| + \|v_{xx}(t)\|) \leq C_2, \tag{5.3.36}$$

$$\|v_x(t)\|_{L^\infty}^2 \leq C\|v_x(t)\|\|v_{xx}(t)\| \leq C_2, \quad \|w_x(t)\|_{L^\infty}^2 \leq C\|w_x(t)\|\|w_{xx}(t)\| \leq C_2, \tag{5.3.37}$$

$$\|\theta_x(t)\|_{L^\infty}^2 \leq C\|\theta_x(t)\|\|\theta_{xx}(t)\| \leq C_2, \tag{5.3.38}$$

$$\|\eta_x(t)\|_{L^\infty}^2 \leq C\|\eta_x(t)\|\|\eta_{xx}(t)\| \leq C_2. \tag{5.3.39}$$

Applying Lemma 5.3.4 to (5.3.30)–(5.3.33) and using Lemmas 5.3.1–5.3.3, we obtain that, as $t \to +\infty$,

$$\|v_t(t)\| \to 0, \quad \|w_t(t)\| \to 0, \quad \|\theta_t(t)\| \to 0, \quad \|\eta_{xx}(t)\| \to 0 \tag{5.3.40}$$

which, together with (5.1.1), (5.1.7) and (5.3.34)–(5.3.39), implies that as $t \to +\infty$,

$$\|v_{xx}(t)\| + \|w_{xx}(t)\| + \|\theta_{xx}(t)\| + \|\eta_t(t)\|_{H^1} \to 0, \tag{5.3.41}$$

$$\|\eta_t(t)\|_{L^\infty} + \|(\eta_x(t), v_x(t), w_x(t), \theta_x(t))\|_{L^\infty} \to 0. \tag{5.3.42}$$

Thus (5.1.12)–(5.1.13) follows from (5.1.1) and (5.3.40)–(5.3.42). The proof is complete. □

Proof of Theorem 5.1.2. The theorem is an immediate consequence of Lemmas 5.3.1–5.3.4 and Lemma 5.3.5. □

5.4 Global Existence and Asymptotic Behavior in $H^4(\mathbb{R})$

In this section, we derive estimates in $H^4(\mathbb{R})$ and complete the proof of Theorem 5.1.3. The following several lemmas concern the estimates in $H^4(\mathbb{R})$.

Lemma 5.4.1. *Under the assumptions of Theorem 5.1.3, the following estimates hold for any $t > 0$:*

$$\|v_{tx}(x,0)\| + \|w_{tx}(x,0)\| + \|\theta_{tx}(x,0)\| \le C_3, \tag{5.4.1}$$

$$\|v_{tt}(x,0)\| + \|\theta_{tt}(x,0)\| + \|w_{tt}(x,0)\|$$
$$+ \|v_{txx}(x,0)\| + \|w_{txx}(x,0)\| + \|\theta_{txx}(x,0)\| \le C_4, \tag{5.4.2}$$

$$\|v_{tt}(t)\|^2 + \int_0^t \|v_{ttx}(s)\|^2 ds \le C_4 + C_4 \int_0^t \|\theta_{txx}(s)\|^2 ds, \tag{5.4.3}$$

$$\|w_{tt}(t)\|^2 + \int_0^t (\|w_{ttx}\|^2 + \|w_{tt}\|^2)(s)ds \le C_4, \tag{5.4.4}$$

$$\|\theta_{tt}(t)\|^2 + \int_0^t \|\theta_{ttx}(s)\|^2 ds \le C_4 + C_4 \int_0^t (\|\theta_{txx}\|^2 + \|v_{txx}\|^2$$
$$+ \|w_{xt}\|^3 \|w_{txx}\|)(s)ds. \tag{5.4.5}$$

Here $C_3 > 0$ is a constant depending on C_1, C_2 and the H^3 norm of the initial data $(\eta_0(x), v_0(x), w_0(x), \theta_0(x))$.

Proof. We easily infer from (5.1.2) and Lemmas 5.3.1–5.3.3 that

$$\|v_t(t)\| \le C_2 \|v_x(t)\|_{H^1} + \|\eta_x(t)\| + \|\theta_x(t)\|). \tag{5.4.6}$$

Differentiating (5.1.2) with respect to x and using Lemmas 5.3.1–5.3.2, we have

$$\|v_{tx}(t)\| \le C_2(\|v_x(t)\| + \|v_{xxx}(t)\| + \|\theta_x(t)\|_{H^1} + \|\eta_x(t)\|_{H^1}) \tag{5.4.7}$$

and

$$\|v_{xxx}(t)\| \le C_2(\|v_x(t)\| + \|\eta_x(t)\|_{H^1} + \|\theta_x(t)\|_{H^1} + \|v_{tx}(t)\|). \tag{5.4.8}$$

Next, differentiating (5.1.2) with respect to x twice and using Lemmas 5.3.1–5.3.3 and the interpolation inequalities, we have

$$\|v_{txx}(t)\| \le C_2(\|\eta_x(t)\|_{H^2} + \|v_x(t)\|_{H^3} + \|\theta_x(t)\|_{H^2}) \tag{5.4.9}$$

or

$$\|v_{xxxx}(t)\| \leq C_2(\|\eta_x(t)\|_{H^2} + \|v_x(t)\|_{H^2} + \|\theta_x(t)\|_{H^2} + \|v_{txx}(t)\|). \quad (5.4.10)$$

In the same manner, we deduce from (5.1.3) and (5.1.4) that

$$\|w_t(t)\| \leq C_2(\|\eta_x(t)\| + \|w(t)\|_{H^2}), \quad (5.4.11)$$

$$\|w_{tx}(t)\| \leq C_2(\|\eta_x(t)\|_{H^1} + \|w_x(t)\|_{H^2}), \quad (5.4.12)$$

$$\|\theta_t(t)\| \leq C_2(\|\theta_x(t)\|_{H^1} + \|v_x(t)\| + \|\eta_x(t)\| + \|w(t)\|_{H^2}), \quad (5.4.13)$$

$$\|\theta_{tx}(t)\| \leq C_2(\|\theta_x(t)\|_{H^2} + \|v_x(t)\|_{H^1} + \|\eta_{xx}(t)\| + \|w_x(t)\|_{H^1}) \quad (5.4.14)$$

or

$$\|w_{xxx}(t)\| \leq C_2(\|w_{tx}(t)\| + \|\eta_x(t)\|_{H^1} + \|w_x(t)\|_{H^1}), \quad (5.4.15)$$

$$\|\theta_{xxx}(t)\| \leq C_2(\|\theta_x(t)\|_{H^1} + \|v_x(t)\|_{H^1} + \|\eta_{xx}(t)\| + \|\theta_{tx}(t)\| + \|w_x(t)\|_{H^1}) \quad (5.4.16)$$

and

$$\|w_{txx}(t)\| \leq C_2(\|w_{xxxx}(t)\| + \|\eta_x(t)\|_{H^2} + \|w_x(t)\|_{H^2}), \quad (5.4.17)$$

$$\|\theta_{txx}(t)\| \leq C_2(\|\eta_x(t)\|_{H^2} + \|v_x(t)\|_{H^2} + \|\theta_x(t)\|_{H^3} + \|w_x(t)\|_{H^2}) \quad (5.4.18)$$

or

$$\|w_{xxxx}(t)\| \leq C_2(\|w_{txx}(t)\| + \|\eta_x(t)\|_{H^2} + \|w_x(t)\|_{H^2} + \|w_x(t)\|_{H^2}), \quad (5.4.19)$$

$$\|\theta_{xxxx}(t)\| \leq C_2(\|\eta_x(t)\|_{H^2} + \|v_x(t)\|_{H^2} + \|\theta_x(t)\|_{H^2} + \|\theta_{txx}(t)\| + \|w_x(t)\|_{H^2}). \quad (5.4.20)$$

Now differentiating (5.1.2) with respect to t, and using Lemmas 5.3.1–5.3.3 and (5.1.1), we have

$$\|v_{tt}(t)\| \leq C_2\big(\|\theta_x(t)\| + \|\eta_x(t)\| + \|v_{xx}(t)\| + \|v_{tx}(t)\|_{H^1}$$
$$+ \|\theta_{xt}(t)\| + \|\theta_t(t)\| + \|w_x(t)\|_{H^2}\big) \quad (5.4.21)$$

which, together with (5.4.7), (5.4.9) and (5.4.12), implies

$$\|v_{tt}(t)\| \leq C_2(\|\theta_x(t)\|_{H^2} + \|v_x(t)\|_{H^3} + \|\eta_x(t)\|_{H^2} + \|w_x(t)\|_{H^2}). \quad (5.4.22)$$

Analogously, we derive from (5.1.3), (5.1.4) and Lemmas 5.3.1–5.3.3 that

$$\|w_{tt}(t)\| \leq C_2(\|w_{txx}(t)\| + \|w_{xt}(t)\| + \|w_x x(t)\| + \|w_x(t)\| + \|v_{xx}(t)\|$$
$$+ \|\eta_x(t)\| + \|v_x(t)\|), \quad (5.4.23)$$

$$\|\theta_{tt}(t)\| \leq C_2(\|\theta_t(t)\| + \|\theta_x(t)\| + \|\theta_{tx}(t)\|_{H^1} + \|\theta_{txx}(t)\| + \|v_x(t)\|$$
$$+ \|v_{xt}(t)\| + \|w_x(t)\|_{H^2}) \quad (5.4.24)$$

which, combined with (5.4.11)–(5.4.14), (5.4.17), (5.4.18) and (5.4.7), gives

$$\|w_{tt}(t)\| \leq C_2(\|w_x(t)\|_{H^3} + \|v_x(t)\|_{H^2} + \|\eta_x\|), \quad (5.4.25)$$

$$\|\theta_{tt}(t)\| \le C_2(\|\theta_x(t)\|_{H^3} + \|v_x(t)\|_{H^2} + \|\eta_x\|_{H^2} + \|w_x(t)\|_{H^2}). \qquad (5.4.26)$$

Thus estimates (5.4.1)–(5.4.2) follow from (5.4.7), (5.4.9), (5.4.12), (5.4.14), (5.4.17), (5.4.18), (5.4.22), (5.4.25) and (5.4.26).

Further, differentiating (5.1.2) with respect to t twice, multiplying the resulting equation by v_{tt} in $L^2(\mathbb{R})$, and using (5.1.1) and Lemmas 5.3.1–5.3.3, we deduce that

$$
\begin{aligned}
\frac{1}{2}\frac{d}{dt}\|v_{tt}(t)\|^2 &= -\int_{\mathbb{R}}\sigma_{tt}v_{ttx}dx - \int_{\mathbb{R}}\frac{v_{ttx}^2}{\eta}dx \\
&\quad + C_2\|v_{ttx}(t)\|(\|\theta_{tt}(t)\| + \|v_{tx}(t)\| + \|\theta_t(t)\| + \|v_x(t)\|) \\
&\le -(2C_1)^{-1}\|v_{ttx}(t)\|^2 \\
&\quad + C_2(\|\theta_{tt}(t)\|^2 + \|v_{tx}(t)\|^2 + \|\theta_t(t)\|^2 + \|v_x(t)\|^2)
\end{aligned}
$$

which, along with (5.4.24), implies

$$
\begin{aligned}
\frac{d}{dt}\|v_{tt}(t)\|^2 + C_1^{-1}\|v_{ttx}(t)\|^2 &\le C_2(\|\theta_{txx}(t)\|^2 + \|\theta_{xx}(t)\|^2 + \|\theta_{tx}(t)\|^2 + \|v_x(t)\|_{H^1}^2 \\
&\quad + \|v_{tx}(t)\|^2 + \|\theta_t(t)\|^2 + \|\eta_x(t)\|^2). \qquad (5.4.27)
\end{aligned}
$$

Thus estimate (5.4.3) follows from Lemmas 5.3.1–5.3.3, (5.4.2) and (5.4.27). Analogously, we obtain from (5.1.3) that

$$\frac{1}{2}\frac{d}{dt}\|w_{tt}(t)\|^2 = -A\int_{\mathbb{R}}\frac{w_{ttx}^2}{\eta}dx - A\int_{\mathbb{R}}w_{tt}^2\eta dx - A\int_{\mathbb{R}}(w_tv_xw_{tt} + wv_{tx}w_{tt})dx, \qquad (5.4.28)$$

which, integrated over $(0,t)$, $t > 0$, implies

$$
\begin{aligned}
\|w_{tt}(t)\|^2 &+ C_1^{-1}\int_0^t\|w_{ttx}(s)\|^2ds + C_1^{-1}\int_0^t\|w_{tt}(s)\|^2ds \\
&\le C_4 + 2C_1^{-1}\int_0^t\|w_{tt}(s)\|^2ds + C_1\int_0^t\Big(\|v_{tx}(s)\|^2 + \|w_t(s)\|^2\Big)ds.
\end{aligned}
$$

Hence

$$\|w_{tt}(t)\|^2 + \int_0^t\Big(\|w_{ttx}\|^2 + \|w_{tt}\|^2\Big)(s)ds \le C_4, \qquad (5.4.29)$$

which gives (5.4.4). In the same way, we obtain from (5.1.4)

$$
\begin{aligned}
\frac{1}{2}\frac{d}{dt}\|\theta_{tt}(t)\|^2 &\le -\int_{\mathbb{R}}\theta_{ttx}^2dx + C_2\|\theta_{ttx}(t)\|(\|\theta_{tx}(t)\| + \|v_{tx}(t)\| + \|v_x(t)\| \\
&\quad + C_2\|\theta_{tt}(t)\|(\|\sigma_{tt}(t)\| + \|\sigma_t(t)\|\|v_{tx}(t)\|_{L^\infty} + \|v_{ttx}(t)\|) \\
&\quad + C_2(\|\theta_{tt}\|^2 + \|w_{ttx}\|^2 + \|w_{xt}\|_{L^4}^4 + \|w_{tt}\|^2 \\
&\quad + \|w_{tx}\|^2 + \|v_{xt}\|^2) + C_2. \qquad (5.4.30)
\end{aligned}
$$

By Lemmas 5.3.1–5.3.3 and the interpolation inequality, we derive

$$\|\sigma_t(t)\| \le C_2(\|v_{tx}(t)\| + \|\theta_t(t)\| + \|v_x(t)\|), \tag{5.4.31}$$

$$\|\sigma_{tt}(t)\| \le C_2(\|v_{ttx}(t)\| + \|\theta_{tt}(t)\| + \|v_{tx}(t)\| + \|\theta_t(t)\| + \|v_x(t)\|), \tag{5.4.32}$$

$$\|v_{tx}(t)\|_{L^\infty}^2 \le C\|v_{tx}(t)\|\|v_{txx}(t)\|, \tag{5.4.33}$$

and

$$\|w_{tx}(t)\|_{L^4}^4 \le \|w_{txx}(t)\|\|w_{tx}(t)\|^3. \tag{5.4.34}$$

By virtue of (5.4.31)–(5.4.34), we infer from (5.4.30) that

$$\frac{d}{dt}\|\theta_{tt}(t)\|^2 + C_1^{-1}\|\theta_{ttx}(t)\|^2$$

$$\le C_2 + C_2(\|\theta_{tx}(t)\|^2 + \|v_x(t)\|^2 + \|v_{tx}(t)\|^2 + \|\theta_t(t)\|^2 + \|v_{ttx}(t)\|^2 + \|\theta_{tt}(t)\|^2)$$

$$+ C_2\|\theta_{tt}(t)\|(\|v_{tx}(t)\| + \|\theta_t(t)\| + \|v_x(t)\|)(\|v_{tx}(t)\| + \|v_{txx}\|)$$

$$+ C_2(\|w_{tt}\|^2 + \|w_{tx}\|^2 + \|v_{xt}\|^2 + \|w_{xt}\|^3\|w_{txx}\|), \tag{5.4.35}$$

which, together with (5.4.2)–(5.4.3), (5.4.24) and Lemmas 5.3.1–5.3.3, yields

$$\|\theta_{tt}(t)\|^2 + C_1^{-1}\int_0^t \|\theta_{ttx}(s)\|^2 ds$$

$$\le C_4 + C_4 \int_0^t (\|\theta_{tt}\|^2 + \|v_{ttx}\|^2)(s)ds$$

$$+ C_2\left[\int_0^t (\|\theta_{tt}\|^2(\|v_{tx}\|^2 + \|\theta_t\|^2 + \|v_x\|^2)(s)ds)\right]^{1/2}$$

$$\times \left[\int_0^t (\|v_{tx}\|^2 + \|v_{txx}\|^2)(s)ds\right]^{1/2}$$

$$+ C_4 \int_0^t \|w_{xt}(s)\|^3\|w_{txx}(s)\|ds$$

$$\le C_4 + C_4 \int_0^t \|\theta_{txx}(s)\|^2 ds + C_2 \sup_{0 \le s \le t} \|\theta_{tt}(s)\|\left[1 + \left(\int_0^t \|v_{txx}(s)\|^2 ds\right)^{1/2}\right]$$

$$+ C_4 \int_0^t \|w_{xt}(s)\|^3\|w_{txx}(s)\|ds$$

$$\le \frac{1}{2}\sup_{0 \le s \le t}\|\theta_{tt}(s)\|^2 + C_4 + C_4 \int_0^t (\|v_{txx}\|^2 + \|\theta_{txx}\|^2)(s)ds$$

$$+ C_4 \int_0^t \|w_{xt}(s)\|^3\|w_{txx}(s)\|ds. \tag{5.4.36}$$

Hence, taking the supremum on the right-hand side of (5.4.36) gives the required estimate (5.4.5). The proof is complete. $\qquad\square$

Lemma 5.4.2. *Under the assumptions of Theorem 5.1.3, the following estimates hold for any $t > 0$:*

$$\|v_{tx}(t)\|^2 + \int_0^t \|v_{txx}(s)\|^2 ds \le C_3, \tag{5.4.37}$$

$$\|w_{tx}(t)\|^2 + \int_0^t (\|w_{txx}\|^2 + \|w_{tx}\|^2)(s) ds \le C_3, \tag{5.4.38}$$

$$\|\theta_{tx}(t)\|^2 + \int_0^t \|\theta_{txx}(s)\|^2 ds \le C_3, \tag{5.4.39}$$

$$\|\theta_{tt}(t)\|^2 + \|v_{tt}(t)\|^2 + \int_0^t (\|v_{ttx}\|^2 + \|\theta_{ttx}\|^2)(s) ds \le C_4. \tag{5.4.40}$$

Proof. Differentiating (5.1.2) with respect to x and t, multiplying the resulting equation by v_{tx} in $L^2(\mathbb{R})$, and integrating by parts, we deduce that

$$\begin{aligned}
\frac{1}{2}\frac{d}{dt}\|v_{tx}(t)\|^2 \le\ & -\int_{\mathbb{R}} \frac{v_{txx}^2}{\eta} dx + C_2\|v_{txx}(t)\|(\|\theta_{tx}(t)\| + \|v_{tx}(t)\| + \|\theta_t(t)\| \\
& + \|v_{xx}(t)\| + \|\theta_x(t)\| + \|\eta_x(t)\|) \\
\le\ & -(2C_1)^{-1}\|v_{txx}(t)\|^2 + C_2(\|\theta_{tx}(t)\|^2 + \|v_{tx}(t)\|^2 + \|\theta_t(t)\|^2 \\
& + \|v_{xx}(t)\|^2 + \|\theta_x(t)\|^2 + \|\eta_x(t)\|^2) \tag{5.4.41}
\end{aligned}$$

which, combined with Lemmas 5.3.1–5.3.3 and (5.4.2), gives estimate (5.4.37). In the same way, we infer from (5.1.3) that

$$\begin{aligned}
\frac{1}{2}\frac{d}{dt}\|w_{tx}(t)\|^2 \le\ & -C_1^{-1}\int_{\mathbb{R}} \frac{w_{txx}^2}{\eta} dx - C_1^{-1}\int_{\mathbb{R}} \eta w_{tx}^2 dx + \frac{1}{2C_1}\|w_{txx}(t)\|^2 \tag{5.4.42} \\
& + C_2(\|w_{xx}(t)\|^2 + \|w_{tx}(t)\|^2 + \|w_x(t)\|^2 + \|v_{xx}(t)\|^2 + \|w_t(t)\|^2),
\end{aligned}$$

which leads to (5.4.38).

Analogously, from (5.1.4) and (5.4.38) we obtain

$$\begin{aligned}
\frac{1}{2}\frac{d}{dt}\|\theta_{tx}(t)\|^2 \le\ & -\frac{1}{C_1}\int_{\mathbb{R}} \frac{\theta_{txx}^2}{\eta} dx + C_2\|\theta_{txx}(t)\|(\|\theta_{tx}(t)\| + \|\theta_{xx}(t)\| + \|\eta_x(t)\| \\
& + \|v_{xx}(t)\|) + C_2\|\theta_{tx}(t)\|(\|w_{xx}(t)w_{tx}(t)\| + \|w_{txx}(t)\| + \|w_{tx}(t)\| \\
& + \|w_t(t)\| + \|w_{tx}(t)\| + \|w_x(t)\|_{L^4(\mathbb{R})}^2 + \|w_x(t)\| + \|v_{xx}\|) \\
\le\ & -(2C_1)^{-1}\|\theta_{txx}(t)\|^2 + C_2(\|\theta_{tx}(t)\|^2 + \|\theta_{xx}(t)\|^2 + \|v_{xx}(t)\|^2 \\
& + \|\eta_x(t)\|^2 + \|w_{txx}\|^2 + \|w_{tx}\|^2 + \|w_t\|^2), \tag{5.4.43}
\end{aligned}$$

which, combined with Lemmas 5.3.1–5.3.3 and (5.4.38), implies estimate (5.4.39). Inserting (5.4.37)–(5.4.39) into (5.4.3) and (5.4.5) yields estimate (5.4.40). The proof is now complete. $\square$

Lemma 5.4.3. *Under the assumptions of Theorem 5.1.3, the following estimates hold for any $t > 0$:*

$$\|\eta_{xxx}(t)\|_{H^1}^2 + \|\eta_{xx}(t)\|_{W^{1,\infty}}^2 + \int_0^t (\|\eta_{xxx}\|_{H^1}^2 + \|\eta_{xx}\|_{W^{1,\infty}}^2)(s)ds \leq C_4, \quad (5.4.44)$$

$$\|v_{xxx}(t)\|_{H^1}^2 + \|v_{xx}(t)\|_{W^{1,\infty}}^2 + \|w_{xxx}(t)\|_{H^1}^2 + \|w_{xx}(t)\|_{W^{1,\infty}}^2 + \|\theta_{xxx}(t)\|_{H^1}^2$$

$$+ \|\theta_{xx}(t)\|_{W^{1,\infty}}^2 + \|\eta_{txxx}(t)\|^2 + \|v_{txx}(t)\|^2 + \|w_{txx}(t)\|^2 + \|\theta_{txx}(t)\|^2$$

$$+ \int_0^t (\|v_{tt}\|^2 + \|w_{tt}\|^2 + \|\theta_{tt}\|^2 + \|v_{xx}\|_{W^{2,\infty}}^2 + \|w_{xx}\|_{W^{2,\infty}}^2 + \|\theta_{xx}\|_{W^{2,\infty}}^2$$

$$+ \|v_{txx}\|_{H^1}^2 + \|w_{txx}\|_{H^1}^2 + \|\theta_{txx}\|_{H^1}^2 + \|\theta_{tx}\|_{W^{1,\infty}}^2 + \|w_{tx}\|_{W^{1,\infty}}^2$$

$$+ \|v_{tx}\|_{W^{1,\infty}}^2 + \|\eta_{txxx}\|_{H^1}^2)(s)ds \leq C_4, \quad (5.4.45)$$

$$\int_0^t (\|v_{xxxx}\|_{H^1}^2 + \|w_{xxxx}\|_{H^1}^2 + \|\theta_{xxxx}\|_{H^1}^2)(s)ds \leq C_4. \quad (5.4.46)$$

Proof. Differentiating (5.3.23) with respect to x and using (5.1.1) we arrive at

$$\frac{\partial}{\partial t}\left(\frac{\eta_{xxx}}{\eta}\right) + \frac{R\theta\eta_{xxx}}{\eta^2} = E_1(x,t) \quad (5.4.47)$$

with

$$E_1(x,t) = \left[\frac{v_{xxx}\eta_x + \eta_{xx}v_{xx}}{\eta^2} - \frac{2\eta_x\eta_{xx}v_x}{\eta^3}\right] - \frac{\theta_x\eta_{xx}}{\eta^2} + \frac{2R\theta\eta_x\eta_{xx}}{\eta^3} + v_{txx} + E_x(x,t),$$

$$E(x,t) = \frac{R\theta_{xx}}{\eta} + \frac{2\mu v_{xx}\eta_x - 2R\theta_x\eta_x}{\eta^2} + \frac{2R\theta\eta_x^2 - 2v_x\eta_x^2}{\eta^3}.$$

An easy calculation based on Lemmas 5.3.1–5.3.3 and Lemmas 5.4.1–5.4.2 gives

$$\|E_1(t)\| \leq C_2(\|\eta_x(t)\|_{H^1} + \|v_x(t)\|_{H^2} + \|\theta_x(t)\|_{H^2} + \|v_{tx}(t)\|_{H^1}) \quad (5.4.48)$$

and

$$\int_0^t \|E_1(s)\|^2 ds \leq C_4. \quad (5.4.49)$$

Now multiplying (5.4.47) by η_{xxx}/η in $L^2(\mathbb{R})$, we derive

$$\frac{d}{dt}\left\|\frac{\eta_{xxx}}{\eta}(t)\right\|^2 + C_1^{-1}\left\|\frac{\eta_{xxx}}{\eta}(t)\right\|^2 \leq C_1\|E_1(t)\|^2 \quad (5.4.50)$$

which, combined with (5.4.49) and Lemmas 5.3.1–5.3.3 and Lemmas 5.4.1–5.4.2, yields, for $t > 0$,

$$\|\eta_{xxx}(t)\|^2 + \int_0^t \|\eta_{xxx}(s)\|^2 ds \leq C_4. \quad (5.4.51)$$

In view of (5.4.8), (5.4.10), (5.4.16), (5.4.20) and Lemmas 5.3.1–5.3.3 and Lemmas 5.4.1–5.4.2, we get that, for any $t > 0$,

$$\|v_{xxx}(t)\|^2 + \|w_{xxx}(t)\|^2 + \|\theta_{xxx}(t)\|^2$$
$$+ \int_0^t (\|v_{xxx}\|_{H^1}^2 + \|w_{xxx}\|_{H^1}^2 + \|\theta_{xxx}\|_{H^1}^2)(s)ds \leq C_4, \qquad (5.4.52)$$
$$\|v_{xx}(t)\|_{L^\infty}^2 + \|w_{xx}(t)\|_{L^\infty}^2 + \|\theta_{xx}(t)\|_{L^\infty}^2$$
$$+ \int_0^t (\|v_{xx}\|_{W^{1,\infty}}^2 + \|w_{xx}\|_{W^{1,\infty}}^2 + \|\theta_{xx}\|_{W^{1,\infty}}^2)(s)ds \leq C_4. \qquad (5.4.53)$$

Differentiating (5.1.2) with respect to t, we infer that for any $t > 0$,

$$\|v_{txx}(t)\| \leq C_1\|v_{tt}(t)\| + C_2(\|\eta_x(t)\| + \|v_{xx}(t)\| + \|v_{tx}(t)\| + \|\theta_x(t)\|$$
$$+ \|\theta_t(t)\| + \|\theta_{tx}(t)\|) \leq C_4, \qquad (5.4.54)$$

which, together with (5.4.10), gives

$$\|v_{xxxx}(t)\|^2 + \int_0^t (\|v_{txx}\|^2 + \|v_{xxxx}\|^2)(s)ds \leq C_4. \qquad (5.4.55)$$

Similarly, (5.4.17)–(5.4.20) and (5.4.52)–(5.4.53) yield

$$\|w_{txx}(t)\|^2 + \|w_{xxxx}(t)\|^2 + \int_0^t (\|w_{txx}\|^2 + \|w_{xxxx}\|^2)(s)ds \leq C_4, \qquad (5.4.56)$$

$$\|\theta_{txx}(t)\|^2 + \|\theta_{xxxx}(t)\|^2 + \int_0^t (\|\theta_{txx}\|^2 + \|\theta_{xxxx}\|^2)(s)ds \leq C_4, \qquad (5.4.57)$$

which, combined with (5.4.52) and (5.4.55)–(5.4.56), implies

$$\|v_{xxx}(t)\|_{L^\infty}^2 + \|w_{xxx}(t)\|_{L^\infty}^2 + \|\theta_{xxx}(t)\|_{L^\infty}^2$$
$$+ \int_0^t (\|v_{xxx}\|_{L^\infty}^2 + \|w_{xxx}\|_{L^\infty}^2 + \|\theta_{xxx}\|_{L^\infty}^2)(s)ds \leq C_4. \qquad (5.4.58)$$

Differentiating (5.4.47) with respect to x, we see that

$$\frac{\partial}{\partial t}\left(\frac{\eta_{xxxx}}{\eta}\right) + \frac{R\theta\eta_{xxxx}}{\eta^2} = E_2(x,t) \qquad (5.4.59)$$

with

$$E_2(x,t) = \frac{v_{xx}\eta_{xxx} + \eta_x v_{xxxx}}{\eta^2} - \frac{2\eta_x v_x \eta_{xxx}}{\eta^3} + \frac{2R\theta\eta_x\eta_{xxx}}{\eta^3} - \frac{R\theta_x\eta_{xxx}}{\eta^2} + E_{1x}(x,t).$$

Using Lemmas 5.3.1–5.3.3 and Lemmas 5.4.1–5.4.2, we can deduce that

$$\|E_{xx}(t)\| \leq C_4(\|\theta_x(t)\|_{H^3} + \|\eta_x(t)\|_{H^2} + \|v_x(t)\|_{H^3}), \qquad (5.4.60)$$

$$\|E_{1x}(t)\| \le C_4(\|v_x(t)\|_{H^3} + \|\eta_x(t)\|_{H^2} + \|v_{tx}(t)\|_{H^2} + \|\theta_x(t)\|_{H^3}), \quad (5.4.61)$$
$$\|E_2(t)\| \le C_4(\|v_x(t)\|_{H^3} + \|\eta_x(t)\|_{H^2} + \|v_{tx}(t)\|_{H^2} + \|\theta_x(t)\|_{H^3}). \quad (5.4.62)$$

On the other hand, differentiating (5.1.2) with respect to t and x, we infer

$$\|v_{txxx}(t)\| \le C_1\|v_{ttx}(t)\| + C_2(\|v_{xx}\|_{H^1} + \|\theta_x(t)\|_{H^1} + \|\eta_x(t)\|_{H^1}$$
$$+ \|\theta_{tx}(t)\|_{H^1} + \|\theta_t(t)\| + \|v_{tx}(t)\|_{H^1}). \quad (5.4.63)$$

Similarly, we have

$$\|w_{txxx}(t)\| \le C_1\|w_{ttx}(t)\| + C_2(\|w_{tx}(t)\|_{H^1} + \|w_x\|_{H^1}$$
$$+ \|v_x\|_{H^1} + \|\eta_x(t)\|_{H^1}), \quad (5.4.64)$$
$$\|\theta_{txxx}(t)\| \le C_1\|\theta_{ttx}(t)\| + C_2(\|\eta_x(t)\| + \|v_{xx}\|_{H^1} + \|\theta_x(t)\|_{H^2} + \|\theta_{tx}(t)\|_{H^1}$$
$$+ \|\theta_t(t)\| + \|v_{tx}(t)\|_{H^1} + \|w_{tx}\|_{H^1} + \|w_x\|_{H^1}). \quad (5.4.65)$$

Thus it follows from Lemmas 5.3.1–5.3.3, Lemmas 5.4.1–5.4.2 and (5.4.63)–(5.4.65) that

$$\int_0^t (\|v_{txxx}\|^2 + \|w_{txxx}\|^2 + \|\theta_{txxx}\|^2)(s)ds \le C_4, \ \forall t > 0. \quad (5.4.66)$$

By virtue of (5.4.51), (5.4.55)–(5.4.57), (5.4.62)–(5.4.63), Lemmas 5.3.1–5.3.3 and Lemmas 5.4.1–5.4.2, we have

$$\int_0^t \|E_2(s)\|^2 ds \le C_4, \ \ \forall t > 0. \quad (5.4.67)$$

Next, multiplying (5.4.59) by η_{xxxx}/η in $L^2(\mathbb{R})$, we get

$$\frac{d}{dt}\left\|\frac{\eta_{xxxx}}{\eta}(t)\right\|^2 + C_1^{-1}\left\|\frac{\eta_{xxxx}}{\eta}(t)\right\|^2 \le C_1\|E_2(t)\|^2 \quad (5.4.68)$$

which, combined with (5.4.67), implies

$$\|\eta_{xxxx}(t)\|^2 + \int_0^t \|\eta_{xxxx}(s)\|^2 ds \le C_4, \ \forall t > 0. \quad (5.4.69)$$

From (5.4.21)–(5.4.26), Lemmas 5.3.1–5.3.3, Lemmas 5.4.1–5.4.2 and (5.4.51)–(5.4.58), we obtain

$$\int_0^t (\|v_{tt}\|^2 + \|w_{tt}\|^2 + \|\theta_{tt}\|^2)(s)ds \le C_4, \ \forall t > 0. \quad (5.4.70)$$

Differentiating (5.1.2) with respect to x three times and using the following estimates

$$\|\sigma_x(t)\| \le C_2(\|v_{xx}(t)\| + \|\theta_x(t)\| + \|\eta_x(t)\|),$$

$$\|\sigma_{xx}(t)\| \leq C_2(\|v_x(t)\|_{H^2} + \|\theta_x(t)\|_{H^1} + \|\eta_x(t)\|_{H^1}),$$
$$\|\sigma_{xxx}(t)\| \leq C_2(\|v_x(t)\|_{H^3} + \|\theta_x(t)\|_{H^2} + \|\eta_x(t)\|_{H^2}),$$

we deduce that

$$\|v_{xxxxx}(t)\| \leq C_1\|v_{txxx}(t)\| + C_2(\|\eta_x(t)\|_{H^3} + \|v_x(t)\|_{H^3} + \|\theta_x(t)\|_{H^3}). \quad (5.4.71)$$

Thus we conclude from (5.1.1), (5.4.55), (5.4.57), (5.4.66), (5.4.69), (5.4.71), Lemmas 5.3.1–5.3.2 and Lemmas 5.4.1–5.4.2 that

$$\int_0^t (\|v_{xxxxx}\|^2 + \|\eta_{txxx}\|_{H^1}^2)(s)ds \leq C_4, \ \forall t > 0. \quad (5.4.72)$$

Similarly, we can deduce that for any $t > 0$,

$$\int_0^t \|w_{xxxxx}(s)\|^2 ds \leq C_4, \quad (5.4.73)$$

$$\int_0^t \|\theta_{xxxxx}(s)\|^2 ds \leq C_4, \quad (5.4.74)$$

$$\int_0^t (\|v_{xx}\|_{W^{2,\infty}}^2 + \|w_{xx}\|_{W^{2,\infty}}^2 + \|\theta_{xx}\|_{W^{2,\infty}}^2)(s)ds \leq C_4. \quad (5.4.75)$$

Thus employing (5.1.1), (5.4.51)–(5.4.58), (5.4.66), (5.4.69)–(5.4.70), (5.4.73)–(5.4.74) and the interpolation inequality, we can derive the desired estimates (5.4.44)–(5.4.46). The proof is complete. $\qquad\square$

Lemma 5.4.4. *Under the assumptions of Theorem 5.1.3, the following estimates hold for any $t > 0$:*

$$\|\eta(t) - 1\|_{H^4}^2 + \|\eta_t(t)\|_{H^3}^2 + \|\eta_{tt}(t)\|_{H^1}^2 + \|v(t)\|_{H^4}^2 + \|v_t(t)\|_{H^2}^2 + \|v_{tt}(t)\|^2$$
$$+ \|w(t)\|_{H^4}^2 + \|w_t(t)\|_{H^2}^2 + \|w_{tt}(t)\|^2 + \|\theta(t) - 1\|_{H^4}^2 + \|\theta_t(t)\|_{H^2}^2 + \|\theta_{tt}(t)\|^2$$
$$+ \int_0^t \Big(\|\eta_x\|_{H^3}^2 + \|v_x\|_{H^4}^2 + \|v_t\|_{H^3}^2 + \|v_{tt}\|_{H^1}^2 + \|w_x\|_{H^4}^2 + \|w_t\|_{H^3}^2 + \|w_{tt}\|_{H^1}^2$$
$$+ \|\theta_x\|_{H^4}^2 + \|\theta_t\|_{H^3}^2 + \|\theta_{tt}\|_{H^1}^2 \Big)(s)ds \leq C_4, \quad (5.4.76)$$

$$\int_0^t (\|\eta_t\|_{H^4}^2 + \|\eta_{tt}\|_{H^2}^2 + \|\eta_{ttt}\|^2)(s)ds \leq C_4. \quad (5.4.77)$$

Proof. Using (5.1.1), Lemmas 5.3.1–5.3.3 and Lemmas 5.4.1–5.4.3, we can derive estimates (5.4.76)–(5.4.77). The proof is complete. $\qquad\square$

We now prove the large-time behavior of H^4-global solution $(\eta(t), v(t), w(t), \theta(t))$.

Lemma 5.4.5. *Under the assumptions of Theorem 5.1.3, the H^4-global solution $(\eta(t), v(t), w(t), \theta(t))$ obtained in Lemmas 5.4.1–5.4.4 to the Cauchy problem (5.1.1)–(5.1.6) satisfies the relations (5.1.19) and (5.1.20).*

Proof. Similarly to (5.4.27)–(5.4.28), (5.4.35), (5.4.41)–(5.4.43), (5.4.50), (5.4.68) and using (5.1.18), we obtain

$$\frac{d}{dt}\|v_{tt}(t)\|^2 + (2C_1)^{-1}\|v_{ttx}(t)\|^2 \leq C_2(\|\theta_{xx}(t)\|^2 + \|\theta_{tx}(t)\|_{H^1}^2 + \|v_x(t)\|_{H^1}^2$$
$$+ \|v_{tx}(t)\|^2 + \|\theta_t(t)\|^2 + \|\eta_x(t)\|^2), \tag{5.4.78}$$

$$\frac{d}{dt}\|w_{tt}(t)\|^2 + C_1^{-1}\|w_{ttx}(t)\|^2 + (2C_1)^{-1}\|w_{tt}(t)\|^2 \leq C_4(\|v_{tx}\|^2 + \|w_t(t)\|^2), \tag{5.4.79}$$

$$\frac{d}{dt}\|\theta_{tt}(t)\|^2 + C_1^{-1}\|\theta_{ttx}(t)\|^2 \leq C_4(\|\theta_{tx}(t)\|^2 + \|v_{tx}(t)\|_{H^1}^2 + \|v_x(t)\|^2 + \|v_{tx}(t)\|^2$$
$$+ \|\theta_t(t)\|^2 + \|v_{ttx}(t)\|^2 + \|\theta_{tt}(t)\|^2 + \|w_{tx}(t)\|^2 + \|w_{txx}(t)\|^2$$
$$+ \|w_{tt}(t)\|^2 + \|v_x(t)\|^2 + \|w_t(t)\|^2), \tag{5.4.80}$$

$$\frac{d}{dt}\|v_{tx}(t)\|^2 + C_1^{-1}\|v_{txx}(t)\|^2 \leq C_2(\|\theta_{tx}(t)\|^2 + \|v_{tx}(t)\|^2 + \|\theta_t(t)\|^2$$
$$+ \|v_{xx}(t)\|^2 + \|\theta_x(t)\|^2 + \|\eta_x(t)\|^2), \tag{5.4.81}$$

$$\frac{d}{dt}\|w_{tx}(t)\|^2 + C_1^{-1}\|w_{txx}(t)\| + C_1^{-1}\|w_{tx}\|^2 \leq C_2(\|w_{txx}(t)\|^2 + \|w_{tx}(t)\|^2$$
$$+ \|w_x(t)\|^2 + \|v_{xx}(t)\|^2 + \|w_t(t)\|^2), \tag{5.4.82}$$

$$\frac{d}{dt}\|\theta_{tx}(t)\|^2 + C_1^{-1}\|\theta_{txx}(t)\|^2 \leq C_2(\|\theta_{tx}(t)\|^2 + \|\theta_{xx}(t)\|^2 + \|v_{xx}(t)\|^2 + \|\eta_x(t)\|^2$$
$$+ \|w_{tx}(t)\|^2 + \|w_{txx}(t)\|^2 + \|w_{xx}(t)\|^2 + \|v_{xx}(t)\|^2 + \|w_t(t)\|^2$$
$$+ \|w_{tx}(t)\|^2 + \|w_x(t)\|^2), \tag{5.4.83}$$

$$\frac{d}{dt}\|\frac{\eta_{xxx}}{\eta}(t)\|^2 + C_1^{-1}\|\frac{\eta_{xxx}}{\eta}(t)\|^2 \leq C_1\|E_1(t)\|^2, \tag{5.4.84}$$

$$\frac{d}{dt}\|\frac{\eta_{xxxx}}{\eta}(t)\|^2 + C_1^{-1}\|\frac{\eta_{xxxx}}{\eta}(t)\|^2 \leq C_1\|E_2(t)\|^2 \tag{5.4.85}$$

where, by (5.1.18), (5.4.49) and (5.4.67),

$$\int_0^t (\|E_1\|^2 + \|E_2\|^2)(s)ds \leq C_4, \quad \forall t > 0. \tag{5.4.86}$$

Applying Lemma 5.3.4 to (5.4.78)–(5.4.89) and using estimates (5.1.15) and (5.4.86), we infer that, as $t \to +\infty$,

$$\|v_{tt}(t)\| \to 0, \quad \|w_{tt}(t)\| \to 0, \quad \|\theta_{tt}(t)\| \to 0, \quad \|v_{tx}(t)\| \to 0, \tag{5.4.87}$$
$$\|w_{tx}(t)\| \to 0, \quad \|\theta_{tx}(t)\| \to 0, \quad \|\eta_{xxx}(t)\| \to 0, \quad \|\eta_{xxxx}(t)\| \to 0. \tag{5.4.88}$$

In the same manner as for (5.4.8), (5.4.10), (5.4.54) and using the interpolation inequality, we obtain

$$\|v_{xxx}(t)\| \leq C_2(\|v_x(t)\| + \|\eta_x(t)\|_{H^1} + \|\theta_x(t)\|_{H^1} + \|v_{tx}(t)\|), \tag{5.4.89}$$

$$\|v_{txx}(t)\| \le C_1\|v_{tt}(t)\| + C_2(\|v_{xx}(t)\| + \|u_x(t)\| + \|v_{tx}(t)\| + \|\theta_x(t)\|$$
$$+ \|\theta_t(t)\| + \|\theta_{tx}(t)\|), \tag{5.4.90}$$

$$\|v_{xxxx}(t)\| \le C_2(\|v_x(t)\|_{H^2} + \|u_x(t)\|_{H^2} + \|\theta_x(t)\|_{H^2} + \|v_{txx}(t)\|), \tag{5.4.91}$$

$$\|v_{tx}(t)\|_{L^\infty}^2 \le C\|v_{tx}(t)\|\|v_{txx}(t)\|, \ \|v_t(t)\|_{L^\infty}^2 \le C\|v_t(t)\|\|v_{tx}(t)\|, \tag{5.4.92}$$

$$\|v_{xx}(t)\|_{L^\infty}^2 \le C\|v_{xx}(t)\|\|v_{xxx}(t)\|, \ \|v_{xxx}(t)\|_{L^\infty}^2 \le C\|v_{xxx}(t)\|\|v_{xxxx}(t)\|, \tag{5.4.93}$$

$$\|u_{xx}(t)\|_{L^\infty}^2 \le C\|u_{xx}(t)\|\|u_{xxx}(t)\|, \ \|u_{xxx}(t)\|_{L^\infty}^2 \le C\|u_{xxx}(t)\|\|u_{xxxx}(t)\|. \tag{5.4.94}$$

Thus it follows from (5.1.1), (5.4.87)–(5.4.94) and Lemma 5.3.5 that, as $t \to +\infty$,

$$\|(\eta_x(t), v_x(t))\|_{H^3} + \|v_t(t)\|_{H^2} + \|\eta_t(t)\|_{H^3} + \|\eta_t(t)\|_{W^{2,\infty}}$$
$$+ \|\eta_{tt}(t)\|_{H^1} + \|(\eta_x(t), v_x(t))\|_{W^{2,\infty}} \to 0. \tag{5.4.95}$$

Analogously, we can show that, as $t \to +\infty$,

$$\|w_x(t)\|_{H^3} + \|w_t(t)\|_{H^2} + \|w_t(t)\|_{W^{1,\infty}} + \|w_x(t)\|_{W^{2,\infty}} \to 0, \tag{5.4.96}$$
$$\|\theta_x(t)\|_{H^3} + \|\theta_t(t)\|_{H^2} + \|\theta_t(t)\|_{W^{1,\infty}} + \|\theta_x(t)\|_{W^{2,\infty}} \to 0, \tag{5.4.97}$$

which, together with Lemma 5.3.5 and (5.4.95), implies estimates (5.1.16) and (5.1.17). The proof is complete. ☐

Proof of Theorem 5.1.3. Lemmas 5.4.1–5.4.5 establish the global existence of the H^4-solution to problem (5.1.1)–(5.1.6). ☐

Proof of Corollary 5.1.1. Employing the Sobolev embedding theorem with the estimates (5.1.16) and (5.1.17), we can get the desired conclusion immediately. ☐

Proof of Theorem 5.1.4.
Case I. The proof proceeds as in the verification of Theorems 5.1.1–5.1.3.

Case II. The proof of global the existence and asymptotic behavior of H^1 solutions to the problem (5.1.1)–(5.1.6) can be found in [82] and the proof for H^i ($i = 2, 4$) is actually the same as Theorems 5.1.2 and 5.1.3. ☐

5.5 Bibliographic Comments

The 1D compressible Navier-Stokes system has attracted the interest of physicists and mathematicians for a relatively long time. A crucial observation in the research of this field is that in the 1D case treated in Lagrangian coordinates, the specific volume η can be expressed in terms of other unknown variable, and so positive upper and lower bounds can be obtained. This fact was used in significant manner, in particular, in the results by Kazhikhov and Shelukhin [61, 63]. For initial-boundary value problems, there are numerous works concerning global existence, regularity and asymptotic behavior of solutions, etc., in various cases (see, e.g., [1, 8, 9, 30, 43, 100, 101, 105, 138, 140, 149, 152]). Recently, Feireisl [37, 38] has studied the stabilizability of weak solutions besides global energy of viscous, compressible, and heat conducting fluid in the 3D case, which is more general than the 1D case. Concerning the Cauchy problem, Kazhikhov and Shelukhin [62, 63] also proved that given $\eta_0 \bar{\eta}$, ν_0, and $\theta_0 \bar{\theta}$ with some positive constants $\bar{\eta}$, $\bar{\theta}$ and initial data η_0, $\theta_0 > 0$, there exist a there exists a unique global H^1 solution $(\eta(t), \nu(t), \theta(t))$ on $[0, T)$ $(T > 0)$. Actually, in their work the domain is unbounded and hence the Poincaré inequality is not available, and consequently the properties of the initial data $(\eta_0, \nu_0, \theta_0)$ must be given with the help of some stationary state data like $\eta_0 - \bar{\eta}$, etc., which is known as the small initial data proposed in Kanel's paper [56].

Under the small initial data hypothesis, the global existence and large-time behavior of smooth solutions have been obtained for the Cauchy problem, including the 2D- or 3D-dimensional cases (see, e.g., [1, 46, 53–55, 74–77, 93, 150]). Although these results are based on the same assumption, namely, small initial data, the specific hypotheses are quite different. For instance, in [54] the small data hypothesis in the 1D case is expressed in the weighted form weighted

$$e_0^2 := \|\eta - \eta_0\|_{L^\infty}^2 + \int_{\mathbb{R}} (1+x)^\gamma \left\{ (\eta_0 - \bar{\eta})^2 + \nu_0^2 + (\theta_0 - \bar{\theta})^2 + \nu_0^4 \right\} dx \le \varepsilon$$

where $(1+x)^\gamma$ $(\gamma > 1/2)$ is a weight function, ε is a sufficiently small positive constant. In another work, by Okada et al. [93], the corresponding condition takes on the form $E_0 E_1 \le \varepsilon$, where $E_i := \|(\log(\eta_0/\bar{\eta}), \log(\nu_0), \log(\theta_0/\bar{\theta})\|_{H^i}$ $(i = 0, 1)$. Then, in Qin et al.'s work [112], the two cases have been treated together to establish further regularity properties. Moreover, Mujakovic recently considered the Cauchy problem in the latter case for micropolar fluid dynamics (see, e.g., [80–83]). In the present chapter we only begin to utilize Jiang's approach to prove the existence of global solutions in H^1 for the 1D micropolar fluid system (5.1.1)–(5.1.6) and then obtain the regularity of H^2 and H^4 global solutions. Since we are dealing with a system that has one extra equation compared to that considered in [54], we need more precise estimates to treat the more complex terms. Interpolation inequalities are our powerful tools for establishing the desired estimates. In [80, 81] it was shown that, for any length of time $T > 0$, the solution $\eta, \nu, w, \theta)$ enjoys the following properties:

$$1/\eta - 1 \in L^\infty(0, T; H^1(\mathbf{R})) \cap H^1(\mathbb{R} \times (0, T)), \tag{5.5.1}$$

$$\nu, w, \theta - 1 \in L^\infty((0, T); H^1(\mathbb{R})) \cap H^1(\mathbb{R} \times (0, T)) \cap L^2(0, T; H^2(\mathbb{R})), \tag{5.5.2}$$

and

$$1/\eta, \theta > 0. \tag{5.5.3}$$

Chapter 6

Global Existence and Exponential Stability for a 1D Compressible Viscous Micropolar Fluid Model

6.1 Introduction

In this chapter, we shall study the global existence and exponential stability of H^i-global solutions ($i = 1, 2, 4$) to a classical kind of Navier-Stokes equations describing the motion of a one-dimensional compressible viscous heat-conducting micropolar fluids, which belong to a class of fluids with nonsymmetric stress tensor called polar fluids (see, e.g., [32], [73]). Precisely, in Lagrangian coordinates, such a system can be represented as follows:

$$\eta_t = v_x, \tag{6.1.1}$$

$$v_t = \sigma_x, \tag{6.1.2}$$

$$w_t = A\left(\frac{w_x}{\eta}\right)_x - A\eta w, \tag{6.1.3}$$

$$e_t = \pi_x + \sigma v_x + \frac{w_x^2}{\eta} + \eta w^2, \tag{6.1.4}$$

where $(x, t) \in [0, 1] \times \mathbb{R}_+$ are the Lagrangian mass and time coordinates and

$$\sigma = \frac{\nu v_x}{\eta} - P, \quad \pi = \frac{\hat{\kappa}\theta_x}{\eta}, \quad P(\eta, \theta) = \frac{R\theta}{\eta}, \quad e(\eta, \theta) = \theta \tag{6.1.5}$$

denote stress, heat flux, pressure and internal energy, respectively, and $A > 0$ is a constant. In addition,

$$\rho = \frac{1}{\eta} \tag{6.1.6}$$

denotes the density of the fluid.

Here we consider (6.1.1)–(6.1.5) subject to the initial conditions

$$(\eta(x,0), v(x,0), w(x,0), \theta(x,0)) = (\eta_0(x), v_0(x), w_0(x), \theta_0(x)), \quad \text{for all} \quad x \in [0,1],$$
(6.1.7)

and the boundary conditions

$$v(0,t) = v(1,t) = 0, \quad w(0,t) = w(1,t) = 0, \quad \theta_x(0,t) = \theta_x(1,t) = 0, \quad \forall t \geq 0,$$
(6.1.8)

or

$$v(0,t) = v(1,t) = 0, \quad w(0,t) = w(1,t) = 0, \quad \theta(0,t) = \theta(1,t) = 1, \quad \forall t \geq 0.$$
(6.1.9)

The unknown quantities η, v, w, θ denote the specific volume, velocity, angular momentum and absolute temperature, respectively. The function $\hat{\kappa}(x,t)$ is the heat conductivity, satisfying the Fourier law for heat flux $-\pi$

$$-\pi(\eta, \theta, \theta_x) = -\frac{\hat{\kappa}(\eta, \theta)}{\eta}\theta_x.$$
(6.1.10)

For simplicity, in the present chapter, the heat conductivity is assumed to be as a positive constant κ and the total mass of the fluid is taken to be 1. The remaining symbols like ν, R, etc., denote physical constants, representing the viscosity and the Boltzmann constants, etc., with ν simplified as $\nu = 1$ in this chapter as in [110].

In this chapter, $\Omega = (0,1)$, $\mathbb{R}_+ = [0, +\infty)$, $H^i = W^{i,2}$ ($i = 1, 2, 4$) and we use $\| \cdot \|$ and $C^{k,\alpha}(\Omega)$ to denote the norm in $L^2(\Omega)$, and the space of functions whose derivatives are Hölder continuous with exponent α and order of differentiability from 0 to k, respectively. C_i ($i = 1, 2, 3, 4$) denote universal constants depending on $\min_{x\in\Omega} \eta_0(x)$, $\min_{x\in\Omega} \theta_0(x)$, the $H^i(\Omega)$ ($i = 1, 2, 3, 4$) spatial norms of $(\eta_0, v_0, w_0, \theta_0)$, but independent of time $T > 0$. To facilitate our analysis, we also define three function classes:

$$H_+^1 = \Big\{(\eta, v, w, \theta) \in \left(H^1(\Omega)\right)^4 : \eta(x) > 0, \ \theta(x) > 0,$$

$$x \in \Omega, \ v(0) = v(1) = 0, \ w(0) = w(1) = 0, \ \theta(0) = \theta(1) = 1 \text{ for } (6.1.9)\Big\},$$

and

$$H_+^i = \Big\{(\eta, v, w, \theta) \in \left(H^i(\Omega)\right)^4 : \eta(x) > 0, \ \theta(x) > 0,$$

$$x \in \Omega, \ v(0) = v(1) = 0, w(0) = w(1) = 0,$$

$$\theta'(0) = \theta'(1) = 0 \text{ for } (6.1.8), \quad \text{or } \theta(0) = \theta(1) = 1 \text{ for } (6.1.9)\Big\}, \quad i = 2, 4.$$

The Cauchy problem for (6.1.1)–(6.1.6) was considered in Chapter 5. In this chapter, we shall consider the initial-boundary value problem (6.1.1)–(6.1.9).

We now state our main results in this chapter.

Theorem 6.1.1. *Assume that* $(\eta_0, v_0, w_0, \theta_0) \in H_+^1$. *Then the problem* (6.1.1)–(6.1.8) *or the problem* (6.1.1)–(6.1.7), (6.1.9) *have a unique* H^1-*global solution* $(\eta(t), v(t), w(t), \theta(t))$ *and the following inequalities hold:*

$$0 < C_1^{-1} \le \eta(t,x) \le C_1^{-1}, \quad \text{on } \Omega \times \mathbb{R}_+, \tag{6.1.11}$$

$$0 < C_1^{-1} \le \theta(t,x) \le C_1^{-1}, \quad \text{on } \Omega \times \mathbb{R}_+, \tag{6.1.12}$$

$$\|v(t)\|^2 + \|\theta(t)\|^2 + \|w(t)\|^2 + \|\eta_x\|^2 + \|v_x(t)\|^2 + \|\theta_x(t)\|^2$$

$$+ \|w_x(t)\|^2 + \int_0^t \Big[\|\eta_x\|^2 + \|v_x\|^2 + \|\theta_x\|^2 + \|w_x\|^2 + \|v_{xx}\|^2 + \|\theta_{xx}\|^2$$

$$+ \|w_{xx}\|^2 + \|v_t\|^2 + \|\theta_t\|^2 + \|w_t\|^2 \Big](s)ds \le C_1, \quad \forall t > 0. \tag{6.1.13}$$

Moreover, there exists a positive constant $\gamma_1 = \gamma_1(C_1)$ *such that, for any fixed constant* $\gamma \in (0, \gamma_1]$, *the following estimate holds for any* $t > 0$:

$$\left\| \Big(\eta(t) - \bar{\eta},\, v(t),\, w(t),\, \theta(t) - \bar{\theta} \Big) \right\|_{H_+^1}^2 \le C_1 e^{-\gamma t}, \tag{6.1.14}$$

where

$$\bar{\eta} = \int_\Omega \eta_0(x)dx, \quad \bar{\theta} = 1 \text{ for } (6.1.9), \tag{6.1.15}$$

or, for (6.1.8), $\bar{\theta} > 0$ *is uniquely determined by the energy form*

$$e(\bar{\eta}, \bar{\theta}) := \int_\Omega \left(\frac{v_0^2}{2} + \frac{w_0^2}{2A} + \theta_0 \right)(x)dx. \tag{6.1.16}$$

Theorem 6.1.2. *Assume that* $(\eta_0, v_0, w_0, \theta_0) \in H_+^2$. *Then the problem* (6.1.1)–(6.1.8) *or the problem* (6.1.1)–(6.1.7), (6.1.9) *has a unique* H^2-*global solution* $(\eta(t), v(t), w(t), \theta(t)) \in H_+^2$ *for any* $t > 0$, *and the following estimates hold:*

$$\|\eta(t)\|_{H^2}^2 + \|\eta(t)\|_{W^{1,\infty}}^2 + \|\eta_t(t)\|_{H^1}^2 + \|v(t)\|_{H^2}^2 + \|v(t)\|_{W^{1,\infty}}^2 + \|w(t)\|_{H^2}^2$$

$$+ \|w(t)\|_{W^{1,\infty}}^2 + \|v_t(t)\|^2 + \|\theta(t)\|_{H^2}^2 + \|\theta(t)\|_{W^{1,\infty}}^2 + \|\theta_t(t)\|^2 + \int_0^t \Big[\|\eta_x\|_{H^1}^2$$

$$+ \|\eta_x\|_{L^\infty}^2 + \|\eta_t\|_{H^2}^2 + \|v_x\|_{H^2}^2 + \|v_x\|_{W^{1,\infty}}^2 + \|v_t\|_{H^1}^2 + \|w_x\|_{H^2}^2 + \|w_x\|_{W^{1,\infty}}^2$$

$$+ \|w_t\|_{H^1}^2 + \|\theta_x\|_{H^2}^2 + \|\theta_x\|_{W^{1,\infty}}^2 + \|\theta_t\|_{H^1}^2 \Big](s)ds \le C_2. \tag{6.1.17}$$

Moreover, there exists a positive constant $\gamma_2 = \gamma_2(C_2) \le \gamma_1(C_1)$ *such that, for any fixed* $\gamma \in (0, \gamma_2]$, *the following estimate holds for any* $t > 0$:

$$\left\| \Big(\eta(t) - \bar{\eta},\, v(t),\, w(t),\, \theta(t) - \bar{\theta} \Big) \right\|_{H_+^2} \le C_2 e^{-\gamma t}. \tag{6.1.18}$$

Theorem 6.1.3. *Assume that* $(\eta_0, v_0, w_0, \theta_0) \in H_+^4$. *Then the problem* (6.1.1)–(6.1.8) *or the problem* (6.1.1)–(6.1.7), (6.1.9) *admits a unique* H^4-*global solution* $(\eta(t), v(t), w(t), \theta(t)) \in H_+^4$ *for any* $t > 0$ *and the following estimates hold:*

$$\|\eta(t)\|_{H^4}^2 + \|\eta(t)\|_{W^{3,\infty}}^2 + \|\eta_t(t)\|_{H^3}^2 + \|\eta_{tt}(t)\|_{H^1}^2 + \|v(t)\|_{H^4}^2 + \|v(t)\|_{W^{3,\infty}}^2$$
$$+ \|w(t)\|_{H^4}^2 + \|w(t)\|_{W^{3,\infty}}^2 + \|w_t(t)\|_{H^2}^2 + \|w_{tt}(t)\|^2 + \|w_t(t)\|_{H^2}^2 + \|w_{tt}(t)\|^2$$
$$+ \|\theta(t) - 1\|_{H^4}^2 + \|\theta(t) - 1\|_{W^{3,\infty}}^2 + \|\theta_t(t)\|_{H^2}^2 + \|\theta_{tt}(t)\|^2 \leq C_4, \qquad (6.1.19)$$

$$\int_0^t \Big[\|\eta_x\|_{H^3}^2 + \|\eta_t\|_{H^4}^2 + \|\eta_{tt}\|_{H^2}^2 + \|\eta_{ttt}\|^2 + \|\eta_x\|_{W^{2,\infty}}^2 + \|v_x\|_{H^4}^2 + \|v_t\|_{H^3}^2$$
$$+ \|v_{tt}\|_{H^1}^2 + \|v_x\|_{W^{3,\infty}}^2 + \|w_x\|_{H^4}^2 + \|w_t\|_{H^3}^2 + \|w_{tt}\|_{H^1}^2 + \|w_x\|_{W^{3,\infty}}^2$$
$$+ \|\theta_x\|_{H^4}^2 + \|\theta_t\|_{H^3}^2 + \|\theta_{tt}\|_{H^1}^2 + \|\theta_x\|_{W^{3,\infty}}^2 \Big](s)ds \leq C_4. \qquad (6.1.20)$$

Moreover, there exists a positive constant $\gamma_4 = \gamma_4(C_4) \leq \gamma_2(C_2)$ *such that, for any fixed* $\gamma \in (0, \gamma_4]$, *the following estimate holds for any* $t > 0$,:

$$\left\| \Big(\eta(t) - \bar{\eta}, \, v(t), \, w(t), \, \theta(t) - \bar{\theta} \Big) \right\|_{H_+^4}^2 \leq C_4 e^{-\gamma t}. \qquad (6.1.21)$$

Corollary 6.1.1. *The* H^4-*global solution* $(\eta(t), v(t), w(t), \theta(t))$ *obtained in Theorem 6.1.3 is actually a classical one when the compatibility conditions hold. Precisely,* $(\eta(t), v(t), w(t), \theta(t)) \in C^{3,\frac{1}{2}}(\Omega)$. *Moreover, for any* $t > 0$,

$$\left\| \Big(\eta(t) - \bar{\eta}, \, v(t), \, w(t), \, \theta(t) - \bar{\theta} \Big) \right\|_{C^{3,\frac{1}{2}}(\Omega)}^2 \leq C_4 e^{-\gamma t}, \qquad (6.1.22)$$

for any $\gamma \in (0, \gamma_4]$.

6.2   Global Existence and Exponential Stability in H_+^1

In this section, we mainly follow Mujakoć's basic idea in [84] to establish global H^1 estimates for (η, v, w, θ) and Qin's method to obtain the exponential stability of solutions in [99].

Lemma 6.2.1. *For all* $t > 0$,

$$\int_0^1 \eta(x)dx = \int_0^1 \eta_0(x)dx \equiv \bar{\eta}, \qquad (6.2.1)$$

$$\frac{1}{2}\int_0^1 v^2 dx + \frac{1}{2A}\int_0^1 w^2 dx + \int_0^1 (\theta - \log\theta - 1)dx + R\int_0^1 (\eta - \log\eta - 1)dx$$
$$+ \int_0^t \int_0^1 \left(\frac{v_x^2}{\theta\eta} + \frac{w_x^2}{\theta\eta} + \frac{\eta w^2}{\theta} + \hat{k}\frac{\theta_x^2}{\eta\theta^2} \right) dx ds := E_1, \qquad (6.2.2)$$

$$0 < \alpha \leq \int_0^1 \theta(x)dx \leq \beta, \tag{6.2.3}$$

where α and β are constants, and

$$E_1 := \int_0^1 \left(v_0^2 dx + \frac{1}{2A}w_0^2 + (\theta_0 - \log\theta_0 - 1) + R(\eta_0 - \log\eta_0 - 1)\right)dx. \tag{6.2.4}$$

Proof. See, e.g., [84], Lemmas 3.1–3.2. $\qquad\square$

Now we borrow the related results for the upper and lower bounds of η from [101]; the version we use here is just a simpler case of [101]. In fact, these results are the improved ones of [63].

Lemma 6.2.2. *There holds that*

$$0 < C_1^{-1} \leq \eta(x,t) \leq C_1, \text{ for all } (x,t) \in [0,1] \times \mathbb{R}_+, \tag{6.2.5}$$

$$0 \leq C_1^{-1} \leq \rho(x,t) \leq C_1, \text{ for all } (x,t) \in [0,1] \times \mathbb{R}_+. \tag{6.2.6}$$

Proof. See, e.g., [101], pp. 53–55, Lemma 2.1.5. $\qquad\square$

As it stands in [84] combined with Lemmas 6.2.1–6.2.2, we can immediately obtain (6.1.11)–(6.1.13). $\qquad\square$

Now let us establish the verification of the exponential stability of the solution $(\eta(t), v(t), w(t), \theta(t))$. We introduce some thermodynamic quantities and describe their properties. First, the entropy $S(\rho, \theta)$ satisfies

$$\frac{\partial S}{\partial\rho} = -\frac{P_\theta}{\rho^2}, \quad \frac{\partial S}{\partial\theta} = \frac{e_\theta}{\theta}. \tag{6.2.7}$$

Now we take η and S as the basic independent variables and make the coordinate transformation

$$\mathfrak{T}: (\rho, \theta) \in \mathfrak{D}_{\rho,\theta} = \{(\rho, \theta): \rho, \theta > 0\} \ni (\rho, \theta) \mapsto (\eta, S) \in \mathfrak{TD}_{\rho,\theta}. \tag{6.2.8}$$

Since the Jacobian $|\partial(\eta, S)/\partial(\rho, \theta)| = -e_\theta\eta^2/\theta < 0$ on $\mathfrak{D}_{\rho,\theta}$, the transformation $\mathfrak{T}$ has a unique inverse and there exist $\theta = \theta(\eta, S)$, $e(\eta, S)$ and $P(\eta, S)$ which are smooth functions of (η, S). Note that $e = e(\eta, S) := e(\eta, \theta(\eta, S)) = e(\eta, \theta)$, $P = P(\eta, S) := P(\eta, \theta(\eta, S)) = P(\eta, \theta)$, and

$$e_\eta = -P, \quad e_S = \theta, \quad P_\eta = -\left(\frac{P_\rho}{\eta^2} + \frac{\theta P_\theta^2}{e_\theta}\right), \quad P_S = \frac{\theta P_\theta}{e_\theta}, \quad \theta_\eta = -\frac{\theta P_\theta}{e_\theta}, \quad \theta_S = \frac{\theta}{e_\theta}. \tag{6.2.9}$$

We define the energy form

$$\mathfrak{E}(\eta, v, w, S) := \frac{v^2}{2} + \frac{w^2}{2A} + e(\eta, S) - e(\bar\eta, \bar S) - \frac{\partial e}{\partial\eta}(\bar\eta, \bar S)(\eta - \bar\eta) - \frac{\partial e}{\partial S}(\bar\eta, \bar S)(S - \bar S), \tag{6.2.10}$$

where $\bar S = S(\bar\eta, \bar\theta)$.

Lemma 6.2.3. *The following estimate holds for any $t > 0$:*

$$\frac{v^2}{2} + \frac{w^2}{2A} + C_1^{-1}\left(|\eta - \overline{\eta}|^2 + |S - \overline{S}|^2\right) \leq \mathfrak{E}(\eta, v, w, S) \tag{6.2.11}$$

$$\leq \frac{v^2}{2} + \frac{w^2}{2A} + C_1\left(|\eta - \overline{\eta}|^2 + |S - \overline{S}|^2\right).$$

Proof. By the mean value theorem, there exists a point (ξ, ζ) between (η, S) and $(\overline{\eta}, \overline{S})$ such that

$$\mathfrak{E}(\eta, v, w, S) := \frac{v^2}{2} + \frac{w^2}{2A} + \frac{1}{2}\left[\frac{\partial^2 e}{\partial \eta^2}(\xi, \zeta)(\eta - \overline{\eta})^2 + 2\frac{\partial^2 e}{\partial \eta \partial S}(\xi, \zeta)(\eta - \overline{\eta})(S - \overline{S})\right.$$

$$\left. + \frac{\partial^2 e}{\partial \eta^2}(\xi, \zeta)(S - \overline{S})^2\right], \tag{6.2.12}$$

where $\xi = \lambda\overline{\eta} + (1-\lambda)\eta$, $\zeta = \lambda\overline{S} + (1-\lambda)S$, with a constant $\lambda \in [0, 1]$. (6.2.5) and (6.1.12) yield

$$\left|\frac{\partial^2 e}{\partial \eta^2}(\xi, \zeta)\right| + \left|\frac{\partial^2 e}{\partial \eta \partial S}(\xi, \zeta)\right| + \left|\frac{\partial^2 e}{\partial \eta^2}(\xi, \zeta)\right| \leq C_1, \tag{6.2.13}$$

which, combined with (6.2.12), gives

$$\mathfrak{E}(\eta, v, w, S) \leq \frac{v^2}{2} + \frac{w^2}{2A} + C_1\left[(\eta - \overline{\eta})^2 + (S - \overline{S})^2\right]. \tag{6.2.14}$$

On the other hand, it follows from (6.2.9) that

$$e_{\eta\eta} = \frac{P_\rho}{\eta^2} + \frac{\theta P_\theta^2}{e_\theta}, \quad e_{\eta S} = \frac{-\theta P_\theta}{e_\theta}, \quad e_{SS} = \frac{\theta}{e_\theta},$$

and so the Hessian of e is positive definite for $\eta, \theta > 0$. Therefore,

$$\mathfrak{E}(\eta, v, w, S) \geq \frac{v^2}{2} + \frac{w^2}{2A} + C_1^{-1}\left[(\eta - \overline{\eta})^2 + (S - \overline{S})^2\right]. \tag{6.2.15}$$

Combining with (6.2.14)–(6.2.15), we arrive at (6.2.11). □

Lemma 6.2.4. *There exists a positive constant $\gamma_1^{(1)} = \gamma_1^{(1)}(C_1) > 0$ such that, for any fixed $\gamma \in (0, \gamma_1^{(1)}]$, the following estimate holds for any $t > 0$:*

$$e^{\gamma t}\left(\|\eta(t) - \overline{\eta}\|^2 + \|v(t)\|^2 + \|w(t)\|^2 + \|\theta(t) - \overline{\theta}\|^2 + \|\eta_x(t)\|^2\right)$$

$$+ \int_0^t e^{\gamma s}\left(\|\eta_x\|^2 + \|v_x\|^2 + \|w_x\|^2 + \|\theta_x\|^2\right)(s)ds \leq C_1. \tag{6.2.16}$$

Proof. We can see from (6.1.1)–(6.1.4) that

$$\left(\frac{v^2}{2} + \frac{w^2}{2A} + e\right)_t = \left[\sigma v + \frac{\kappa\theta_x}{\eta}\right]_x + \frac{w_x^2}{\eta} + \left(\frac{w_x}{\eta}\right)_x w, \qquad (6.2.17)$$

$$S_t = -P_\theta\rho_t\eta^2 + \frac{e_\theta}{\theta}\theta_t = \left(\frac{\kappa\eta\theta_x}{\theta}\right)_x + \kappa\frac{1}{\eta}\left(\frac{\theta_x}{\theta}\right)^2 + \frac{v_x^2}{\eta\theta} + \frac{w_x^2}{\eta\theta} + \frac{\eta w^2}{\theta}. \qquad (6.2.18)$$

Since $\overline{\eta}_t \equiv 0$, $\overline{\theta}_t \equiv 0$, we infer from (6.2.20), (6.2.21), (6.1.1) and (6.1.2) that

$$\mathfrak{E}_t(\eta, v, w, S) + \left(\frac{\overline{\theta}}{\theta}\right)\left[\frac{v_x^2}{\eta} + \frac{w_x^2}{\eta} + \eta w^2 + \frac{\kappa\theta_x^2}{\eta\theta}\right]$$

$$= \left[\frac{vv_x}{\eta} + \kappa\left(1 - \frac{\overline{\theta}}{\theta}\right)\frac{\theta_x}{\eta} - (P - P(\overline{\eta}, \overline{S}))v\right]_x + \frac{w_x^2}{\eta} + \left(\frac{w_x}{\eta}\right)_x w, \qquad (6.2.19)$$

$$\left[\frac{1}{2}\left(\frac{\rho_x}{\rho}\right)^2 + \rho_x v\eta\right]_t + P_\rho\rho_x^2\eta = -P_\theta\rho_x\theta_x\eta - \left(\frac{vv_x}{\eta}\right)_x + \frac{v_x^2}{\eta}. \qquad (6.2.20)$$

Multiplying (6.2.19) and (6.2.20) by $e^{\gamma t}$ and $\delta e^{\gamma t}$, integrating over $[0, 1] \times [0, t]$ ($\forall t > 0$), and using the Young inequality and the Poincaré inequality, we derive that for small $\delta > 0$,

$$e^{\gamma t}\left[\|\eta(t) - \overline{\eta}\|^2 + \|v(t)\|^2 + \|w(t)\|^2 + \|\eta_x(t)\|^2\right]$$

$$+ \int_0^t e^{\gamma s}\left[\|\eta_x\|^2 + \|v_x\|^2 + \|w_x\|^2 + \|w\|^2 + \|\theta_x\|^2\right](s)ds \leq C_1. \qquad (6.2.21)$$

By virtue of (6.1.8) and the Poincaré inequality,

$$\|\theta - \overline{\theta}\|_{L^\infty} \leq \|\int_0^x \theta_y(y, t)dy\|_{L^\infty} \leq \|\theta_x(t)\|, \qquad (6.2.22)$$

$$\|\eta(t) - \overline{\eta}\| \leq C_1\|\eta_x(t)\|, \quad \|v(t)\| \leq C_1\|v_x(t)\|, \quad \|w(t)\| \leq C_1\|w_x(t)\|.$$

By (6.1.9),

$$\int_0^1 \left(\frac{v^2}{2} + \frac{w^2}{2A} + e(\eta, \theta)\right)dx \equiv e(\overline{\eta}, \overline{\theta}),$$

whence

$$\|e(\eta, \theta) - e(\overline{\eta}, \overline{\theta})\| \leq \|e(\eta, \theta) - \int_0^1 e(\eta, \theta)dx\| + \frac{1}{2}\|v(t)\|^2 + \frac{1}{2A}\|w(t)\|^2$$

$$\leq C_1(\|\eta_x(t)\| + \|v_x(t)\| + \|w_x(t)\| + \|\theta_x(t)\|).$$

By the mean value theorem, we get

$$\|\theta - \overline{\theta}\| \leq C_1(\|e(\eta, \theta) - e(\overline{\eta}, \overline{\theta})\| + \|\eta(t) - \overline{\eta}\|)$$

$$\leq C_1(\|\eta_x(t)\| + \|v_x(t)\| + \|w_x(t)\| + \|\theta_x(t)\|). \tag{6.2.23}$$

As a result, (6.2.22) and (6.2.23) lead to

$$e^{\gamma t}\|\theta(t) - \overline{\theta}\|^2 \leq C_1.$$

This completes the proof of (6.2.16). ☐

Lemma 6.2.5. *There is a positive constant* $\gamma_1 = \gamma_1(C_1) \leq \gamma_1^{(1)}$ *such that, for any fixed* $\gamma \in (0, \gamma_1]$, *the following estimate holds for any* $t > 0$:

$$e^{\gamma t}\left(\|v_x(t)\|^2 + \|w_x(t)\|^2 + \|\theta_x(t)\|^2\right) \tag{6.2.24}$$

$$+ \int_0^t e^{\gamma s}\left(\|v_{xx}\|^2 + \|w_{xx}\|^2 + \|\theta_{xx}\|^2 + \|v_t\|^2 + \|w_t\|^2 + \|\theta_t\|^2\right)(s)ds \leq C_1.$$

Proof. Multiplying (6.1.2) by $-e^{\gamma t}v_{xx}$, integrating over $[0,1] \times [0,t]$ for any $t > 0$ and using the Poincaré inequality, the interpolation inequalities, the Young inequality and Lemma 6.2.4, we derive that

$$e^{\gamma t}\|v_x(t)\|^2 + \int_0^t e^{\gamma s}\|v_{xx}(s)\|^2 ds \leq C_1 + C_1 \int_0^t e^{\gamma s}(\|\eta_x\|^2 + \|v_x\|^2 + \|\theta_x\|^2)(s)ds$$

$$\leq C_1, \tag{6.2.25}$$

which, combined with (6.1.2), gives

$$\int_0^t e^{\gamma s}\|v_t(s)\|^2 ds \leq C_1 + C_1 \int_0^t e^{\gamma s}\|v_{xx}(s)\|^2 ds \leq C_1. \tag{6.2.26}$$

Similarly to (6.1.3) and (6.1.4), we can derive (6.2.24). ☐

Proof of Theorem 6.1.1. The global existence and the exponential stability of solution $(\eta(t), v(t), w(t), \theta(t))$ asserted in Theorem 6.1.1 follow from the preceding lemmas. ☐

6.3   Global Existence and Exponential Stability in H_+^2

In this section, we shall complete the proof of Theorem 6.1.2. We begin to prove the global existence of H^2 solutions with the following lemma on estimates in $H^1(\Omega)$.

Lemma 6.3.1. *Under the assumptions of Theorem 6.1.1, the* H^1 *global solution* $(\eta(t), v(t), w(t), \theta(t))$ *to the problem* (6.1.1)–(6.1.8) *or problem* (6.1.1)–(6.1.7), (6.1.9) *exists, and, for any* $t > 0$,

$$\|\eta(t) - \overline{\eta}\|_{H^1}^2 + \|v(t)\|_{H^1}^2 + \|w(t)\|_{H^1}^2 + \|\theta(t) - \overline{\theta}\|_{H^1}^2 + \|\eta_t(t)\|^2$$

$$+ \int_0^t \left(\|v_x\|_{H^1}^2 + \|w_x\|_{H^1}^2 + \|\theta_x\|_{H^1}^2 + \|\eta_x\|^2 + \|v_t\|^2 + \|w_t\|^2 + \|\theta_t\|^2 \right)(s)ds$$

$$\leq C_1, \tag{6.3.1}$$

$$\|\eta(t) - \overline{\eta}\|_{L^\infty}^2 + \|v(t)\|_{L^\infty}^2 + \|w(t)\|_{L^\infty}^2 + \|\theta(t) - \overline{\theta}\|_{L^\infty}^2$$

$$+ \int_0^t \left(\|\eta_t\|_{H^1}^2 + \|v_x\|_{L^\infty}^2 + \|w_x\|_{L^\infty}^2 + \|\theta_x\|_{L^\infty}^2 \right)(s)ds \leq C_1. \tag{6.3.2}$$

Proof. The proof is identical to that of Lemma 5.3.1. □

Lemma 6.3.2. *Under the assumptions of Theorem 6.1.2, the following estimates hold for any $t > 0$:*

$$\|\theta_t(t)\|^2 + \|v_t(t)\|^2 + \|w_t(t)\|^2 + \int_0^t (\|v_{xt}\|^2 + \|w_{xt}\|^2 + \|\theta_{xt}\|^2)(s)ds \leq C_2,$$

$$\tag{6.3.3}$$

$$\|v_x(t)\|_{L^\infty}^2 + \|v_{xx}(t)\|^2 + \|w_x(t)\|_{L^\infty}^2 + \|w_{xx}(t)\|^2 + \|\theta_x(t)\|_{L^\infty}^2 + \|\theta_{xx}(t)\|^2 \leq C_2, \tag{6.3.4}$$

$$\|v(t)\|_{H^2}^2 + \|w(t)\|_{H^2}^2 + \|\theta(t) - \overline{\theta}\|_{H^2}^2 + \|\eta_t(t)\|_{H^1}^2 \leq C_2. \tag{6.3.5}$$

Proof. The proof is identical to that of Lemma 5.3.2. □

Lemma 6.3.3. *Under the assumptions of Theorem 6.1.2, the following estimates hold for any $t > 0$:*

$$\|\eta_{xx}(t)\|^2 + \|\eta_x(t)\|_{L^\infty}^2 + \int_0^t (\|\eta_{xx}\|^2 + \|\eta_x\|_{L^\infty}^2)(s)ds \leq C_2, \tag{6.3.6}$$

$$\int_0^t (\|v_{xxx}\|^2 + \|w_{xxx}\|^2 + \|\theta_{xxx}\|^2)(s)ds \leq C_2. \tag{6.3.7}$$

Proof. The proof is identical to that of Lemma 5.3.3. □

Next, we shall deal with the exponential stability of the global H^2 solution. In fact, the argument is similar to that for the global existence, with the difference that the weight function $\exp(\gamma t)$ accompanies in the estimates.

Lemma 6.3.4. *There exists a positive constant $\gamma_2^{(1)} = \gamma_2^{(1)}(C_2)$ such that, for any fixed $\gamma \in (0, \gamma_2^{(1)}]$, the following estimate holds for any $t > 0$:*

$$e^{\gamma t}(\|v(t)\|_{H^2}^2 + \|w(t)\|_{H^2}^2 + \|\theta(t) - \overline{\theta}\|_{H^2}^2) \tag{6.3.8}$$

$$+ \int_0^t e^{\gamma s} \left(\|v_{xxx}\|^2 + \|v_{tx}\|^2 + \|w_{xxx}\|^2 + \|w_{tx}\|^2 + \|\theta_{xxx}\|^2 + \|\theta_{tx}\|^2 \right)(s)ds \leq C_2.$$

Proof. By Lemma 3.5 in [99], one can show that

$$e^{\gamma t}\|v(t)\|_{H^2}^2 + \int_0^t e^{\gamma s}(\|v_{xxx}\|^2 + \|v_{xt}\|^2)(s)ds \leq C_2.$$

Similarly, multiplying (6.1.3) and (6.1.4) by $w_t e^{\gamma t}$ and $\theta_t e^{\gamma t}$, respectively, then integrating the results over $[0,1] \times [0,t]$, one can obtain the remaining estimates. This completes the proof. $\qquad\square$

Lemma 6.3.5. *There exists a positive constant $\gamma_2 = \gamma_2(C_2) \leq \gamma_2^{(1)}(C_2)$ such that, for any fixed $\gamma \in (0, \gamma_2]$, the following estimate holds for any $t > 0$:*

$$e^{\gamma t}\|\eta(t) - \overline{\eta}\|_{H^2}^2 + \int_0^t e^{\gamma s}\|\eta_{xx}(s)\|^2 ds \leq C_2. \qquad (6.3.9)$$

Proof. See, e.g., Lemma 3.6 in [99]. $\qquad\square$

Proof of Theorem 6.1.2. Combining with Lemmas 6.3.1–6.3.5, we complete the proof of Theorem 6.1.2 with no difficulty. $\qquad\square$

6.4 Global Existence and Exponential Stability in H_+^4

Proof of Theorem 6.1.3. Similarly to [112] and similarly to the proof of Theorem 2.1.3, we differentiate the system of equations more than twice to get the result of Theorem 6.1.3. $\qquad\square$

Proof of Corollary 6.1.1. Employing the Sobolev embedding theorem, the desired conclusion follows immediately. $\qquad\square$

6.5 Bibliographic Comments

The mathematical research of the 1D compressible Navier-Stokes system has made great steps since the original breakthrough work by Kazhikhov et al. [1, 61, 63]. A crucial specific feature in this research is that in the 1D case, treated in Lagrangian coordinates, the specific volume can be expressed in terms of other unknown variables, and positive upper and lower bounds can be derived in the respective initial-boundary value problems. Moreover, this feature was exploited and improved further, with slight differences, in [30, 43], etc. For the Cauchy problem, since one is dealing with an unbounded domain, the Poincaré inequality is not available. To overcome this we consider the problem in the case of small initial data, as advocated in [56, 93] and done in [112]. A similar situation for the micropolar model has been studied in recent years (see, e.g., [82, 110]). For initial-boundary value problems, there are numerous works concerning global existence, regularity and asymptotic behavior, etc., in various cases, etc. (see, e.g., [1, 8, 9, 15, 16, 30, 43, 54, 55, 84, 86–89, 96, 97, 99, 100, 113, 138, 140]) and many cases are rigorously investigated under various conditions on the pressure P and the heat conductivity $\widetilde{\kappa}$ (see, e.g., [53, 54]). Actually Okada's work [93] provides us with the main significant ideas for attacking the problem. On the contrary, in the case of the Cauchy problem, considering a bounded underlying domain enables the application of Poincaré's inequality and consequently exponential decay rates can be obtained, which differ considerably from those on the unbounded domain. Qin [99] has flexibly modified Okada's subtle method and obtained a series of properties concerning global existence, regularity, exponential stability and existence of attractors. Recently, in the case of micropolar fluids, Mujaković has extended the study to case of non-homogeneous boundary value problems (see, e.g., [85]). Earlier also in [84] and [79], she derived similar results, including results on the large-time behavior of solutions. However, exponential stability has not been established (see, e.g., [79]) till the results in the present chapter. In fact, here we improve such results and in particular establish the exponential stability under less restrictive boundary conditions.

Note that in Chapter 5 we have proved the global existence and large-time behavior of solutions in H^i ($i = 1, 2, 4$) for the Cauchy problem corresponding to problem (6.1.1)–(6.1.8) (or problem (6.1.1)–(6.1.7), (6.1.9)).

Chapter 7

Global Existence and Exponential Stability of Solutions to the Equations of 1D Full Non-Newtonian Fluids

7.1 Introduction

In this chapter, we shall prove the global existence and exponential stability of solutions to the following full non-Newtonian fluid model:

$$\rho_t + (\rho u)_x = 0, \tag{7.1.1}$$

$$(\rho u)_t + (\rho u^2)_x - u_{xx} + (R\rho\theta)_x = 0, \tag{7.1.2}$$

$$\left(\rho\theta + \frac{1}{2}\rho u^2\right)_t + \left[\left(\rho\theta + \frac{1}{2}\rho u^2\right) + (R\rho\theta)u - uu_x\right]_x - \left[(\theta_x^2 + \mu_0)^{\frac{p-2}{2}}\theta_x\right]_x = 0. \tag{7.1.3}$$

Here subscripts indicate partial differentiations, ρ, u and θ denote the unknown density, velocity and absolute temperature, respectively. R, $p > 2$, $\mu_0 > 0$ are given constants. The initial and boundary conditions are given by

$$(\rho, u, \theta)|_{t=0} = (\rho_0, u_0, \theta_0), \tag{7.1.4}$$

$$(u, \theta_x)|_{x=0} = (u, \theta_x)|_{x=1} = 0. \tag{7.1.5}$$

This full non-Newtonian fluid model ($p > 2$) is more complicated than the models considered in Chapters 3–5 in Qin and Huang [102], which are Newtonian fluid models ($p = 2$). Therefore, we need to more delicate estimates to deal with this non-Newtonian model. For convenience, we introduce the Lagrangian coordinates (y, t), defined by

$$y = \int_0^x \rho(\xi, t)d\xi.$$

By (7.1.1) and (7.1.5), we get

$$\int_0^1 \rho(x,t)dx = \int_0^1 \rho_0(x)dx,$$

and without loss of generality we assume $\int_0^1 \rho_0(x)dx = 1$.

Thus problem (7.1.1)–(7.1.5) in Euler coordinates (x,t) is transformed into the following problem in Lagrangian coordinates (y,t) for $y \in \Omega = [0,1]$, $t > 0$:

$$v_t - u_y = 0, \tag{7.1.6}$$

$$u_t - \sigma_y = 0, \tag{7.1.7}$$

$$\left(e + \frac{u^2}{2}\right)_t - (\sigma u)_y + Q_y = 0 \tag{7.1.8}$$

with initial and boundary data

$$(v,u,\theta)|_{t=0} = (v_0, u_0, \theta_0)(y), \quad y \in \Omega, \tag{7.1.9}$$

$$(u,\theta_y)|_{\partial\Omega} = 0, \quad t > 0, \tag{7.1.10}$$

where $v = 1/\rho$ is the specific volume, P, e, σ and Q are the pressure, internal energy, stress and heat flux, respectively, which have the following expressions

$$P = \frac{R\theta}{v}, \quad e = \theta, \quad \sigma = -P + \frac{u_y}{v}, \quad Q = -\left(\left(\frac{\theta_y}{v}\right)^2 + \mu_0\right)^{\frac{p-2}{2}} \frac{\theta_y}{v}.$$

In this chapter, we study the non-Newtonian models (7.1.6)–(7.1.10) to establish the global existence and exponential stability of solutions in H^i ($i = 1,2,4$) for one-dimensional full compressible non-Newtonian fluids with large initial data, which were not studied in [3–5, 20, 44, 90, 94, 114, 122, 133, 141, 142, 146, 147].

We define three function classes as follows:

$$H^1_+ = \Big\{(v,u,\theta) \in (H^1[0,1])^3 : v(y) > 0, \theta(y) > 0, y \in [0,1],$$

$$u|_{y=0} = u|_{y=1} = 0, \theta_y|_{y=0} = \theta_y|_{y=1} = 0\Big\},$$

$$H^i_+ = \Big\{(v,u,\theta) \in (H^i[0,1])^3 : v(y) > 0, \theta(y) > 0, y \in [0,1],$$

$$u|_{y=0} = u|_{y=1} = 0, \theta_y|_{y=0} = \theta_y|_{y=1} = 0\Big\}, i = 2,4.$$

We will use the following notations:

L^p, $1 \leq p \leq +\infty$, $W^{m,p}$, $m \in \mathbb{N}$, $H^1 = W^{1,2}$, $H^1_0 = W^{1,2}_0$ denote the usual (Sobolev) spaces on $[0,1]$. In addition, $\|\cdot\|_B$ denotes the norm in the space B, we also put $\|\cdot\| = \|\cdot\|_{L^2[0,1]}$. $\overline{f}(t) = \int_0^1 f(y,t)dy$, $f^*(y,t) =$

$f(y,t) - \overline{f}(t)$, $(f(y,t), g(y,t)) = \int_0^1 f(y,t)g(y,t)dy$. Letters C_i $(i = 1,2,3,4)$ will denote universal constants not depending on t, but possibly on the norm of the initial data in H^i $(i = 1,2,3,4)$.

Now we can state our main result.

Theorem 7.1.1. *Assume that the initial data $(v_0, u_0, \theta_0) \in H_+^1$ and compatibility conditions are satisfied. Then the problem (7.1.6)–(7.1.10) admits a unique global solution $(v(t), u(t), \theta(t)) \in H_+^1$ verifying*

$$0 < C_1^{-1} \leq v(x,t) \leq C_1, \quad 0 < C_1^{-1} \leq \theta(x,t) \leq C_1, \quad \forall(x,t) \in [0,1] \times [0,+\infty) \tag{7.1.11}$$

and

$$\|v(t) - \overline{v}\|_{H^1}^2 + \|u(t)\|_{H^1}^2 + \|\theta(t) - \overline{\theta}\|_{H^1}^2 + \|\theta_y(t)\|_{L^p}^p$$
$$+ \int_0^t \left(\|v - \overline{v}\|_{H^1}^2 + \|\theta_y(t)\|_{L^p}^p + \|u\|_{H^2}^2 + \|\theta - \overline{\theta}\|_{H^2}^2 + \|u_t\|^2 + \|\theta_t\|^2 \right)(s)ds$$
$$+ \int_0^t \||\theta_y|^{\frac{p-2}{2}} + \theta_y)\theta_{yy}(s)\|^2 ds \leq C_1, \quad \forall t > 0, \tag{7.1.12}$$

where $\overline{v} = \int_0^1 v dy = \int_0^1 v_0 dy$ and $\overline{\theta} > 0$ is determined by $e(\overline{v}, \overline{\theta}) = \int_0^1 (\frac{u_0^2}{2} + e(v_0, \theta_0))dy$.

Moreover, there are constants $C_1 > 0$ and $\gamma_1 = \gamma_1(C_1) > 0$ such that, for any fixed $\gamma \in (0, \gamma_1]$, the following estimate holds for any $t > 0$:

$$e^{\gamma t}(\|v(t) - \overline{v}\|_{H^1}^2 + \|u(t)\|_{H^1}^2 + \|\theta(t) - \overline{\theta}\|_{H^1}^2 + \|\theta_y(t)\|_{L^p}^p)$$
$$+ \int_0^t e^{\gamma s} \left(\|v - \overline{v}\|_{H^1}^2 + \|\theta_y(t)\|_{L^p}^p + \|u\|_{H^2}^2 + \|\theta - \overline{\theta}\|_{H^2}^2 + \|u_t\|^2 + \|\theta_t\|^2 \right)(s)ds$$
$$+ \int_0^t e^{\gamma s}\||\theta_y|^{\frac{p-2}{2}} + \theta_y)\theta_{yy}(s)\|^2 ds \leq C_1. \tag{7.1.13}$$

Theorem 7.1.2. *Assume that the initial data $(v_0, u_0, \theta_0) \in H_+^2$ and compatibility conditions are satisfied. Then the problem (7.1.6)–(7.1.10) admits a unique global solution $(v(t), u(t), \theta(t)) \in H_+^2$ verifying*

$$\|v(t) - \overline{v}\|_{H^2}^2 + \|u(t)\|_{H^2}^2 + \|\theta(t) - \overline{\theta}\|_{H^2}^2 + \|u_t(t)\|^2 + \|\theta_t(t) - \overline{\theta}\|^2 \tag{7.1.14}$$
$$+ \int_0^t \left(\|v - \overline{v}\|_{H^2}^2 + \|u\|_{H^3}^2 + \|\theta - \overline{\theta}\|_{H^3}^2 + \|u_{ty}\|^2 + \|\theta_{ty}\|^2 \right)(s)ds \leq C_2, \quad \forall t > 0.$$

Moreover, there are constants $C_2 > 0$ and $\gamma_2 = \gamma_2(C_2) > 0$ such that, for any fixed $\gamma \in (0, \gamma_2]$, the following estimate holds for any $t > 0$:

$$e^{\gamma t}(\|v(t) - \overline{v}\|_{H^2}^2 + \|u(t)\|_{H^2}^2 + \|\theta(t) - \overline{\theta}\|_{H^2}^2 + \|u_t(t)\|^2 + \|\theta_t(t) - \overline{\theta}\|^2) \tag{7.1.15}$$
$$+ \int_0^t e^{\gamma s} \left(\|v - \overline{v}\|_{H^2}^2 + \|u\|_{H^3}^2 + \|\theta - \overline{\theta}\|_{H^3}^2 + \|u_{ty}\|^2 + \|\theta_{ty}\|^2 \right)(s)ds \leq C_2.$$

Theorem 7.1.3. *Assume that the initial data* $(v_0, u_0, \theta_0) \in H_+^4$ *and compatibility conditions are satisfied. Then the problem (7.1.6)–(7.1.10) admits a unique global solution* $(v(t), u(t), \theta(t)) \in H_+^4$ *verifying*

$$\|v(t) - \overline{v}\|_{H^4}^2 + \|u(t)\|_{H^4}^2 + \|\theta(t) - \overline{\theta}\|_{H^4}^2 + \|v_t(t)\|_{H^3}^2 + \|v_{tt}(t)\|_{H^1}^2 + \|u_t(t)\|_{H^2}^2$$

$$+ \|\theta_t(t)\|_{H^2}^2 + \|u_{tt}(t)\|^2 + \|\theta_{tt}(t)\|^2 + \int_0^t \Big(\|v - \overline{v}\|_{H^4}^2 + \|u\|_{H^5}^2 + \|\theta - \overline{\theta}\|_{H^5}^2$$

$$+ \|u_t\|_{H^3}^2 + \|\theta_t\|_{H^3}^2 + \|v_t\|_{H^4}^2 + \|v_{tt}\|_{H^2}^2 + \|v_{ttt}\|_{H^4}^2 \Big)(s)ds \leq C_4, \quad \forall t > 0.$$

$$(7.1.16)$$

Moreover, there are constants $C_4 > 0$ *and* $\gamma_4 = \gamma_1(C_4) > 0$ *such that, for any fixed* $\gamma \in (0, \gamma_4]$, *the following estimate holds for any* $t > 0$:

$$e^{\gamma t}(\|v(t) - \overline{v}\|_{H^4}^2 + \|u(t)\|_{H^4}^2 + \|\theta(t) - \overline{\theta}\|_{H^4}^2 + \|v_t(t)\|_{H^3}^2 + \|v_{tt}(t)\|_{H^1}^2 + \|u_t(t)\|_{H^2}^2$$

$$+ \|\theta_t(t)\|_{H^2}^2 + \|u_{tt}(t)\|^2 + \|\theta_{tt}(t)\|^2) + \int_0^t e^{\gamma s}\Big(\|v - \overline{v}\|_{H^4}^2 + \|u\|_{H^5}^2 + \|\theta - \overline{\theta}\|_{H^5}^2$$

$$+ \|u_t\|_{H^3}^2 + \|\theta_t\|_{H^3}^2 + \|v_t\|_{H^4}^2 + \|v_{tt}\|_{H^2}^2 + \|v_{ttt}\|_{H^4}^2 \Big)(s)ds \leq C_4. \qquad (7.1.17)$$

Remark 7.1.1. If we replace (7.1.10) by the conditions

$$u(y,t)|_{y=0} = u(y,t)|_{y=1} = 0, \quad \theta(y,t)|_{y=0} = \theta(y,t)|_{y=1} = T_0, \qquad t > 0$$

with $T_0 = \text{const.} > 0$, then estimates (7.1.11)–(7.1.17) also hold with $\overline{\theta} = T_0$.

Corollary 7.1.1. *The* H^4-*global solution* $(v(t), u(t), \theta(t)) \in H_+^4$ *obtained in Theorem 7.1.3 is actually a classical solution* $(v(t), u(t), \theta(t)) \in C^{3,/2}(0,1)$ *and as* $t \to +\infty$,

$$\|(v_x(t), u_x(t), \theta_x(t))\|_{C^{2,\frac{1}{2}}(0,1)} + \|v_t(t)\|_{C^{2,\frac{1}{2}}(0,1)} \qquad (7.1.18)$$

$$+ \|(u_t(t), \theta_t(t))\|_{C^{1,\frac{1}{2}}} + \|v_{tt}(t)\|_{C^{\frac{1}{2}}(0,1)} \to 0.$$

7.2   Proof of Theorem 7.1.1

In this section, we shall give the proof of Theorem 7.1.1. We assume throughout that its hypotheses hold.

Lemma 7.2.1. *The following estimates hold:*

$$\theta(y,t) > 0, \quad \forall (y,t) \in [0,1] \times [0,+\infty), \qquad (7.2.1)$$

$$\int_0^1 v(y,t)dy = \int_0^1 v_0(y)dy = \overline{v}_0, \quad \forall t > 0, \qquad (7.2.2)$$

$$\int_0^1 \left(e + \frac{u^2}{2} \right)(y,t)dy = \int_0^1 \left(e_0 + \frac{u_0^2}{2} \right)(y)dy = E_0, \quad \forall t > 0, \tag{7.2.3}$$

$$\int_0^1 \left[(\theta - \log\theta - 1) + (v - \log v - 1) \right](y,t)dy$$

$$+ \int_0^t \int_0^1 \left[\frac{u_y^2}{v\theta} + \frac{|\theta_y|^p}{v^{p-1}\theta^2} + \frac{\theta_y^2}{v\theta^2} \right](y,t)dyds \le C_1, \quad \forall t > 0. \tag{7.2.4}$$

Proof. Equation (7.1.8) can be rewritten as

$$\theta_t - \frac{R\theta u_y}{v} = \frac{u_y^2}{v} + \left[\left(\left(\frac{\theta_y}{v} \right)^2 + \mu_0 \right)^{\frac{p-2}{2}} \frac{\theta_y}{v} \right]_y. \tag{7.2.5}$$

Applying the compatibility conditions, the positivity of θ_0 and a generalized maximum principle to (7.2.5), we can get (7.2.1). Estimate (7.2.2) is a direct consequence of (7.1.6) and (7.1.10).

Integrating (7.1.8) over $Q_t := [0,1] \times [0,t]$ and noting (7.1.10), we have (7.2.3), the conservation law of total energy.

Multiplying (7.2.5) by θ^{-1}, and integrating the resulting equation over Q_t, we can get

$$\int_0^1 (\log\theta + \log v)(y,t)dy + \int_0^t \int_0^1 \left[\frac{u_y^2}{v\theta} + \frac{\theta_y^2((\frac{\theta_y}{v})^2 + \mu_0)^{\frac{p-2}{2}}}{v\theta^2} \right](y,s)dyds \le C_1,$$

which, along with (7.2.2) and (7.2.3), gives (7.2.4). $\square$

Lemma 7.2.2. *For any $t \ge 0$, there exist a point $y_1 = y_1(t) \in [0,1]$ such that the specific volume $v(y,t)$ in problem (7.1.6)–(7.1.10) can be expressed as*

$$v(y,t) = D(y,t)Z(t) \left[1 + \int_0^t D^{-1}(y,s)Z^{-1}(s)v(y,s)P(y,s)ds \right], \tag{7.2.6}$$

where

$$D(y,t) = v_0(y)\exp\left[\int_{y_1(t)}^y u(x,t)dx - \int_0^y u_0(x)dx + \bar{v}_0^{-1}\int_0^1 v_0(y)\int_0^y u_0 dxdy) \right],$$

$$Z(t) = \exp\left[-\frac{1}{\bar{v}_0}\int_0^t \int_0^1 (u^2 + vP)dyds \right].$$

Proof. The proof is identical to that of Lemma 2.1.3 in Qin [101], pp. 51–52. $\square$

Lemma 7.2.3. *It holds that*

$$0 < C_1^{-1} \le v(y,t) \le C_1, \quad \forall (y,t) \in [0,1] \times [0,+\infty). \tag{7.2.7}$$

Proof. The proof is identical to that of Lemma 2.1.5 in Qin [101], pp. 53–54.   □

Corollary 7.2.1. *It holds that for any $(y, t) \in [0, 1] \times [0, +\infty)$,*

$$C_1^{-1} - C_1 V(t) \leq \theta^{2m}(y, t) \leq C_1 + C_1 V(t) \tag{7.2.8}$$

with $0 \leq 2m \leq 1$ and $V(t) = \int_0^1 \frac{\theta_y^2}{v\theta^2} dy$ satisfying $\int_0^{+\infty} V(s)ds < +\infty$.

Lemma 7.2.4. *The following estimates hold for any $t > 0$:*

$$\int_0^t \|u(s)\|_{L^\infty}^2 ds \leq C_1, \tag{7.2.9}$$

$$\int_0^t \int_0^1 (1 + \theta)^{2m} u^2(y, s)dyds \leq C_1, \tag{7.2.10}$$

$$\|v_y(t)\|^2 + \int_0^t \int_0^1 \theta v_y^2(y, s)dyds \leq C_1 + C_1 A \tag{7.2.11}$$

with $A(t) = \sup_{0 \leq s \leq t} \|\theta(s)\|_{L^\infty}$.

Proof. The proof is the same as that of Lemma 2.1.6 in Qin [101], pp. 56–57.   □

Corollary 7.2.2. *It holds that, for any $t > 0$,*

$$\int_0^t \int_0^1 (1 + \theta)^{2m} v_y^2(y, s)dyds \leq C_1 + C_1 A. \tag{7.2.12}$$

with $0 \leq 2m \leq 1$.

Lemma 7.2.5. *The following estimates hold for any $t > 0$:*

$$\int_0^t \|u_y(s)\|^2 ds \leq C_1 + C_1 A^{\frac{1}{2}}, \tag{7.2.13}$$

$$\|u_y(t)\|^2 + \int_0^t \|u_t(s)\|^2 ds \leq C_1 + C_1 A^2, \tag{7.2.14}$$

$$\int_0^t \|u_{yy}(s)\|^2 ds \leq C_1 + C_1 A^{\frac{5}{2}}. \tag{7.2.15}$$

Proof. Multiplying (7.1.7) by u, u_t, and u_{yy}, respectively, and then integrating the results over Q_t, using Lemmas 7.2.3–7.2.4 and Corollary 7.2.2, we get, for any $\varepsilon > 0$, that

$$\|u(t)\|^2 + \int_0^t \int_0^1 \frac{u_y^2}{v} dyds$$

$$\leq C_1 + C_1 \left| \int_0^t \int_0^1 (v_y \theta + \theta_y) u\, dyds \right|$$

$$\le C_1 + C_1 \left(\int_0^t \int_0^1 v_y^2 \theta dy ds \right)^{\frac{1}{2}} \left(\int_0^t \int_0^1 u^2 \theta dy ds \right)^{\frac{1}{2}}$$

$$+ C_1 \left(\int_0^t \int_0^1 \frac{\theta_y^2}{\theta^2} dy ds \right)^{\frac{1}{2}} \left(\int_0^t \int_0^1 u^2 \theta^2 dy ds \right)^{\frac{1}{2}}$$

$$\le C_1 + C_1 A^{\frac{1}{2}}, \qquad\qquad (7.2.16)$$

$$\|u_y(t)\|^2 + \int_0^t \|u_t(s)\|^2 ds$$

$$\le C_1 + C_1 \left| \int_0^t \int_0^1 (v_y \theta + \theta_y) u_t dy ds \right|$$

$$\le C_1 + C_1 \varepsilon \int_0^t \|u_t(s)\|^2 ds + C_1 A \int_0^t \int_0^1 \theta v_y^2 dy ds$$

$$+ C_1 A^2 \int_0^t \int_0^1 \frac{\theta_y^2}{\theta^2} dy ds$$

$$\le C_1 + C_1 A^2 + C \varepsilon \int_0^t \|u_t(s)\|^2 ds, \qquad\qquad (7.2.17)$$

$$\|u_y(t)\|^2 + \int_0^t \|u_{yy}(s)\|^2 ds$$

$$\le C_1 + C_1 \varepsilon \int_0^t \|u_{yy}(s)\|^2 ds + C_1 \int_0^t \int_0^1 (v_y^2 u_y^2 + v_y^2 \theta^2 + \theta_y^2) dy ds$$

$$\le C_1 + C_1 \varepsilon \int_0^t \|u_{yy}(s)\|^2 ds + C_1 A^2 + C_1 \int_0^t \|u_y\|_{L^\infty}^2 \int_0^1 v_y^2 dy ds$$

$$\le C_1 + C_1 A^2 + C_1 \varepsilon \int_0^t \|u_{yy}(s)\|^2 ds + C_1 A \int_0^t \|u_y(s)\|_{L^\infty}^2 ds$$

$$\le C_1 + C_1 A^2 + C_1 \varepsilon \int_0^t \|u_{yy}(s)\|^2 ds$$

$$+ C_1 A \left(\int_0^t \|u_y(s)\|^2 ds \right)^{\frac{1}{2}} \left(\int_0^t \|u_{yy}(s)\|^2 ds \right)^{\frac{1}{2}}$$

$$\le C_1 + C_1 A^2 + C_1 \varepsilon \int_0^t \|u_{yy}(s)\|^2 ds + C_1 A^2 \int_0^t \|u_y(s)\|^2 ds$$

$$\le C_1 + C_1 A^{\frac{5}{2}} + C_1 \varepsilon \int_0^t \|u_{yy}(s)\|^2 ds. \qquad\qquad (7.2.18)$$

Now (7.2.13) follows from (7.2.16), and (7.2.14)–(7.2.15) from (7.2.17)–(7.2.18) for $\varepsilon > 0$ small enough. $\qquad\qquad\qquad\qquad\qquad\qquad\qquad\qquad\qquad\qquad\qquad$ □

Corollary 7.2.3. *It holds that, for any $t > 0$,*

$$\int_0^t \int_0^1 (1 + \theta)^{2m} u_y^2(y, s) dy ds \leq C_1 + C_1 A^2 \tag{7.2.19}$$

with $0 \leq 2m \leq 1$.

Lemma 7.2.6. *It holds that, for any $t > 0$,*

$$\int_0^t \int_0^1 (|\theta_y|^p + \theta_y^2)(y, s) dy ds \leq C_1 + C_1 A^{\frac{3}{2}}. \tag{7.2.20}$$

Proof. Multiplying (7.2.5) by θ, integrating the resulting equation over Q_t, and using (7.1.10) and Lemmas 7.2.1–7.2.5, we derive that

$$\|\theta(t)\|^2 + \int_0^t \int_0^1 (\theta_y^2 + |\theta_y|^p)(y, s) dy ds$$

$$\leq C_1 + C_1 \left| \int_0^t \int_0^1 (u_y^2 \theta + v_y \theta^2 u + \theta \theta_y u)(y, s) dy ds \right|$$

$$\leq C_1 + C_1 \int_0^t \|u_y\|_{L^\infty}^2 \int_0^1 \theta dy ds + C_1 \left(\int_0^t \int_0^1 \theta v_y^2 dy ds \right)^{\frac{1}{2}} \left(\int_0^t \int_0^1 \theta^3 u^2 dy ds \right)^{\frac{1}{2}}$$

$$+ C_1 \left(\int_0^t \int_0^1 \frac{\theta_y^2}{\theta^2} dy ds \right)^{\frac{1}{2}} \left(\int_0^t \int_0^1 u^2 \theta^4 dy ds \right)^{\frac{1}{2}}$$

$$\leq C_1 + C_1 A^{\frac{3}{2}} + C_1 \left(\int_0^t \|u_y\|^2 ds \right)^{\frac{1}{2}} \left(\int_0^t \|u_{yy}\|^2 ds \right)^{\frac{1}{2}}$$

$$+ C_1 A^{\frac{3}{2}} \left(\int_0^t \|u\|_{L^\infty}^2 \int_0^1 \theta dy ds \right)^{\frac{1}{2}}$$

$$\leq C_1 + C_1 A^{\frac{3}{2}}. \qquad \square$$

Lemma 7.2.7. *It holds that, for any $t > 0$,*

$$\|\theta_y(t)\|^2 + \|\theta_y(t)\|_{L^p}^p + \int_0^t \|\theta_t(s)\|^2 ds \leq C_1 + C_1 A^{\frac{5}{2}}. \tag{7.2.21}$$

Proof. Multiplying (7.2.5) by θ_t and integrating the resulting equation over Q_t, we derive that

$$\|\theta_y(t)\|^2 + \|\theta_y(t)\|_{L^p}^p + \int_0^t \|\theta_t(s)\|^2 ds$$

$$\leq C_1 + C_1 \int_0^t \int_0^1 (|u_y^2 \theta_t| + |\theta u_y \theta_t|)(y, s) dy ds. \tag{7.2.22}$$

Using Lemmas 7.2.5–7.2.6, we get for any $\varepsilon > 0$,

$$\int_0^t \int_0^1 u_y^2 \theta_t(y, s) dy ds \leq \int_0^t \|u_y(s)\|_{L^\infty} \|u_y(s)\| \|\theta_t(s)\| ds$$

$$\leq C_1 \sup_{0 \leq s \leq t} \|u_y(s)\|_{L^\infty} \left(\int_0^t \|u_y(s)\|^2 ds\right)^{\frac{1}{2}} \left(\int_0^t \|\theta_t(s)\|^2 ds\right)^{\frac{1}{2}}$$

$$\leq C_1 A^{\frac{5}{4}} \left(\int_0^t \|\theta_t(s)\|^2 ds\right)^{\frac{1}{2}} \leq C_1 A^{\frac{5}{2}} + C\varepsilon \int_0^t \|\theta_t(s)\|^2 ds, \qquad (7.2.23)$$

$$\int_0^t \int_0^1 \theta u_y \theta_t(y, s) dy ds \leq A \int_0^t \|u_y(s)\| \|\theta_t(s)\| ds$$

$$\leq A \left(\int_0^t \|u_y(s)\|^2 ds\right)^{\frac{1}{2}} \left(\int_0^t \|\theta_t(s)\|^2 ds\right)^{\frac{1}{2}}$$

$$\leq C_1 A^{\frac{5}{2}} + C_1 \varepsilon \int_0^t \|\theta_t(s)\|^2 ds. \qquad (7.2.24)$$

Taking $\varepsilon > 0$ small enough and inserting (7.2.23)–(7.2.24) into (7.2.22), we obtain (7.2.21). $\qquad \square$

Lemma 7.2.8. *It holds that, for any $t > 0$,*

$$\|\theta(t)\|_{L^\infty} \leq C_1, \qquad (7.2.25)$$

$$\int_0^1 \left(\theta_y^2 + v_y^2 + u_y^2\right) dy + \int_0^t \int_0^1 \left(v_y^2 + u_y^2 + u_t^2 + u_{yy}^2 + \theta_t^2 + \theta_y^2\right) dy ds \leq C_1. \qquad (7.2.26)$$

Proof. By the Nirenberg interpolation inequality, we have

$$\|\theta(t)\|_{L^\infty} \leq C_1 \|\theta_y(t)\|^{\frac{2}{3}} \|\theta(t)\|_{L^1}^{\frac{1}{3}} + C\|\theta(t)\|_{L^1} \leq C_1 + C_1 \|\theta_y(t)\|^{\frac{2}{3}}. \qquad (7.2.27)$$

From (7.2.21), (7.2.27) and Young's inequality, we derive that, for any $\varepsilon > 0$,

$$\|\theta_y(t)\|^2 + \|\theta_y(t)\|_{L^p}^p + \int_0^t \|\theta_t(s)\|^2 ds \leq C_1 + C_1 \varepsilon \|\theta_y(t)\|^2,$$

i.e., for $\varepsilon > 0$ small enough,

$$\|\theta_y(t)\|^2 + \|\theta_y(t)\|_{L^p}^p + \int_0^t \|\theta_t(s)\|^2 ds \leq C_1. \qquad (7.2.28)$$

Estimate (7.2.25) follows from (7.2.27)–(7.2.28), and estimate (7.2.26) from (7.2.25) and Lemmas 7.2.1–7.2.7. $\qquad \square$

Lemma 7.2.9. *It holds that, for any $t > 0$,*

$$\int_0^t (\|P^*\|^2 + \|\sigma^*\|^2)(s)ds \leq C_1, \qquad \forall t > 0, \qquad (7.2.29)$$

$$\frac{d}{dt}\|P^*(t)\|^2 \leq C_1(\|u_t(t)\|^2 + \|\theta_t(t)\|^2 + 1), \forall t > 0, \qquad (7.2.30)$$

$$\frac{d}{dt}\|\sigma^*(t)\|^2 \leq C_1(\|u_t(t)\|^2 + \|\theta_t(t)\|^2 + 1), \forall t > 0. \qquad (7.2.31)$$

Proof. The proof is the same as that of Lemma 2.1.12 in Qin [101], pp. 66–68. □

Lemma 7.2.10. *It holds that for any $t > 0$,*

$$\frac{d}{dt}\|v_y(t)\|^2 \leq C_1 \left(\|u_{yy}(t)\|^2 + \|v_y(t)\|^2\right), \ \forall t > 0, \qquad (7.2.32)$$

$$\frac{d}{dt}\|\theta_y(t)\|^2 + \int_0^1 \left(|\theta_y|^{p-2} + \theta_y^2 + 1\right)\theta_{yy}^2 dy \leq C_1 \left(\|u_{yy}(t)\|^2 + 1\right), \ \forall t > 0, \quad (7.2.33)$$

$$\|\theta_y(t)\|^2 + \int_0^t \int_0^1 \left(|\theta_y|^{p-2} + \theta_y^2 + 1\right)\theta_{yy}^2 dyds \leq C_1, \ \forall t > 0. \qquad (7.2.34)$$

Proof. The proof is the same as that of Lemma 2.1.13 in Qin [101], pp. 68–69. □

Lemma 7.2.11. *As $t \to +\infty$, we have*

$$\|v(t) - \overline{v}_0\|_{H^1}^2 \to 0, \quad \|v_y(t)\|^2 \to 0, \qquad (7.2.35)$$

$$\|u(t)\|_{H^1}^2 \to 0, \qquad (7.2.36)$$

$$\|P^*(t)\|^2 \to 0, \quad \|\sigma^*(t)\|^2 \to 0, \quad \|\theta_y(t)\|^2 \to 0, \qquad (7.2.37)$$

$$\|\theta(t) - \overline{\theta}\|_{H^1}^2 \to 0, \quad \|\theta(t) - \overline{\theta}\|_{L^\infty}^2 \to 0. \qquad (7.2.38)$$

Proof. The proof is identical to that of Lemma 2.1.14 in Qin [101], pp. 69–70. □

Lemma 7.2.12. *The following estimate holds:*

$$\theta(y,t) \geq C_1^{-1} > 0, \quad \forall (y,t) \in [0,1] \times [0,+\infty). \qquad (7.2.39)$$

Proof. We prove (7.2.39) by contradiction. If (7.2.39) is not true, that is,

$$\inf_{(y,t)\in[0,1]\times[0,+\infty)} \theta(y,t) = 0,$$

then there exists a sequence $(y_n, t_n) \in [0,1] \times [0,+\infty)$ such that, as $n \to +\infty$,

$$\theta(y_n, t_n) \to 0. \qquad (7.2.40)$$

If the sequence $\{t_n\}$ has a subsequence, denoted also by t_n, converging to $+\infty$, then by the asymptotic behavior results in Lemma 7.2.12, we know that as $n \to +\infty$,

$$\theta(y_n, t_n) \to \overline{\theta} > 0$$

which contradicts (7.2.40). If the sequence $\{t_n\}$ is bounded, i.e., there exists a constant $M > 0$, independent of n, such that for any $n = 1, 2, \ldots,\ 0 < t_n \leq M$. Thus there exists a point $(y^*, t^*) \in [0, 1] \times [0, M]$ such that $(y_n, t_n) \to (y^*, t^*)$ as $n \to +\infty$. On the other hand, by (7.2.40) and the continuity of solutions in Lemmas 7.2.1–7.2.12, we conclude that $\theta(y_n, t_n) \to \theta(y^*, t^*) = 0$ as $n \to +\infty$, which contradicts (7.2.1). Thus the proof is complete. $\qquad\square$

In what follows, we shall establish the exponential stability of the solution in H^1. Let us introduce the flow density $\rho = \frac{1}{v}$. Then we easily get that the specific entropy

$$\eta = \eta(v, \theta) = \eta(\rho, \theta) = R \log v + \log \theta, \qquad (7.2.41)$$

satisfies

$$\frac{\partial \eta}{\partial \rho} = -\frac{R}{\rho^2}, \quad \frac{\partial \eta}{\partial \theta} = \frac{1}{\theta}. \qquad (7.2.42)$$

We consider the transform

$$\mathcal{A} :\in \mathcal{D}_{\rho,\theta} = \{(\rho, \theta) : \rho > 0, \theta > 0\} \ni (\rho, \theta)) \mapsto (v, \eta) \in \mathcal{A}\mathcal{D}_{\rho,\theta},$$

where $v = 1/\rho$ and $\eta = \eta(1/\rho, \theta)$. Since the Jacobian

$$\frac{\partial(v, \eta)}{\partial(\rho, \theta)} = -\frac{1}{\rho} < 0 \quad \text{on} \quad \mathcal{D}_{\rho,\theta},$$

there exists a unique inverse function $\theta = \theta(v, \eta)$, which is a smooth function of $(v, \eta) \in \mathcal{A}\mathcal{D}_{\rho,\theta}$. Thus the functions e, p can be regarded as smooth functions of (v, η). Denote

$$e = e(v, \eta) := e(v, \theta(v, \eta)) = e(\rho^{-1}, \theta), \quad p = p(v, \eta) := p(v, \theta(v, \eta)) = p(\rho^{-1}, \theta).$$

Thus we derive that e, p satisfy

$$e_v = -\frac{R\theta}{v}, \quad e_\eta = 0, \quad P_v = -\frac{R\theta + R^2\theta}{v^2}, \quad P_\eta = \frac{R\theta}{v}. \qquad (7.2.43)$$

Let

$$\mathcal{E}(v, u, \eta) = \frac{u^2}{2} + e(v, \eta) - e(\overline{v}, \overline{\eta}) - \frac{\partial e}{\partial v}(\overline{v}, \overline{\eta})(v - \overline{v}) - \frac{\partial e}{\partial \eta}(\overline{v}, \overline{\eta})(\eta - \overline{\eta}), \qquad (7.2.44)$$

where $\overline{v} = \int_0^1 v_0 dx$ and $\overline{\eta} = \eta(\overline{v}, \overline{\theta})$.

Lemma 7.2.13. *The unique generalized global solution* $(v(t), u(t), \theta(t))$ *to problem* (7.1.6)–(7.1.10) *satisfies the estimate*

$$\frac{u^2}{2} + C_1^{-1}(|v - \overline{v}|^2 + |\eta - \overline{\eta}|^2) \leq \mathcal{E}(v, u, \eta) \leq \frac{u^2}{2} + C_1(|v - \overline{v}|^2 + |\eta - \overline{\eta}|^2). \qquad (7.2.45)$$

Proof. The proof is the same as that of Lemma 2.3.4 in Qin [101], pp. 86–87. $\square$

Lemma 7.2.14. *Under the assumptions of Theorem 7.1.1, there are positive constants $C_1 > 0$ and $\gamma_1' = \gamma_1'(C_1) > 0$ such that for any fixed $\gamma \in (0, \gamma_1']$ and for any $t > 0$, there holds*

$$e^{\gamma t}\left(\|v(t) - \overline{v}\|^2 + \|u(t)\|^2 + \|\theta(t) - \overline{\theta}\|^2 + \|v_y\|^2 + \|\rho_y\|^2\right)$$

$$+ \int_0^t e^{\gamma s}(\|\rho_y\|^2 + \|u_y\|^2 + \|\theta_y\|^2 + \|v_y\|^2)(s)ds \leq C_1. \tag{7.2.46}$$

Proof. By equations (7.1.6)–(7.1.8), it is easy to verify the following

$$\left(e + \frac{u^2}{2}\right)_t = \left[-Pu + \rho u u_y + \rho\theta_y(\rho^2\theta_y^2 + \mu_0)^{\frac{p-2}{2}}\right]_y, \tag{7.2.47}$$

$$\eta_t = \left(\frac{(\rho^2\theta_y^2 + \mu_0)^{\frac{p-2}{2}}\rho\theta_y}{\theta}\right)_y + (\rho^2\theta_y^2 + \mu_0)^{\frac{p-2}{2}}\rho\left(\frac{\theta_y}{\theta}\right)^2 + \frac{\rho u_y^2}{\theta}. \tag{7.2.48}$$

Since $\overline{v}_t = 0$, $\overline{\theta}_t = 0$, we infer from (7.2.47) and (7.2.48) that

$$\mathcal{E}_t(1/\rho, u, \eta) + \frac{\overline{\theta}}{\theta}\left[\rho u_y^2 + (\rho^2\theta_y^2 + \mu_0)^{\frac{p-2}{2}}\rho\theta_y^2/\theta\right]$$

$$= \left[\rho u u_y + (1 - \frac{\overline{\theta}}{\theta})(\rho^2\theta_y^2 + \mu_0)^{\frac{p-2}{2}}\rho\theta_y - (p - \overline{p})u\right]_y, \tag{7.2.49}$$

$$\left[(\rho_y/\rho)^2/2 + \rho_y u/\rho\right]_t + R\theta\rho_y^2/\rho = -R\rho_y\theta_y - (\rho u u_y)_y + \rho u_y^2. \tag{7.2.50}$$

Multiplying (7.2.49), (7.2.50) by $e^{\gamma t}$, $\beta e^{\gamma t}$, respectively, and adding the results, we get

$$\frac{\partial}{\partial t}G(t) + e^{\gamma t}\left[\frac{\overline{\theta}}{\theta}\left(\rho u_y^2 + (\rho^2\theta_y^2 + \mu_0)^{\frac{p-2}{2}}\rho\theta_y^2/\theta\right) + \beta\left(R\theta\rho_y^2/\rho + R\rho_y\theta_y - \rho u_y^2\right)\right]$$

$$= \gamma e^{\gamma t}\left[\mathcal{E}(1/\rho, u, \eta) + \beta\left((\rho_y/\rho)^2/2 + \rho_y u/\rho\right)\right]$$

$$+ e^{\gamma t}\left[(1 - \beta)\rho u u_y - (p - \overline{p})u + (\rho^2\theta_y^2 + \mu_0)^{\frac{p-2}{2}}\left(1 - \frac{\overline{\theta}}{\theta}\right)\rho\theta_y\right]_y, \tag{7.2.51}$$

where $G(t) = e^{\gamma t}\left[\mathcal{E}(1/\rho, u, \eta) + \beta((\rho_y/\rho)^2/2 + \rho_y u/\rho)\right]$.

Next, integrating (7.2.51) over Q_t and using Lemmas 7.2.1–7.2.12, Cauchy's inequality and Poincaré's inequality, we deduce that for small $\beta > 0$ and for any $\gamma > 0$,

$$e^{\gamma t}\left[\|\rho(t) - \overline{\rho}\|^2 + \|\eta(t) - \overline{\eta}\|^2 + \|u(t)\|^2 + \|\rho_y(t)\|^2\right]$$

$$+ \int_0^t e^{\gamma s} \Big[\|\rho_y\|^2 + \|u_y\|^2 + \|\theta_y\|^2 \Big](s) ds$$

$$\leq C_1 + C_1 \gamma \int_0^t e^{\gamma s} \Big[\|\rho - \overline{\rho}\|^2 + \|\theta - \overline{\theta}\|^2 + \|u\|^2 + \|\rho_y\|^2 \Big](s) ds. \qquad (7.2.52)$$

By Lemmas 7.2.1–7.2.12, the mean value theorem and the Poincaré inequality, we have

$$\|v(t) - \overline{v}\| \leq C_1 \|v_y(t)\|,$$
$$\|\theta(t) - \overline{\theta}\| \leq C_1(\|e(v, \theta) - e(\overline{v}, \overline{\theta})\| + \|v(t) - \overline{v}\|),$$
$$\leq C_1(\|e(v, \theta) - e(\overline{v}, \overline{\theta})\| + \|v_y(t)\|),$$
$$\leq C_1(\|\theta_y(t)\| + \|u_y(t)\| + \|v_y(t)\|). \qquad (7.2.53)$$

Similarly, we infer that

$$C_1^{-1} \|v(t) - \overline{v}\| \leq \|\rho(t) - \overline{\rho}\| \leq C_1 \|v(t) - \overline{v}\|, \qquad (7.2.54)$$
$$\|\theta(t) - \overline{\theta}\| \leq C_1(\|\eta(t) - \overline{\eta}\| + \|v(t) - \overline{v}\|). \qquad (7.2.55)$$

It follows from (7.2.53)–(7.2.55) that there exists a constant $\gamma_1' = \gamma_1'(C_1) > 0$ such that, for any fixed $\gamma \in (0, \gamma_1']$, (7.2.46) holds. The proof is complete. □

Lemma 7.2.15. *There exists a positive constant* $\gamma_1 = \gamma_1(C_1) \leq \gamma_1'$ *such that for any* $t > 0$ *and any fixed* $\gamma \in (0, \gamma_1']$, *the following estimate holds*

$$e^{\gamma t} \Big(\|u_y(t)\|^2 + \|\theta_y(t)\|^2 + \|\theta_y(t)\|_{L^p}^p \Big)$$
$$+ \int_0^t e^{\gamma s} \Big(\|u_{yy}\|^2 + \|\theta_{yy}\|^2 + \|u_t\|^2 + \|\theta_t\|^2 \Big)(s) ds$$
$$+ \int_0^t \int_0^1 e^{\gamma s} \Big(|\theta_y|^{p-2} + \theta_y^2 \Big) \theta_{yy}^2 dy ds \leq C_1. \qquad (7.2.56)$$

Proof. By (7.1.6)–(7.1.8), Lemmas 7.2.1–7.2.12 and the Poincaré inequality, we have

$$\|u_t\| \leq C_1 (\|v_y\| + \|\theta_y\| + \|u_{yy}\|), \qquad \|u_y\| \leq C_1 \|u_{yy}\|, \qquad (7.2.57)$$
$$\|\theta_t\| \leq C_1 \Big(\|(|\theta_y|^{\frac{p-2}{2}} + \theta_y + 1)\theta_{yy}\| + \|u_{yy}\| \Big), \qquad \|\theta_y\| \leq C_1 \|\theta_{yy}\|. \qquad (7.2.58)$$

Multiplying (7.1.7), (7.1.8) by $-e^{\gamma t} u_{yy}$, $-e^{\gamma t} \theta_{yy}$, respectively, integrating the results over Q_t, and adding them, and then using Young's inequality, the embedding theorem, Lemmas 7.2.1–7.2.12 and Lemma 7.2.14, we obtain

$$e^{\gamma t} \Big(\|u_y(t)\|^2 + \|\theta_y(t)\|^2 \Big) + \int_0^t \int_0^1 e^{\gamma s} \Big((|\theta_y|^{p-2} + \theta_y^2 + 1)\theta_{yy}^2 + u_{yy}^2 \Big)(s) ds$$

$$\leq C_1 + C_1 \int_0^t e^{\gamma s} \Big\{ \big(\|\theta_y\| + \|v_y\| + \|v_y\| \|u_y\|^{1/2} \|u_y\|^{1/2} \big) \|u_{yy}\| + \big(\|\theta_y\| + \|u_y\|$$

$$+ \|u_y\|^{3/2} \|u_{yy}\|^{1/2} \big) + \big(\|\theta_y\|_{L^\infty}^{p-1} + \|\theta_y\|_{L^\infty}^3 + \|\theta_y\|_{L^\infty} \big) \|\theta_{yy}\|(s) \Big\} ds$$

$$+ C_1 \int_0^t e^{\gamma s} \Big\{ \|\theta_y\|^2 + \|v_y\|^2 + \|u_y\|^2 + \big(\|\theta_t\| + \|u_y\| \big) \|\theta_y\|^{1/2} \|\theta_{yy}\|^{1/2} \Big\}(s) ds$$

$$\leq C_1 + 1/(2C) \int_0^t \int_0^1 e^{\gamma s} \Big(\big(|\theta_y|^{p-2} + \theta_y^2 + 1 \big) \theta_{yy}^2 + u_{yy}^2 \Big)(s) ds, \qquad (7.2.59)$$

which, together with Lemmas 7.2.1–7.2.12 and equations (7.1.6)–(7.1.8), gives (7.2.56).   □

This completes the proof of Theorem 7.1.1.   □

7.3   Proof of Theorem 7.1.2

In this section, we shall study the global existence and exponential stability of solutions to problem (7.1.6)–(7.1.10) in H_+^2. We begin with the following lemmas.

Lemma 7.3.1. *Under the assumptions of Theorem 7.1.2, the following estimate holds:*

$$\|u_t(t)\|^2 + \|\theta_t(t)\|^2 + \int_0^t \Big(\|u_{ty}\|^2 + \||\theta_y|^{\frac{p-2}{2}} \theta_{ty}\|^2 + \|\theta_y \theta_{ty}\|^2 + \|\theta_{ty}\|^2 \Big)(s) ds \leq C_2. \tag{7.3.1}$$

Proof. Differentiating (7.1.7) with respect to t, multiplying the resulting equation by u_t, and integrating over $(0,1)$, we infer that

$$\frac{d}{dt} \|u_t(t)\|^2 + \|u_{ty}(t)\|^2 \leq \frac{1}{2} \|u_{ty}(t)\|^2 + C_1 \big(\|u_y(t)\|^2 + \|\theta_t(t)\|^2 + \|u_y(t)\|_{L^4}^4 \big)$$

$$\leq \frac{1}{2} \|u_{ty}(t)\|^2 + C_2 \big(\|u_{yy}(t)\|^2 + \|\theta_t(t)\|^2 \big),$$

which, together with Theorem 7.1.1, gives

$$\|u_t(t)\|^2 + \int_0^t \|u_{ty}(s)\|^2 ds \leq C_2. \tag{7.3.2}$$

Analogously, we have for any $\varepsilon > 0$,

$$\frac{d}{dt} \|\theta_t(t)\|^2 + C_1^{-1} \Big(\||\theta_y|^{\frac{p-2}{2}} \theta_{ty}(t)\|^2 + \|\theta_y \theta_{ty}(t)\|^2 + \|\theta_{ty}(t)\|^2 \Big)$$

$$\leq \varepsilon \Big(|\theta_y|^{\frac{p-2}{2}} \theta_{ty}(t)\|^2 + \|\theta_y \theta_{ty}(t)\|^2 \Big)$$

$$+ C_1 \Big(\|\theta_t(t)\|^2 + \|u_{ty}(t)\|^2 + \|u_{yy}(t)\|^2 + \|\theta_y(t)\|^2 + \|\theta_y(t)\|_{L^p}^p \Big). \tag{7.3.3}$$

Integrating (7.3.3) over $[0, t]$ and applying Gronwall's inequality, we get for $\varepsilon > 0$ small enough,

$$\|\theta_t(t)\|^2 + \int_0^t \left(\||\theta_y|^{\frac{p-2}{2}} \theta_{ty}\|^2 + \|\theta_y \theta_{ty}\|^2 + \|\theta_{ty}\|^2 \right)(s)ds \leq C_2$$

which, together with (7.3.2), gives (7.3.1). $\qquad\square$

Lemma 7.3.2. *Under the assumptions of Theorem 7.1.2, the following estimate holds:*

$$\|u_{yy}(t)\|^2 + \|\theta_{yy}(t)\|^2 + \int_0^t \left(\|u_{yyy}\|^2 + \|\theta_{yyy}\|^2 \right)(s)ds \leq C_2. \qquad (7.3.4)$$

Proof. The proof is identical to that of Lemma 2.3.8 in Qin [101], pp. 90–91. $\qquad\square$

Lemma 7.3.3. *Under the assumptions of Theorem 7.1.2, the following estimate holds:*

$$\|v_{yy}(t)\|^2 + \int_0^t \|v_{yy}(s)\|^2 ds \leq C_2. \qquad (7.3.5)$$

Proof. The proof is identical to that of Lemma 2.3.9 in Qin [101], pp. 91–92. $\qquad\square$

Lemma 7.3.4. *Under the assumptions of Theorem 7.1.2, for any $(v_0, u_0, \theta_0) \in H_+^2$, there exists a positive constant $\gamma_2' = \gamma_2'(C_2) \leq \gamma_1$ such that, for any fixed $\gamma \in (0, \gamma_2']$, the following estimate holds for any $t > 0$:*

$$e^{\gamma t}\left(\|u_t(t)\|^2 + \|\theta_t(t)\|^2 + \|u_{yy}(t)\|^2 + \|\theta_{yy}(t)\|^2 \right)$$
$$+ \int_0^t e^{\gamma s}\left(\|u_{ty}\|^2 + \|\theta_{ty}\|^2 \right)(s)ds \leq C_2. \qquad (7.3.6)$$

Proof. The proof is identical to that of Lemma 2.3.11 in Qin [101], p. 96. $\qquad\square$

Lemma 7.3.5. *There exists a positive constant $\gamma_2 = \gamma_2(C_2) \leq \gamma_2'$ such that, for any fixed $\gamma \in (0, \gamma_2]$, the following estimate holds:*

$$\|v(t) - \bar{v}\| \leq C_2 e^{-\gamma t}, \quad \forall t > 0. \qquad (7.3.7)$$

Proof. The proof is identical to that of Lemma 2.1.12 in Qin [101], pp. 69–70. $\qquad\square$

This completes the proof of Theorem 7.1.2. $\qquad\square$

7.4 Proof of Theorem 7.1.3

In this section, we shall study the global existence and exponential stability of solutions to problem (7.1.6)–(7.1.10) in H_+^4. We begin with the following lemmas.

Lemma 7.4.1. *Under the assumptions of Theorem 7.1.3, the following estimates hold:*

$$\|u_{ty}(y,0)\| + \|\theta_{ty}(y,0)\| \le C_3, \tag{7.4.1}$$

$$\|u_{tt}(y,0)\| + \|\theta_{tt}(y,0)\| + \|u_{tyy}(y,0)\| + \|\theta_{tyy}(y,0)\| \le C_4, \tag{7.4.2}$$

$$\|u_{tt}(t)\|^2 + \int_0^t \|u_{tty}\|^2(s)ds \le C_3 + C_3 \int_0^t \|\theta_{tyy}(s)\|^2 ds, \tag{7.4.3}$$

$$\|\theta_{tt}(t)\|^2 + \int_0^t \|\theta_{tty}(s)\|^2 ds$$

$$\le C_3 + C_2\varepsilon^{-1}\int_0^t \|\theta_{tyy}\|^2(s)ds + C_1\varepsilon\int_0^t (\|u_{tyy}\|^2 + \|u_{tty}\|^2)(s)ds. \tag{7.4.4}$$

Proof. The proof is the same as that of Lemma 2.4.1 in Qin [101], pp. 100–104. □

Lemma 7.4.2. *Under the assumptions of Theorem 7.1.3, the following estimates hold for any $\varepsilon > 0$:*

$$\|u_{ty}(t)\|^2 + \int_0^t \|u_{tyy}(s)\|^2 ds \le C_3\varepsilon^{-6} + C_1\varepsilon^2 \int_0^t (\|\theta_{tyy}\|^2 + \|u_{tty}\|^2)(s)ds, \tag{7.4.5}$$

$$\|\theta_{ty}(t)\|^2 + \int_0^t \|\theta_{tyy}(s)\|^2 ds \le C_3\varepsilon^{-6} + C_2\varepsilon^2 \int_0^t (\|u_{tyy}\|^2 + \|\theta_{tty}\|^2$$

$$+ \|\theta_{yyy}\|^2 \|\theta_{tyy}\|^2)(s)ds. \tag{7.4.6}$$

Proof. The proof is the same as that of Lemma 2.4.2 in Qin [101], pp. 104–107. □

Lemma 7.4.3. *Under the assumptions of Theorem 7.1.3, the following estimates hold:*

$$\|u_{tt}(t)\|^2 + \|u_{ty}(t)\|^2 + \|\theta_{tt}(t)\|^2 + \|\theta_{ty}(t)\|^2 + \int_0^t \Big(\|u_{tty}\|^2 + \|u_{tyy}\|^2$$

$$+ \|\theta_{tty}\|^2 + \|\theta_{tyy}\|^2 \Big)(s)ds \le C_4, \tag{7.4.7}$$

$$\|v_{yyy}(t)\|_{H^1}^2 + \|v_{yy}(t)\|_{W^{1,\infty}}^2 + \int_0^t (\|v_{yyy}\|_{H^1}^2 + \|v_{yy}\|_{W^{1,\infty}}^2)(s)ds \le C_4, \tag{7.4.8}$$

$$\|u_{yyy}(t)\|_{H^1}^2 + \|u_{yy}(t)\|_{W^{1,\infty}}^2 + \|\theta_{yyy}(t)\|_{H^1}^2 + \|\theta_{yy}(t)\|_{W^{1,\infty}}^2 + \|v_{tyyy}(t)\|^2$$

$$+ \|u_{tyy}(t)\|^2 + \|\theta_{tyy}(t)\|^2 + \int_0^1 \Big(\|u_{tt}\|^2 + \|\theta_{tt}\|^2 + \|u_{yy}\|_{W^{2,\infty}}^2 + \|\theta_{yy}\|_{W^{2,\infty}}^2$$

$$+ \|\theta_{tyy}\|_{H^1}^2 + \|u_{tyy}\|_{H^1}^2 + \|\theta_{ty}\|_{W^{1,\infty}}^2 + \|u_{ty}\|_{W^{1,\infty}}^2 + \|v_{tyyy}\|_{H^1}^2 \Big)(s)ds \le C_4, \tag{7.4.9}$$

$$\int_0^t (\|u_{yyyy}\|_{H^1}^2 + \|\theta_{yyyy}\|_{H^1}^2)(s)ds \le C_4. \tag{7.4.10}$$

Proof. The proof is identical to that of Lemma 2.4.3 in Qin [101], pp. 107–110. □

Lemma 7.4.4. *Under the assumptions of Theorem 7.1.3, for any* $(v_0, u_0, \theta_0) \in H^4_+$, *there exists a positive constant* $\gamma_4^{(1)} = \gamma_4^{(1)}(C_4) \leq \gamma_2(C_2)$ *such that, for any fixed* $\gamma \in (0, \gamma_4^{(1)}]$, *the following estimates hold* $\varepsilon \in (0, 1)$ *small enough:*

$$e^{\gamma t}\|u_{tt}(t)\|^2 + \int_0^t e^{\gamma s}\|u_{tty}(s)\|^2 ds \leq C_3 + C_3 \int_0^t e^{\gamma s}\|\theta_{tyy}(s)\|^2 ds, \qquad (7.4.11)$$

$$e^{\gamma t}\|\theta_{tt}(t)\|^2 + \int_0^t e^{\gamma s}\|\theta_{tty}(s)\|^2 ds \leq C_3 + C_2\varepsilon^{-1} \int_0^t e^{\gamma s}\|\theta_{tyy}(s)\|^2 ds$$

$$+ C_1\varepsilon \int_0^t e^{\gamma s}\Big(\|u_{tyy}\|^2 + \|u_{tty}\|^2\Big)(s)ds, \quad \forall t > 0. \quad (7.4.12)$$

Proof. The proof is identical to that of Lemma 2.4.6 in Qin [101], pp. 119–120. □

Lemma 7.4.5. *Under the assumptions of Theorem 7.1.3, for any* $(v_0, u_0, \theta_0) \in H^4_+$, *there exists a positive constant* $\gamma_4^{(2)} \leq \gamma_4^{(1)}$ *such that, for any fixed* $\gamma \in (0, \gamma_4^{(2)}]$, *the following estimates hold* $\varepsilon \in (0, 1)$ *small enough:*

$$e^{\gamma t}\|u_{ty}(t)\|^2 + \int_0^t e^{\gamma s}\|u_{tyy}(s)\|^2 ds$$

$$\leq C_3\varepsilon^{-6} + C_2\varepsilon^2 \int_0^t e^{\gamma s}\Big(\|\theta_{tyy}\|^2 + \|u_{tty}\|^2\Big)(s)ds, \quad \forall t > 0, \qquad (7.4.13)$$

$$e^{\gamma t}\|\theta_{ty}(t)\|^2 + \int_0^t e^{\gamma s}\|\theta_{tyy}(s)\|^2 ds$$

$$\leq C_3\varepsilon^{-6} + C_2\varepsilon^2 \int_0^t e^{\gamma s}\Big(\|u_{tyy}\|^2 + \|\theta_{tty}\|^2\Big)(s)ds, \quad \forall t > 0. \qquad (7.4.14)$$

Proof. The proof is the same as that of Lemma 2.4.7 in Qin [101], pp. 120–121. □

Lemma 7.4.6. *Under the assumptions of Theorem 7.1.3, for any* $(v_0, u_0, \theta_0) \in H^4_+$, *there exists a positive constant* $\gamma_4 \leq \gamma_4^{(2)}$ *such that, for any fixed* $\gamma \in (0, \gamma_4]$, *the following estimates hold for any* $t > 0$:

$$e^{\gamma t}\Big(\|u_{tt}(t)\|^2 + \|u_{ty}(t)\|^2 + \|\theta_{tt}(t)\|^2 + \|\theta_{ty}(t)\|^2\Big) + \int_0^t e^{\gamma s}\Big(\|u_{tty}\|^2 + \|u_{tyy}\|^2$$

$$+ \|\theta_{tty}\|^2 + \|\theta_{tyy}\|^2\Big)(s)ds \leq C_4, \qquad (7.4.15)$$

$$e^{\gamma t}\Big(\|v_{yyy}(t)\|^2_{H^1} + \|v_{yy}(t)\|^2_{W^{1,\infty}}\Big) + \int_0^t e^{\gamma s}\Big(\|v_{yyy}\|^2_{H^1} + \|v_{yy}\|^2_{W^{1,\infty}}\Big)(s)ds \leq C_4, \qquad (7.4.16)$$

$$e^{\gamma t}\Big(\|u_{yyy}(t)\|^2_{H^1} + \|u_{yy}(t)\|^2_{W^{1,\infty}} + \|\theta_{yyy}(t)\|^2_{H^1} + \|\theta_{yy}(t)\|^2_{W^{1,\infty}} + \|v_{tyyy}(t)\|^2$$

$$+ \|u_{tyy}(t)\|^2 + \|\theta_{tyy}(t)\|^2 \big) + \int_0^t e^{\gamma s}\Big(\|u_{tt}\|^2 + \|\theta_{tt}\|^2 + \|u_{yy}\|_{W^{2,\infty}}^2 + \|\theta_{yy}\|_{W^{2,\infty}}^2$$

$$+ \|\theta_{tyy}\|_{H^1}^2 + \|u_{tyy}\|_{H^1}^2 + \|\theta_{ty}\|_{W^{1,\infty}}^2 + \|u_{ty}\|_{W^{1,\infty}}^2 + \|v_{tyyy}\|_{H^1}^2\Big)(s)ds \le C_4,$$

$$\tag{7.4.17}$$

$$\int_0^t e^{\gamma s}\left(\|u_{yyyy}\|_{H^1}^2 + \|\theta_{yyyy}\|_{H^1}^2\right)(s)ds \le C_4. \tag{7.4.18}$$

Proof. The proof is identical to that of Lemma 2.4.8 in Qin [101], pp. 121–123. □

This completes the proof of Theorem 7.1.3. □

7.5 Bibliographic Comments

The motion of a viscous, heat-conducting fluid in a domain $\Omega \subseteq \mathbb{R}^3$ can be described by the system of equations, in the Newtonian form, known as the Navier-Stokes equations. These fundamental equations in fluid mechanics have been studied by many mathematicians and a lot of results have been obtained (the interested reader is referred to the monographs [35, 36, 71, 91, 101] and the references cited therein). In the following we only recall results in the literature that are closely related to ours. For the one-dimensional linearly viscous gas (or Newtonian fluid) satisfying Fourier's law of heat flux and standard thermodynamical relations, Kawohl [59] and Jiang [50] obtained the existence of global solutions for 1D viscous heat-conductive real gas with different growth assumptions on the pressure p, internal energy e and heat conductivity κ in terms of temperature. Qin [97, 99–101] established the regularity and asymptotic behavior of global solutions under more general growth assumptions on p, e, κ than those in [50, 59].

Ladyzhenskaya [66, 67] proposed a new model to study some kinds of non-Newtonian fluids which is of interest to us. Since then there has been a remarkable research in the field of non-Newtonian flows, both theoretically and experimentally (cf. [3–5, 20, 44, 90, 94, 114, 122, 133, 141, 142, 146, 147] and references therein). Let us briefly recall related results in the literature. Bellout, Bloom and Nečas [4] studied the non-Newtonian fluids for space periodic problems and showed that there exist Young measure-valued solutions under some conditions. Dong and Li [20] established the large time behavior for weak solutions for a certain class of incompressible non-Newtonian fluids in $\mathbb{R}^2$. Guo and Zhu [44] investigated the partial regularity of the generalized solutions (which are called suitable weak solutions) to the modified Navier-Stokes equations which describe the dynamics of incompressible monopolar non-Newtonian fluids. Nećasová and Penel [90] proved the L^2 decay for weak solution to the equations of non-Newtonian incompressible fluids in whole space under some assumptions. Zhao and Li [146] studied the long-time behavior of a non-Newtonian system in two-dimensional unbounded domains and proved the existence of H^2-compact attractor for the system by showing the corresponding semigroup is asymptotically compact. Zhao, Zhou and Li [147] constructed the trajectory attractor and global attractor for the case of autonomous two-dimensional non-Newtonian fluid flows. Recently, Yuan and Wang [142] proved the local existence and uniqueness of strong solutions for (7.1.1)–(7.1.3) in one space dimension under the boundary conditions

$$u(0,t) = u(1,t) = 0, \quad \theta(0,t) = \theta(1,t) = 0.$$

Wang and Yuan [133] established the global (in time) existence and uniqueness of strong solutions under the assumptions $4/3 < p,\ q < 2$ for the following problem:

$$\rho_t + (\rho u)_x = 0,$$
$$(\rho u)_t + (\rho u^2)_x - (|u_x|^{q-2}u_x)_x + (R\rho\theta)_x = 0,$$
$$(\rho\theta)_t + (\rho u\theta)_x - (|\theta_x|^{p-2}\theta_x)_x + R\rho\theta u_x = (u_x)^2.$$

Chapter 8

Exponential Stability of Spherically Symmetric Solutions to Nonlinear Non-autonomous Compressible Navier-Stokes Equations

8.1 Main Results

This chapter is a continuation of Chapter 1 in Qin and Huang [102]. Here we establish the exponential stability of spherically symmetric solutions to an initial-boundary value problem for nonlinear non-autonomous compressible Navier-Stokes equations with an external force and a heat source in bounded annular domains $G_n = \{x \in \mathbb{R}^n : 0 < a \le |x| \le b\}$ in $\mathbb{R}^n$ $(1 \le n \le 3)$, based on the uniform estimates obtained in Chapter 1 of Qin and Huang [102]. In Eulerian coordinates, the equations under consideration read

$$\partial_t \rho + \partial_r(\rho v) + \frac{(n-1)}{r}\rho v = 0, \tag{8.1.1}$$

$$C_v \rho(\partial_t v + v\partial_r v) - (\lambda + 2\mu)\left[\partial_r^2 v + \frac{(n-1)}{r}\partial_r v - \frac{(n-1)}{r^2}v\right] + R\partial_r(\rho v) = f(r,t), \tag{8.1.2}$$

$$C_v \rho(\partial_t \theta + v\partial_r \theta) - \kappa\partial_r^2 \theta - k\frac{(n-1)}{r}\partial_r \theta + R\rho\theta\left[\partial_r v + \frac{(n-1)}{r}v\right]$$
$$- \lambda\left[\partial_r v + \frac{(n-1)}{r}v\right]^2 - 2\mu(\partial_r v)^2 - 2\mu\frac{(n-1)}{r^2}v^2 = g(r,t), \tag{8.1.3}$$

subject to the initial and boundary conditions

$$\rho(r,0) = \rho_0(r), \quad v(r,0) = v_0(r), \quad \theta(r,0) = \theta_0(r), \quad r \in G_n, 1 \le n \le 3, \tag{8.1.4}$$

$$v(a,t) = v(b,t) = 0, \quad \theta_r(a,t) = \theta_r(b,t) = 0, \, 1 \le n \le 3. \tag{8.1.5}$$

When $n = 2, 3$, equations (8.1.1)–(8.1.3) describe the spherically symmetric motion of a viscous polytropic ideal gas with a non-autonomous external force f and a heat source g. The unknown functions ρ, v, θ are the density, velocity, and absolute temperature, respectively, λ and μ are the constant viscosity coefficients, R, C_v, and κ are the gas constant, specific heat capacity and thermal conductivity, respectively, where one assumes constant $R, C_v, \kappa, \mu > 0$ and $\beta = \lambda + 2\mu$.

It is convenient to rewrite system (8.1.1)–(8.1.3) in Lagrangian coordinates. The Eulerian coordinates (r,t) are connected to the Lagrangian coordinates (ζ, t) by the relation

$$r(\zeta, t) = r_0(\zeta) + \int_0^t \tilde{v}(\zeta, \tau)d\tau, \tag{8.1.6}$$

where

$$\tilde{v}(\zeta, t) := v(r(\zeta, t), t), \quad r_0(\zeta) := \eta^{-1}(\zeta), \quad \eta(r) := \int_{d_n}^r s^{n-1}\rho_0(s)ds,$$

$$r \in G_n, \quad d_n = 0 \, (n = 1), \quad d_n = a \, (n = 2, 3).$$

Suppose that $\rho_0(s) > 0, s \in G_n$. Denote $L = \int_a^b s^{n-1}\rho_0(s)ds > 0$. Using equations (8.1.1), (8.1.5) and (8.1.6), we obtain

$$\partial_t \int_{d_n}^{r(\zeta,t)} s^{n-1}\rho(s,t)ds = \delta_{n1}v(0,t)\rho(0,t), \quad \delta_{ij} = 1 \text{ if } i = j, \quad \delta_{ij} = 0 \text{ if } i \ne j.$$

By integration, we derive

$$\int_{d_n}^{r(\zeta,t)} s^{n-1}\rho(s,t)ds = \int_{d_n}^{r_0(\zeta)} s^{n-1}\rho_0(s)ds + \delta_{n1}\int_0^t (v\rho)(0,\tau)d\tau$$

$$= \zeta + \delta_{n1}\int_0^t (v\rho)(0,\tau)d\tau.$$

Thus under the assumption $\inf\{\rho(s,t) : s \in G_n, t \ge 0\} > 0$, G_n is transformed to Ω_n, with $\Omega_n = (0, L)$, $(n = 1, 2, 3)$. Moreover, we have

$$\partial_\zeta r(\zeta, t) = [r(\zeta,t)^{n-1}\rho(r(\zeta,t),t)]^{-1}. \tag{8.1.7}$$

For a function $\varphi(r,t)$, we write $\tilde{\varphi}(\zeta,t) := \varphi(r(\zeta,t),t)$. By virtue of (8.1.6) and (8.1.7), we have

$$\partial_t \tilde{\varphi}(\zeta,t) = \partial_t\varphi(r,t) + v\partial_r\varphi(r,t),$$

$$\partial_\zeta \tilde{\varphi}(\zeta,t) = \partial_r\varphi(r,t)\partial_\zeta r(\zeta,t) = \frac{1}{r^{n-1}\rho(r,t)}\partial_r\varphi(r,t). \tag{8.1.8}$$

If we denote $(\tilde{\rho}, \tilde{v}, \tilde{\theta})$ still by (ρ, v, θ) and (ζ, t) by (x, t). We use $u := \frac{1}{\rho}$ to denote the specific volume. Therefore, by (8.1.7)–(8.1.8), equations (8.1.1)–(8.1.5) in the new variables (x, t) are

$$u_t - (r^{n-1}v)_x = 0, \tag{8.1.9}$$

$$v_t - r^{n-1}\left(\beta \frac{(r^{n-1}v)_x}{u} - R\frac{\theta}{u}\right)_x = f(r(x, t), t), \tag{8.1.10}$$

$$C_v\theta_t - k\left(\frac{r^{2n-2}\theta_x}{u}\right)_x - \frac{1}{u}\left(\beta(r^{n-1}v)_x - R\theta\right)(r^{n-1}v)_x$$

$$+ 2\mu(n-1)(r^{n-2}v^2)_x = g(r(x, t), t), \tag{8.1.11}$$

together with

$$u(x, 0) = u_0(x), \quad v(x, 0) = v_0(x), \quad \theta(x, 0) = \theta_0(x), \quad x \in \Omega_n, \ 1 \le n \le 3, \tag{8.1.12}$$

$$v(0, t) = v(L, t) = 0, \quad \theta_x(0, t) = \theta_x(L, t) = 0, \quad t \ge 0, \ 1 \le n \le 3 \tag{8.1.13}$$

where $\beta = \lambda + 2\mu$. By (8.1.6), we have

$$r(x, t) = r_0(x) + \int_0^t v(x, \tau)d\tau, \quad r_0(x) := \left\{(d_n)^n + n\int_0^x u_0(y)dy\right\}^{\frac{1}{n}}$$

i.e.,

$$r_t = v, \quad r^{n-1}r_x = u, \quad r|_{x=0} = a, \quad r|_{x=L} = b. \tag{8.1.14}$$

When $n = 2, 3$, for constants λ and μ, we assume that

$$n\lambda + 2\mu > 0. \tag{8.1.15}$$

The aim of this chapter is, based on the uniform estimates established in Chapter 1 of [102], to investigate the asymptotic behavior of solutions in H^i ($i = 1, 2, 4$), including the exponential stability of solutions in H^i ($i = 1, 2, 4$). Assume that

$$u_0(x) > 0, \quad \theta_0(x) > 0 \text{ on } [0, L]. \tag{8.1.16}$$

Assume that $f(r, t)$, $g(r, t)$ satisfy the following conditions which will be used in various theorems.

$$f(r, \cdot) \in L^1(\mathbb{R}_+, L^\infty[a, b]) \cap L^2(\mathbb{R}_+, L^2[a, b]), \tag{8.1.17}$$

$$g(r, \cdot) \in L^1(\mathbb{R}_+, L^\infty[a, b]) \cap L^2(\mathbb{R}_+, L^2[a, b]), \quad g(r, t) \ge 0, \tag{8.1.18}$$

$$f(r, \cdot) \in L^\infty(\mathbb{R}_+, L^2[a, b]), \quad f_r(r, \cdot) \in L^2(\mathbb{R}_+, L^2[a, b]),$$

$$f_t(r, \cdot) \in L^2(\mathbb{R}_+, L^2[a, b]), \tag{8.1.19}$$

$$g(r, \cdot) \in L^\infty(\mathbb{R}_+, L^2[a, b]), \quad g_r(r, \cdot) \in L^2(\mathbb{R}_+, L^2[a, b]),$$

$$g_t(r, \cdot) \in L^2(\mathbb{R}_+, L^2[a, b]), \tag{8.1.20}$$

$$f_{rr}(r, \cdot), f_{rt}(r, \cdot), f_{tt}(r, \cdot), f_{rrr}(r, \cdot) \in L^2(\mathbb{R}_+, L^2[a, b]),$$

$$f_r(r, \cdot), f_t(r, \cdot), f_{rr}(r, \cdot) \in L^\infty(\mathbb{R}_+, L^2[a, b]), \tag{8.1.21}$$

$$g_{rr}(r, \cdot), g_{rt}(r, \cdot), g_{tt}(r, \cdot), g_{rrr}(r, \cdot) \in L^2(\mathbb{R}_+, L^2[a, b]),$$

$$g_r(r, \cdot), g_t(r, \cdot), g_{rr}(r, \cdot) \in L^\infty(\mathbb{R}_+, L^2[a, b]). \tag{8.1.22}$$

We will use the following notations: $\|\cdot\|_B$ denotes norm of the space B, $\|\cdot\| = \|\cdot\|_{L^2}$. C_1 stands for a generic positive constant depending only on the H^1 norm of initial data (u_0, v_0, θ_0), $\|f\|_{L^1(\mathbb{R}_+, L^\infty[a,b])}$, $\|f\|_{L^2(\mathbb{R}_+, L^2[a,b])}$, $\|g\|_{L^1(\mathbb{R}_+, L^\infty[a,b])}$, $\|g\|_{L^2(\mathbb{R}_+, L^2[a,b])}$. C_2 stands for a generic positive constant depending only on the H^2 norm of initial data (u_0, v_0, θ_0), i.e., on

$$\|f\|_{L^\infty(\mathbb{R}_+, L^2[a,b])}, \ \|f_r\|_{L^2(\mathbb{R}_+, L^2[a,b])}, \ \|f_t\|_{L^2(\mathbb{R}_+, L^2[a,b])},$$

$$\|g\|_{L^\infty(\mathbb{R}_+, L^2[a,b])}, \ \|g_r\|_{L^2(\mathbb{R}_+, L^2[a,b])}, \ \|g_t\|_{L^2(\mathbb{R}_+, L^2[a,b])}$$

and the constant C_1. Finally, C_4 denotes a generic positive constant depending only on the H^4 norm of the initial data (u_0, v_0, θ_0), i.e., on

$$\|f_{rr}\|_{L^2(\mathbb{R}_+, L^2[a,b])}, \ \|f_{rt}\|_{L^2(\mathbb{R}_+, L^2[a,b])}, \ \|f_{tt}\|_{L^2(\mathbb{R}_+, L^2[a,b])}, \ \|f_{rrr}\|_{L^2(\mathbb{R}_+, L^2[a,b])},$$

$$\|g_{rr}\|_{L^2(\mathbb{R}_+, L^2[a,b])}, \ \|g_{rt}\|_{L^2(\mathbb{R}_+, L^2[a,b])}, \ \|g_{tt}\|_{L^2(\mathbb{R}_+, L^2[a,b])}, \ \|g_{rrr}\|_{L^2(\mathbb{R}_+, L^2[a,b])},$$

and the constants C_1, C_2.

The next theorem concerns the asymptotic behavior of global solutions in H^i $(i = 1, 2, 4)$.

Theorem 8.1.1.

(1) *Assume that* $(u_0, v_0, \theta_0) \in H^1[0, L] \times H_0^1[0, L] \times H^1[0, L]$ *and* (8.1.16)–(8.1.18) *hold. Then the problem* (8.1.9)–(8.1.15) *admits a unique global solution*

$$(u(t), v(t), \theta(t)) \in C([0, +\infty), H^1[0, L] \times H_0^1[0, L] \times H^1[0, L])$$

such that

$$u(t) \equiv u(x, t) > 0, \ \ \theta(t) \equiv \theta(x, t) > 0 \ \ on \ \ [0, L] \times \mathbb{R}_+ \tag{8.1.23}$$

and, as $t \to +\infty$,

$$\|(u(t) - \overline{u}, v(t), \theta(t) - \overline{\theta})\|_{H^1 \times H^1 \times H^1} \to 0, \tag{8.1.24}$$

where

$$\overline{u} = \frac{1}{L} \int_0^L u_0(x) dx, \qquad \overline{\theta} = \frac{1}{C_v L} \int_0^L \left(C_v \theta_0 + \frac{v_0^2}{2} \right)(x) dx.$$

(2) *Assume that $(u_0, v_0, \theta_0) \in H^2[0, L] \times H^2_0[0, L] \times H^2[0, L]$ and (8.1.16)–(8.1.20) hold, then the problem (8.1.9)–(8.1.15) admits a unique global solution*

$$(u(t), v(t), \theta(t)) \in C([0, +\infty), H^2[0, L] \times H^2_0[0, L] \times H^2[0, L])$$

such that (8.1.23) holds and, as $t \to +\infty$,

$$\|(u(t) - \bar{u}, v(t), \theta(t) - \bar{\theta})\|_{H^2 \times H^2 \times H^2} \to 0. \tag{8.1.25}$$

(3) *Assume that $(u_0, v_0, \theta_0) \in H^4[0, L] \times H^4_0[0, L] \times H^4[0, L]$ and (8.1.16)–(8.1.22) hold. Then the problem (8.1.9)–(8.1.15) admits a unique global solution*

$$(u(t), v(t), \theta(t)) \in C([0, +\infty), H^4[0, L] \times H^4_0[0, L] \times H^4[0, L])$$

such that (8.1.23) holds and, as $t \to +\infty$,

$$\|(u(t) - \bar{u}, v(t), \theta(t) - \bar{\theta})\|_{H^4 \times H^4 \times H^4} \to 0. \tag{8.1.26}$$

The following three theorems concern the exponential stability of global solutions in H^i ($i = 1, 2, 4$).

Theorem 8.1.2. *Under assumptions (1) of Theorem 8.1.1, suppose that there exist positive constants α_1 and C_0 such that*

$$\|f(r(x, t), t)\|^2_{L^2[0, L]} + \|g(r(x, t), t)\|^2_{L^2[0, L]} \leq C_0 e^{-\alpha_1 t}. \tag{8.1.27}$$

Then there are constants $C_1 > 0$ and $\gamma_1 = \gamma_1(C_1) > 0$ such that, for any fixed $\gamma \in (0, \gamma_1]$, the global solution $(u(t), v(t), \theta(t))$ obtained in (1) of Theorem 8.1.1 satisfies for any $t > 0$,

$$e^{\gamma t} \left(\|u(t) - \bar{u}\|^2_{H^1} + \|v(t)\|^2_{H^1} + \|\theta(t) - \bar{\theta}\|^2_{H^1} + \|u_t(t)\|^2 \right) \tag{8.1.28}$$

$$+ \int_0^t e^{\gamma s} \left(\|v\|^2_{H^2} + \|u - \bar{u}\|^2_{H^1} + \|\theta - \bar{\theta}\|^2_{H^2} + \|u_t\|^2_{H^1} + \|v_t\|^2 + \|\theta_t\|^2 \right)(s)ds \leq C_1.$$

Theorem 8.1.3. *Under assumptions (2) of Theorem 8.1.1, if there exist positive constants α_2 and $\bar{C}_0$ such that*

$$\|f(r(x, t), t)\|^2_{L^2[0, L]} + \|f_r(r(x, t), t)\|^2_{L^2[0, L]} + \|f_t(r(x, t), t)\|^2_{L^2[0, L]}$$

$$+ \|g_r(r(x, t), t)\|^2_{L^2[0, L]} + \|g_t(r(x, t), t)\|^2_{L^2[0, L]} \leq \bar{C}_0 e^{-\alpha_2 t}, \tag{8.1.29}$$

then there are constants $C_2 > 0$ and $\gamma_2 = \gamma_2(C_2) > 0$ such that for any fixed $\gamma \in (0, \gamma_2]$, the global solution $(u(t), v(t), \theta(t))$ obtained in (2) of Theorem 8.1.1 satisfies for all $t > 0$,

$$e^{\gamma t} \left(\|u(t) - \bar{u}\|^2_{H^2} + \|v(t)\|^2_{H^2} + \|\theta(t) - \bar{\theta}\|^2_{H^2} + \|v_t(t)\|^2 + \|\theta_t(t)\|^2 \right) \tag{8.1.30}$$

$$+ \int_0^t e^{\gamma s} \left(\|v\|^2_{H^3} + \|u - \bar{u}\|^2_{H^2} + \|\theta - \bar{\theta}\|^2_{H^3} + \|v_{tx}\|^2 + \|\theta_{tx}\|^2 \right)(s)ds \leq C_2.$$

Theorem 8.1.4. *Under assumptions* (3) *of Theorem 8.1.1, if there exist positive constants α_4 and $\tilde{C}_0$ such that*

$$\|f_r(r(x,t),t)\|^2_{L^2[0,L]} + \|f_t(r(x,t),t)\|^2_{L^2[0,L]} + \|f_{rr}(r(x,t),t)\|^2_{L^2[0,L]}$$

$$+ \|f_{rt}(r(x,t),t)\|^2_{L^2[0,L]} + \|f_{tt}(r(x,t),t)\|^2_{L^2[0,L]} + \|f_{rrr}(r(x,t),t)\|^2_{L^2[0,L]}$$

$$+ \|g_r(r(x,t),t)\|^2_{L^2[0,L]} + \|g_t(r(x,t),t)\|^2_{L^2[0,L]} + \|g_{rr}(r(x,t),t)\|^2_{L^2[0,L]}$$

$$+ \|g_{rt}(r(x,t),t)\|^2_{L^2[0,L]} + \|g_{tt}(r(x,t),t)\|^2_{L^2[0,L]}$$

$$+ \|g_{rrr}(r(x,t),t)\|^2_{L^2[0,L]} \le \tilde{C}_0 e^{-\alpha_4 t}, \tag{8.1.31}$$

then there are constants $C_4 > 0$ and $\gamma_4 = \gamma_4(C_4) > 0$ such that, for any fixed $\gamma \in (0, \gamma_4]$, the global solution $(u(t), v(t), \theta(t))$ obtained in (3) *of Theorem 8.1.1 satisfies for any $t > 0$,*

$$e^{\gamma t}\Big(\|u(t) - \bar{u}\|^2_{H^4} + \|v(t)\|^2_{H^4} + \|\theta(t) - \bar{\theta}\|^2_{H^4} + \|v_t(t)\|^2_{H^2} + \|\theta_t(t)\|^2_{H^2} + \|u_t(t)\|^2_{H^3}$$

$$+ \|v_{tt}(t)\|^2 + \|\theta_{tt}(t)\|^2\Big) + \int_0^t e^{\gamma s}\Big(\|v\|^2_{H^5} + \|u - \bar{u}\|^2_{H^4} + \|\theta - \bar{\theta}\|^2_{H^5} + \|u_t\|^2_{H^4}$$

$$+ \|v_t\|^2_{H^3} + \|\theta_t\|^2_{H^3} + \|v_{tt}\|^2_{H^2} + \|\theta_{tt}\|^2_{H^2}\Big)(s)ds \le C_4. \tag{8.1.32}$$

8.2   Asymptotic Behavior of Global Solutions

In this section, we shall establish the asymptotic behavior of global solutions in H^i ($i = 1, 2, 4$). We begin with the following lemma.

Lemma 8.2.1. *Assume that* (8.1.16)–(8.1.18) *hold, if $(u_0, v_0, \theta_0) \in H^1[0, L] \times H_0^1[0, L] \times H^1[0, L]$, then problem* (8.1.9)–(8.1.15) *admits a unique global solution $(u(t), v(t), \theta(t)) \in C([0, +\infty), H^1[0, L] \times H_0^1[0, L] \times H^1[0, L])$ satisfying*

$$0 < a \le r(x,t) \le b, \quad \forall (x,t) \in [0, L] \times [0, +\infty), \tag{8.2.1}$$

$$0 < C_1^{-1} \le u(x,t) \le C_1, \quad \forall (x,t) \in [0, L] \times [0, +\infty), \tag{8.2.2}$$

$$\|r(t) - \bar{r}\|^2_{H^2} + \|r_t(t)\|^2_{H^1} + \|u(t) - \bar{u}\|^2_{H^1} + \|v(t)\|^2_{H^1} + \|\theta(t) - \bar{\theta}\|^2_{H^1} + \|u_t(t)\|^2$$

$$+ \int_0^t \Big(\|u - \bar{u}\|^2_{H^1} + \|v\|^2_{H^2} + \|\theta - \bar{\theta}\|^2_{H^2} + \|u_t\|^2_{H^1} + \|v_t\|^2 + \|\theta_t\|^2$$

$$+ \|r - \bar{r}\|^2_{H^2} + \|r_t\|^2_{H^2}\Big)(\tau)d\tau \le C_1, \quad \forall t \ge 0, \tag{8.2.3}$$

where $\bar{r} = (a^n + n\bar{u}x)^{1/n}$.

Proof. See, e.g., [102] and [111].   □

Lemma 8.2.2. *Under assumptions* (1) *of Theorem 8.1.1, it holds that*

$$\lim_{t \to +\infty} \|u(t) - \bar{u}\|_{H^1} = 0. \tag{8.2.4}$$

Proof. Differentiating (8.1.9) with respect to x, multiplying the result by u_x, and then integrating it over $[0, L]$ and applying Young's inequality, we obtain

$$\frac{1}{2}\frac{d}{dt}\|u_x\|^2 \le \|(r^{n-1}v)_{xx}\|^2 + \|u_x\|^2$$

$$\le \frac{1}{2}\|u_x\|^4 + C_1(\|v_{xx}\|^2 + 1),$$

which, together with Lemmas 1.3.4 and 8.2.1, yields

$$\lim_{t \to +\infty} \|u_x(t)\| = 0. \tag{8.2.5}$$

On the other hand, by the embedding theorem, we can deduce

$$\|u(t) - \overline{u}\| \le C_1\|u_x\|,$$

which, together with (8.2.5), gives (8.2.4). $\qquad\square$

Lemma 8.2.3. *Under assumptions* (1) *of Theorem* 8.1.1, *it holds that*

$$\lim_{t \to +\infty} \|v(t)\|_{H^1} = 0, \quad \lim_{t \to +\infty} \|\theta(t) - \overline{\theta}\|_{H^1} = 0. \tag{8.2.6}$$

Proof. Multiplying the result by v_{xx}, then integrating it over $[0, L]$, by Young's inequality, we can deduce

$$\frac{1}{2}\frac{d}{dt}\|v_x\|^2 \le C_1\|v_{xx}\|^2 + C_1\left(\|u_x\|^2 + \|\theta_x\|^2 + \|f\|^2 + \|v_x u_x\|^2\right)$$

$$\le C_1\|v_x\|^4 + C_1\left(\|v_{xx}\|^2 + \|f\|^2 + 1\right),$$

which, together with Lemma 1.3.4 and Lemmas 8.2.1 and 8.2.2, yields

$$\lim_{t \to +\infty} \|v_x(t)\| = 0. \tag{8.2.7}$$

Similarly, we can obtain the second relation in (8.2.6). $\qquad\square$

Proof of Theorem 8.1.1. Combining Lemma 8.2.2 with Lemma 8.2.3, we complete the proof of (1) of Theorem 8.1.1. Similarly, we can derive (8.1.25)–(8.1.26). Till now we have completed the proof of Theorem 8.1.1. $\qquad\square$

8.3 Exponential Stability of Solutions in H^1

Under the assumptions of Theorem 8.1.2, the global existence of solutions was obtained in [102] and [111], which has been stated in Lemma 8.2.1. In this section, we shall establish the exponential stability of solutions in H^1.

To this end, we introduce the flow density $\rho = \frac{1}{v}$, and then we easily get that the specific entropy

$$\eta = \eta(u, \theta) = \eta(\rho, \theta) = R \log u + C_v \log \theta, \tag{8.3.1}$$

satisfies

$$\frac{\partial \eta}{\partial \rho} = -\frac{R}{\rho}, \quad \frac{\partial \eta}{\partial \theta} = \frac{C_v}{\theta}. \tag{8.3.2}$$

We consider the transform

$$\mathcal{A} : \mathcal{D}_{\rho,\theta} = \{(\rho, \theta) : \rho > 0, \theta > 0\} \ni (\rho, \theta) \mapsto (u, \eta) \in \mathcal{A}\mathcal{D}_{\rho,\theta},$$

where $u = 1/\rho$ and $\eta = \eta(1/\rho, \theta)$. Since the Jacobian

$$\frac{\partial(u, \eta)}{\partial(\rho, \theta)} = -\frac{C_v}{\rho^2} < 0 \quad \text{on} \quad \mathcal{D}_{\rho,\theta},$$

there is a unique inverse function $\theta = \theta(u, \eta)$, which is a smooth function of $(u, \eta) \in \mathcal{A}\mathcal{D}_{\rho,\theta}$.

Thus the function e, p can be regarded as the smooth functions of (u, η). We denote by

$$e = e(u, \eta) := e(u, \theta(u, \eta)) = e(\rho^{-1}, \theta), \quad p = p(u, \eta) := p(u, \theta(v, \eta)) = p(\rho^{-1}, \theta).$$

Thus we derive that e, p satisfy

$$e_u = -\frac{R\theta}{u}, \quad e_\eta = \theta, \quad P_u = -\frac{R\theta + \frac{R^2}{C_v}\theta}{u^2}, \quad P_\eta = \frac{R\theta}{C_v u}. \tag{8.3.3}$$

Let

$$\mathcal{E}(u, u, \eta) = \frac{v^2}{2} + e(u, \eta) - e(\overline{u}, \overline{\eta}) - \frac{\partial e}{\partial u}(\overline{u}, \overline{\eta})(u - \overline{u}) - \frac{\partial e}{\partial \eta}(\overline{u}, \overline{\eta})(\eta - \overline{\eta}), \tag{8.3.4}$$

where $\overline{\eta} = \eta(\overline{u}, \overline{\theta})$.

Lemma 8.3.1. *The global solution* $(v(t), u(t), \theta(t))$ *obtained in* (1) *of Theorem 8.1.1 to problem* (8.1.6)–(8.1.10) *satisfies the estimate*

$$\frac{v^2}{2} + C_1^{-1}(|u - \overline{u}|^2 + |\eta - \overline{\eta}|^2) \le \mathcal{E}(v, u, \eta) \le \frac{v^2}{2} + C_1(|u - \overline{u}|^2 + |\eta - \overline{\eta}|^2). \tag{8.3.5}$$

Proof. By the mean value theorem, there exists a point $(\tilde{u}, \tilde{\eta})$ between (u, η) and $(\overline{u}, \overline{\eta})$ such that

$$\mathcal{E}(v, u, \eta) = \frac{1}{2}v^2 + \frac{1}{2}\left[\frac{\partial^2 e}{\partial^2 u}(\tilde{u}, \tilde{\eta})(u - \overline{u})^2 + \frac{\partial^2 e}{\partial^2 \eta}(\tilde{u}, \tilde{\eta})(\eta - \overline{\eta})^2 \right.$$

$$+ \frac{\partial^2 e}{\partial u \partial \eta}(\tilde{u}, \tilde{\eta})(u - \overline{u})(\eta - \overline{\eta})\Big], \qquad (8.3.6)$$

where $\tilde{u} = \lambda_0 \overline{u} + (1 - \lambda_0)u$, $\tilde{\eta} = \lambda_0 \overline{\eta} + (1 - \lambda_0)\eta$, $0 \leq \lambda_0 \leq 1$.

It follows from Lemma 8.2.1 that

$$0 < C_1^{-1} \leq \tilde{u} \leq C_1, \quad |\tilde{\eta}| \leq C_1$$

which implies

$$\left|\frac{\partial^2 e}{\partial^2 u}(\tilde{u}, \tilde{\eta})\right|^2 + \left|\frac{\partial^2 e}{\partial u \partial \eta}(\tilde{u}, \tilde{\eta})\right|^2 + \left|\frac{\partial^2 e}{\partial^2 \eta}(\tilde{u}, \tilde{\eta})\right|^2 \leq C_1. \qquad (8.3.7)$$

Thus, by (8.3.6)–(8.3.7) and the Cauchy inequality, we get

$$\mathcal{E}(v, u, \eta) \leq \frac{1}{2}v^2 + C_1(|u - \overline{u}|^2 + |\eta - \overline{\eta}|^2). \qquad (8.3.8)$$

On the other hand, we infer from (8.3.3) that

$$e_{uu} = \frac{\frac{R^2}{C_v}\theta + R\theta}{u^2}, \quad e_{u\eta} = -\frac{R\theta}{C_v u}, \quad e_{\eta\eta} = \frac{\theta}{C_v},$$

which implies that the Hessian of $e(u, \theta)$ is positive definite for any $u > 0$ and $\theta > 0$. It then follows from (8.3.6) that

$$\mathcal{E}(v, u, \eta) \geq \frac{1}{2}v^2 + C_1^{-1}(|u - \overline{u}|^2 + |\eta - \overline{\eta}|^2),$$

which, together with (8.3.8), gives (8.3.5). $\qquad \square$

Lemma 8.3.2. *Under the assumptions of Theorem 8.1.2, there are positive constants $C_1 > 0$ and $\gamma_1 = \gamma_1(C_1) > 0$ such that, for any fixed $\gamma \in (0, \gamma_1]$, it holds for any $t > 0$*

$$e^{\gamma t}\Big(\|u(t) - \overline{u}\|^2 + \|v(t)\|^2 + \|\theta(t) - \overline{\theta}\|^2 + \|u_x\|^2 + \|\rho_x\|^2\Big)$$

$$+ \int_0^t e^{\gamma s}\Big(\|\rho_x\|^2 + \|v_x\|^2 + \|\theta_x\|^2 + \|u_x\|^2\Big)(s)ds \leq C_1. \qquad (8.3.9)$$

Proof. Using equations (8.1.9)–(8.1.11), it is easy to verify that

$$\left(e + \frac{v^2}{2}\right)_t = \left[r^{n-1}v\sigma + k\rho r^{2n-2}\theta_x - 2\mu(n-1)(r^{n-2}v^2)\right]_x + fv + g, \qquad (8.3.10)$$

$$\eta_t = \left(\frac{k\rho r^{2n-2}\theta_x - 2\mu(n-1)(r^{n-2}v)2}{\theta}\right)_x + \frac{\beta\rho(r^{n-1}v)_x^2}{\theta}$$

$$+ \frac{kr^{2n-2}\theta_x^2}{\theta^2} + \frac{g}{\theta} - \frac{2\mu(n-1)r^{n-2}v^2\theta_x}{\theta^2}, \tag{8.3.11}$$

where $\sigma = p - R\frac{\theta}{u}$, $p = \frac{\beta(r^{n-1}v)_x}{u}$.

Since $\bar{u}_t = 0$, $\bar{\theta}_t = 0$, we infer from (8.3.10) and (8.3.11) that

$$\mathcal{E}_t(1/\rho, u, \eta) + \frac{\bar{\theta}}{\theta}\left[\beta\rho(r^{n-1}v)_x^2 + \frac{k\rho r^{2n-2}\theta_x^2}{\theta}\right] = \left[\beta\rho(r^{n-1}v)(r^{n-1}v)_x\right.$$

$$\left. + k\left(1 - \frac{\bar{\theta}}{\theta}\right)\rho r^{2n-2}\theta_x^2 - (p - \bar{p})(r^{n-1}v) - 2\mu(n-1)r^{n-2}v^2\right]_x, \tag{8.3.12}$$

$$\left[\beta^2(\rho_x/\rho)^2/2 + \rho_x r^{1-n}v/\rho\right]_t + R\beta\theta\rho_x^2/\rho = -\beta R\rho_x\theta_x + \beta r^{1-n}f\rho_x/\rho$$

$$- \beta r^{2-2n}(\rho(r^{n-1}v)(r^{n-1}v)_x)_x + \beta r^{2-2n}\rho(r^{n-1}v)_x^2 + \beta(1-n)r^{-n}v\rho_x/\rho. \tag{8.3.13}$$

Multiplying (8.3.12), (8.3.13) by $e^{\gamma t}$, $\lambda e^{\gamma t}$, respectively, and then adding the results up, we get

$$\frac{\partial}{\partial t}G(t) + e^{\gamma t}\left[\frac{\bar{\theta}}{\theta}\left(\beta\rho(r^{n-1}v)_x^2 + \frac{k\rho r^{2n-2}\theta_x^2}{\theta}\right)\right.$$

$$\left. + \lambda\left(R\beta\theta\rho_x^2/\rho + \beta R\rho_x\theta_x - \beta r^{2-2n}\rho(r^{n-1}v)_x^2 - \beta(1-n)r^{-n}v\rho_x/\rho\right)\right]$$

$$= \gamma e^{\gamma t}\left[\mathcal{E}(1/\rho, u, \eta) + \lambda\beta^2(\rho_x/\rho)^2/2 + \rho_x r^{1-n}v/\rho\right] + e^{\gamma t}\left[k\left(1 - \frac{\bar{\theta}}{\theta}\right)\rho r^{2n-2}\theta_x\right.$$

$$\left. + (1 - \lambda r^{2-2n})\beta\rho(r^{n-1}v)(r^{n-1}v)_x - (p - \bar{p})(r^{n-1}v) - 2\mu(n-1)r^{n-2}v^2\right]_x$$

$$+ e^{\gamma t}\left[fv + g + \lambda\beta r^{1-n}f\rho_x/\rho\right], \tag{8.3.14}$$

where $G(t) = e^{\gamma t}\left[\mathcal{E}(1/\rho, u, \eta) + \lambda\beta\left((\rho_x/\rho)^2/2 + \rho_x r^{1-n}v/\rho\right)\right]$.

Integrating (8.3.14) over Q_t, and using Lemmas 8.2.1 and 8.3.1, Cauchy's inequality and Poincaré's inequality, we deduce that for small $\beta > 0$ and for any $\gamma > 0$,

$$e^{\gamma t}\left[\|\rho(t) - \bar{p}\|^2 + \|\eta(t) - \bar{\eta}\|^2 + \|v(t)\|^2 + \|\rho_x(t)\|^2\right]$$

$$+ \int_0^t e^{\gamma s}\left[\|\rho_x\|^2 + \|v_x\|^2 + \|\theta_x\|^2 + \|u_x\|^2\right](s)ds$$

$$\leq C_1 + C_1\gamma\int_0^t e^{\gamma s}\left[\|\rho - \bar{p}\|^2 + \|\theta - \bar{\theta}\|^2 + \|u\|^2 + \|\rho_x\|^2\right](s)ds$$

$$+ C_1\int_0^t e^{\gamma s}\left[\|f\|_{L^2[0,L]}^2 + \|g\|_{L^2[0,L]}^2\right](s)ds. \tag{8.3.15}$$

By Lemmas 8.2.1 and 8.3.1, the mean value theorem, and the Poincaré inequality, we have

$$\|u(t) - \overline{u}\| \leq C_1 \|u_x(t)\|,$$

$$\|\theta(t) - \overline{\theta}\| \leq C_1 (\|e(v, \theta) - e(\overline{u}, \overline{\theta})\| + \|u(t) - \overline{u}\|)$$

$$\leq C_1 (\|e(u, \theta) - e(\overline{u}, \overline{\theta})\| + \|u_x(t)\|)$$

$$\leq C_1 (\|\theta_x(t)\| + \|u_x(t)\| + \|v_x(t)\|). \qquad (8.3.16)$$

Similarly, we infer that

$$C_1^{-1} \|u(t) - \overline{u}\| \leq \|\rho(t) - \overline{\rho}\| \leq C_1 \|u(t) - \overline{u}\|, \qquad (8.3.17)$$

$$\|\theta(t) - \overline{\theta}\| \leq C_1 (\|\eta(t) - \overline{\eta}\| + \|u(t) - \overline{u}\|). \qquad (8.3.18)$$

It follows from (8.1.25), (8.3.15)–(8.3.18) that there exists a constant $\gamma_1 = \gamma_1(C_1) > 0$ such that, for any fixed $\gamma \in (0, \gamma_1]$, (8.3.9) holds. Then the proof is complete. $\qquad\square$

Lemma 8.3.3. *Under the assumptions of Theorem 8.1.2, there are positive constants $C_1 > 0$ and $\gamma_1 = \gamma_1(C_1) > 0$ such that, for any fixed $\gamma \in (0, \gamma_1]$, i holds for any $t > 0$,*

$$e^{\gamma t} \left(\|v_x(t)\|^2 + \|\theta_x(t)\|^2 \right) + \int_0^t e^{\gamma s} \left(\|v_t\|^2 + \|v_{xx}\|^2 + \|\theta_t\|^2 + \|\theta_{xx}\|^2 \right)(s) ds \leq C_1.$$
$$(8.3.19)$$

Proof. By (8.1.9)–(8.1.11), Lemma 8.2.1 and the Poincaré inequality, we have

$$\|v_x(t)\|^2 \leq C_1 \|v_{xx}(t)\|^2,$$

$$\|v_t(t)\|^2 \leq C_1 \left(\|v_{xx}(t)\|^2 + \|\theta_x(t)\|^2 + \|u_x(t)\|^2 + \|f(t)\|^2 \right), \qquad (8.3.20)$$

$$\|\theta_x(t)\|^2 \leq C_1 \|\theta_{xx}(t)\|^2, \quad \|\theta_t(t)\|^2 \leq C_1 \left(\|\theta_{xx}(t)\|^2 + \|v_{xx}(t)\|^2 + \|g(t)\|^2 \right).$$
$$(8.3.21)$$

Multiplying (8.1.10)–(8.1.11) by $-e^{\gamma t} v_{xx}$, $-e^{\gamma t} \theta_{xx}$, respectively, integrating the results over $[0, 1] \times [0, t]$, and adding them up, using Young's inequality, the embedding theorem, Lemmas 8.2.1 and 8.3.1–8.3.2, we finally deduce that

$$e^{\gamma t} \left(\|v_x(t)\|^2 + \|\theta_x(t)\|^2 \right) + \int_0^t e^{\gamma s} \left(\|v_{xx}\|^2 + \|\theta_{xx}\|^2 \right)(s) ds$$

$$\leq C_1 + C_1 \int_0^t e^{\gamma s} \Big\{ \left(\|v_x\| + \|u_x\| + \|\theta_x\| + \|v_x u_x\| \right) \|v_{xx}\|$$

$$+ \left(\|v_x^2\| + \|u_x \theta_x\| + \|\theta_x\| + \|\theta v_x\| \right) \|\theta_{xx}\| \Big\}(s) ds$$

$$\leq C_1 + \frac{1}{2} \int_0^t e^{\gamma s} \left(\|v_{xx}\|^2 + \|\theta_{xx}\|^2 \right)(s) ds$$

$$+ C_1 \int_0^t e^{\gamma s} \left(\|f\|_{L^2[0,L]}^2 + \|g\|_{L^2[0,L]}^2 \right)(s) ds. \tag{8.3.22}$$

It follows from (8.1.27) that there exists a positive constant $\gamma_1 = \gamma_1(C_1) < \alpha_1$ such that for any fixed $\gamma \in (0, \gamma_1]$,

$$\int_0^t e^{\gamma s} \left(\|f\|_{L^2[0,L]}^2 + \|g\|_{L^2[0,L]}^2 \right)(s) ds \leq C_1$$

which, together with (8.3.22), gives (8.3.19).   □

This completes the proof of Theorem 8.1.2.   □

8.4   Exponential Stability of Solutions in H^2

In this section, we shall establish the exponential stability of solutions in H^2. The proof of Theorem 8.1.3 can be divided into the following several lemmas.

Lemma 8.4.1. *Under the conditions* (8.1.16)–(8.1.20), *if* $(u_0, v_0, \theta_0) \in H^2[0, L] \times H_0^2[0, L] \times H^2[0, L]$, *problem* (8.1.9)–(8.1.15) *admits a unique global solution*

$$(u(t), v(t), \theta(t)) \in C([0, +\infty), H^2[0, L] \times H_0^2[0, L] \times H^2[0, L])$$

satisfying (8.1.23) *and for any* $t > 0$,

$$\|r(t) - \bar{r}\|_{H^3}^2 + \|r_t(t)\|_{H^2}^2 + \|u(t) - \bar{u}\|_{H^2}^2 + \|v(t)\|_{H^2}^2 + \|\theta(t) - \bar{\theta}\|_{H^2}^2 + \|u_t(t)\|_{H^1}^2$$

$$+ \|v_t(t)\|^2 + \|\theta_t(t)\|^2 + \int_0^t \left(\|u - \bar{u}\|_{H^2}^2 + \|u_t\|_{H^2}^2 + \|v\|_{H^3}^2 + \|\theta - \bar{\theta}\|_{H^3}^2 \right.$$

$$+ \|v_t\|_{H^1}^2 + \|\theta_t\|_{H^1}^2 + \|r - \bar{r}\|_{H^3}^2 + \|r_t\|_{H^3}^2 \bigg)(s) ds \leq C_2.$$

Proof. See, e.g., [102] and [111].   □

Lemma 8.4.2. *Under the assumptions of Theorem 8.1.3, there are positive constants* $C_2 > 0$ *and* $\gamma_2'' = \gamma_2''(C_2) > 0$ *such that, for any fixed* $\gamma \in (0, \gamma_2]$, *it holds that, for any* $t > 0$,

$$e^{\gamma t} \left(\|v_t(t)\|^2 + \|v_{xx}(t)\|^2 \right) + \int_0^t e^{\gamma s} \|v_{tx}(s)\|^2 ds \leq C_2, \tag{8.4.1}$$

$$e^{\gamma t} \left(\|\theta_t(t)\|^2 + \|\theta_{xx}(t)\|^2 \right) + \int_0^t e^{\gamma s} \|\theta_{tx}(s)\|^2 ds \leq C_2. \tag{8.4.2}$$

Proof. Differentiating (8.1.10) with respect to t, we have

$$v_{tt} = (n-1)r^{n-2}r_t \left(\beta \frac{(r^{n-1}v)_x}{u} - R\frac{\theta}{u} \right)_x + r^{n-1} \left(\beta \frac{(r^{n-1}v)_x}{u} - R\frac{\theta}{u} \right)_{tx} + \frac{df}{dt}. \tag{8.4.3}$$

Multiplying (8.4.3) by $e^{\gamma t}v_t$, then integrating the result over $[0, L]$, using (8.1.9) and Theorem 8.1.1, and noting that $\frac{df}{dt} = f_r v + f_t$, we derive

$$\int_0^L e^{\gamma t}v_t v_{tt}dx = \frac{1}{2}e^{\gamma t}\|v_t\|^2 = \frac{1}{2}\frac{d}{dt}\left(e^{\gamma t}\|v_t\|^2\right) - \frac{\gamma}{2}e^{\gamma t}\|v_t\|^2, \tag{8.4.4}$$

$$\int_0^L e^{\gamma t}v_t r^{n-1}\left(\beta \frac{(r^{n-1}v)_x}{u} - R\frac{\theta}{u} \right)_{tx}dx$$

$$= e^{\gamma t}\int_0^L v_t r^{n-1}\left(\beta \frac{(r^{n-1}v)_{tx}}{u} - \frac{(r^{n-1}v)_x u_t}{u^2} - R\frac{\theta_t}{u} + R\frac{\theta u_t}{u^2} \right)_x dx$$

$$\leq -C_2^{-1}e^{\gamma t}\|v_{tx}\|^2 + C_2 e^{\gamma t}\left(\|v_x\|_{L^4}^4 + \|\theta_t\|^2 + \|v_x\|^2 \right)$$

$$\leq -C_2^{-1}e^{\gamma t}\|v_{tx}\|^2 + C_2 e^{\gamma t}\left(\|v_{xx}\|^2 + \|\theta_t\|^2 + \|v_x\|^2 \right), \tag{8.4.5}$$

$$\int_0^L e^{\gamma t}(n-1)r^{n-2}vv_t\left(\beta \frac{(r^{n-1}v)_x}{u} - R\frac{\theta}{u} \right)_x dx$$

$$\leq C_2 e^{\gamma t}\left(\|v_t\|^2 + \|v_x u_x\|^2 + \|v_{xx}\|^2 + \|\theta_x u_x\|^2 + \|\theta_x\|^2 \right)$$

$$\leq C_2 e^{\gamma t}\left(\|v_x\|^2 + \|v_{xx}\|^2 + \|\theta_x\|^2 + \|\theta_{xx}\|^2 + \|u_x\|^2 \right) \tag{8.4.6}$$

$$\left| \int_0^L e^{\gamma t}\frac{df}{dt}v_t dx \right| \leq C_2 e^{\gamma t}\|v_t\|^2 + C_2 e^{\gamma t}\left(\|f_r\|_{L^2[0,L]}^2 + \|f_t\|_{L^2[0,L]}^2 \right). \tag{8.4.7}$$

Combining (8.4.3)–(8.4.7) and making use of (8.1.29), we derive that there exists a positive constant $\gamma_2' = \gamma_2'(C_2) \leq \min(\gamma_1, \alpha_2)$ such that, for any fixed $\gamma \in (0, \gamma_2']$,

$$e^{\gamma t}\|v_t(t)\|^2 + \int_0^t e^{\gamma s}\|v_{tx}(s)\|^2 ds \leq C_2, \tag{8.4.8}$$

which, together with (8.1.10), gives for any fixed $\gamma \in (0, \gamma_2']$ that

$$e^{\gamma t}\|v_{xx}(t)\|^2 \leq C_2 e^{\gamma t}\left(\|v_t(t)\|^2 + \|u_x(t)\|^2 + \|v_x(t)\|_{H^1}^2 + \|\theta(t)\|_{H^1}^2 + \|f\|_{L^2[0,L]}^2 \right)$$

$$\leq C_2. \tag{8.4.9}$$

Hence (8.4.1) follows from (8.4.8) by using the embedding theorem. Similarly, differentiating (8.1.11) with respect to t, we get

$$C_v\theta_{tt} = k\left(\frac{r^{2n-2}\theta_x}{u} \right)_{tx} + \left(\beta \frac{(r^{n-1}v)_x}{u} - R\frac{\theta}{u} \right)_t (r^{n-1}v)_x \tag{8.4.10}$$

$$+ \left(\beta \frac{(r^{n-1}v)_x}{u} - R\frac{\theta}{u} \right) (r^{n-1}v)_{tx} - 2\mu(n-1) \left(r^{n-2}v^2 \right)_{tx} + \frac{dg}{dt}.$$

Multiplying (8.4.10) by θ_t, then integrating the result over $[0, L]$, and noting that

$$\frac{dg}{dt} = g_r v + g_t,$$

we have

$$\int_0^L e^{\gamma t} C_v \theta_t \theta_{tt} dx = \frac{C_v}{2} \frac{d}{dt} \left(e^{\gamma t} \|\theta\|_t^2 \right) - \frac{C_v}{2} \gamma e^{\gamma t} \|\theta\|_t^2, \tag{8.4.11}$$

$$\int_0^L k e^{\gamma t} \theta_t \left(\frac{r^{2n-2}\theta_x}{u} \right)_{tx} dx = -e^{\gamma t} \int_0^L k\theta_{tx} \left(\frac{r^{2n-2}\theta_x}{u} \right)_t dx$$
$$\le -C_2^{-1} e^{\gamma t} \|\theta_{tx}\|^2 + C_2 e^{\gamma t} \left(\|u_t\|^2 + \|\theta_x\|^2 \right)$$
$$\le -C_2^{-1} e^{\gamma t} \|\theta_{tx}\|^2 + C_2 e^{\gamma t} \left(\|v_x\|^2 + \|\theta_x\|^2 \right), \tag{8.4.12}$$

$$\int_0^L e^{\gamma t} \left(\beta \frac{(r^{n-1}v)_x}{u} - R\frac{\theta}{u} \right)_t (r^{n-1}v)_x \theta_t dx$$
$$\le C_2 e^{\gamma t} \left(\|v_{tx}\|^2 + \|v_x\|_{L^4}^4 + \|\theta_t\|^2 + \|\theta v_x\|^2 + \|v_x\theta_t\|^2 \right)$$
$$\le C_2 e^{\gamma t} \left(\|v_{tx}\|^2 + \|v_{xx}\|^2 + \|\theta_t\|^2 + \|v_x\|^2 \right). \tag{8.4.13}$$

Next, it follows from (8.1.14) and Lemmas 8.2.1 and 8.3.1 that

$$\left\| \beta \frac{(r^{n-1}v)_x}{u} - R\frac{\theta}{u} \right\|_{L^\infty} \le C_2, \tag{8.4.14}$$

$$\|(r^{n-1}v)_{tx}\| \le C_2 \left(\|v_x\| + \|v_{tx}\| \right), \tag{8.4.15}$$

$$\left| \int_0^L \theta_t dx \right| \le C_1 \left| \int_0^L \frac{1}{u} \left(\beta(r^{n-1}x)_x - R\theta \right) (r^{n-1}v)_x \right|$$
$$\le C_2 \|v_x\| \tag{8.4.16}$$

which, in conjunction with the Poincaré inequality, gives

$$\|\theta_t\| \le \left| \int_0^L \theta_t dx \right| + C\|\theta_{tx}\| \le C_2 \left(\|v_x\| + \|\theta_{tx}\| \right). \tag{8.4.17}$$

Thus, from (8.4.14)–(8.4.17) we derive, for $\varepsilon > 0$ small enough, that

$$\int_0^L e^{\gamma t} \left(\beta \frac{(r^{n-1}v)_x}{u} - R\frac{\theta}{u} \right) (r^{n-1}v)_{tx} \theta_t dx$$
$$\le C_2 e^{\gamma t} \|(r^{n-1}v)_{tx}\| \|\theta_t\|$$
$$\le C_2 e^{\gamma t} (\|v_x\| + \|v_{tx}\|)(\|v_x\| + \|\theta_{tx}\|)$$
$$\le \varepsilon e^{\gamma t} \|\theta_{tx}\|^2 + C_2 e^{\gamma t} (\|v_x\|^2 + \|v_{tx}\|^2), \tag{8.4.18}$$

$$\left| \int_0^L -2\mu(n-1)(r^{n-2}v^2)_{tx} e^{\gamma t} \theta_t dx \right|$$

$$\leq C_2 e^{\gamma t} \left(\|\theta_t\|^2 + \|v_x v_t\|^2 + \|v v_{tx}\|^2 \right)$$

$$\leq C_2 e^{\gamma t} \left(\|\theta_t\|^2 + \|v_t\|_{L^\infty}^2 + \|v_{tx}\|^2 \right)$$

$$\leq C_2 e^{\gamma t} \left(\|\theta_t\|^2 + \|v_{tx}\|^2 \right), \tag{8.4.19}$$

$$\left| \int_0^L e^{\gamma t} \frac{dg}{dt} \theta_t dx \right| \leq C_2 \left(\|g_r\|_{L^2[0,L]}^2 + \|g_t\|_{L^2[0,L]}^2 + \|\theta_t\|^2 \right). \tag{8.4.20}$$

From (8.4.11)–(8.4.20), it follows that, for $\varepsilon > 0$ small enough,

$$e^{\gamma t} \|\theta_t(t)\|^2 + \int_0^t e^{\gamma s} \|\theta_{tx}(s)\|^2 ds$$

$$\leq C_2 + C_2 \int_0^t e^{\gamma s} \left(\|v_x\|_{L^\infty}^2 \|\theta_t\|^2 + \|g_t\|_{L^2[0,L]}^2 + \|g_r\|_{L^2[0,L]}^2 \right)(s) ds.$$

Next, by (8.1.29) and the Gronwall inequality there exists a positive constant $\gamma_2'' \leq \gamma_2'$ such that, for any fixed $\gamma \in (0, \gamma_2'']$,

$$e^{\gamma t} \|\theta_t(t)\|^2 + \int_0^t e^{\gamma s} \|\theta_{tx}(s)\|^2 ds \leq C_2. \tag{8.4.21}$$

Finally, from (8.1.11), we get, for any fixed $\gamma \in (0, \gamma_2'']$, that

$$e^{\gamma t} \|\theta_{xx}(t)\|^2$$

$$\leq C_2 e^{\gamma t} \left(\|\theta_t(t)\|^2 + \|\theta(t)\|_{H^1}^2 + \|u(t)\|_{H^1}^2 + \|v(t)\|_{H^1}^2 + \|g(t)\|_{L^2[0,L]}^2 \right) \leq C_2,$$

which, together with (8.4.21), yields (8.4.2). $\qquad\square$

Lemma 8.4.3. *Under the assumptions of Theorem 8.1.3, there are positive constants $C_2 > 0$ and $\gamma_2''' = \gamma_2'''(C_2) > 0$ such that, for any fixed $\gamma \in (0, \gamma_2''']$, it holds that*

$$e^{\gamma t} \|u_{xx}(t)\|^2 + \int_0^t e^{\gamma s} \|u_{xx}(s)\|^2 ds \leq C_2, \quad \forall t > 0. \tag{8.4.22}$$

Proof. Differentiating (8.1.10) with respect to x, we have

$$\beta \frac{d}{dt} \left(\frac{u_{xx}}{u} \right) + \frac{R\theta u_{xx}}{u^2} = r^{1-n} v_{tx} + (1-n) r^{1-2n} u v_t + 2\beta \frac{(r^{n-1}v)_{xx} u_x}{u^2}$$

$$- 2\beta \frac{(r^{n-1}v)_x u_x^2}{u^3} + R \frac{\theta_{xx}}{u} - 2R \frac{\theta_x u_x}{u^2}$$

$$+ 2R \frac{\theta u_x^2}{u^3} - r^{1-n} f_r u - (1-n) r^{1-2n} u f$$

$$:= M \qquad\qquad (8.4.23)$$

where

$$\|M\| \le C_2(\|v_{tx}\| + \|v_t\| + \|u_x\| + \|v_x\|_{H^1} + \|\theta\|_{H^2} + \|f_r\|_{L^2[0,L]} + \|f\|_{L^2[0,L]}).$$

By Theorem 8.1.2, condition (8.1.19), (8.1.29), (8.3.22), (8.4.7) and Lemma 8.4.2, we get, for any fixed $\gamma \in (0, \gamma_2'']$,

$$\int_0^t e^{\gamma s}\|M(s)\|^2 ds \le C_2, \quad \forall t > 0. \qquad\qquad (8.4.24)$$

Multiplying (8.4.23) by $e^{\gamma t}\dfrac{u_{xx}}{u}$, then integrating the result over $[0, L]$, we arrive at

$$\beta \frac{d}{dt}\left(e^{\gamma t}\|\frac{u_{xx}}{u}\|^2\right) + C_2^{-1}e^{\gamma t}\|\frac{u_{xx}}{u}\|^2 \le \gamma e^{\gamma t}\|\frac{u_{xx}}{u}\|^2 + C_2 e^{\gamma t}\|M\|^2. \qquad (8.4.25)$$

Integrating (8.4.25) over $[0, t]$ and using (8.4.24), we conclude that there exists a positive constant $\gamma_2''' \le \gamma_2''$ such that (8.4.22) holds for any fixed $\gamma \in (0, \gamma_2''']$. $\square$

Lemma 8.4.4. *Under the assumptions of Theorem 8.1.3, there are positive constants $C_2 > 0$ and $\gamma_2 = \gamma_2(C_2) > 0$ such that, for any fixed $\gamma \in (0, \gamma_2]$,*

$$\int_0^t e^{\gamma s}\left(\|v_{xxx}\|^2 + \|\theta_{xxx}\|^2\right)(s)ds \le C_2, \quad \forall t > 0. \qquad (8.4.26)$$

Proof. Differentiating (8.1.10) with respect to x, we get

$$\|v_{xxx}(t)\|^2 \le C_2\left(\|v_{tx}\|^2 + \|v_x\|_{H^1}^2 + \|u_x\|_{H^1}^2 + \|\theta\|_{H^2}^2 + \|f_r\|_{L^2[0,L]}^2\right).$$

By (8.1.29), Theorem 8.1.2, and Lemmas 8.4.1 and 8.4.2, we have for any fixed $\gamma \in (0, \gamma_2''']$,

$$\int_0^t e^{\gamma s}\|v_{xxx}(s)\|^2 ds \le C_2, \forall t > 0. \qquad\qquad (8.4.27)$$

Similarly, we can show that there exists a positive constant $\gamma_2 \le \gamma_2'''$ such that, for any fixed $\gamma \in (0, \gamma_2]$,

$$\int_0^t e^{\gamma s}\|\theta_{xxx}(s)\|^2 ds \le C_2, \forall t > 0,$$

which, together with (8.4.26), yields (8.4.25). $\square$

Proof of Theorem 8.1.3. Combining Lemmas 8.4.1–8.4.4, we can obtain (8.1.30). $\square$

8.5 Exponential Stability of Solutions in H^4

In this section, we shall establish the exponential stability of solutions in H^4. We begin with the following lemma.

Lemma 8.5.1. *Under conditions* (8.1.16)–(8.1.22), *if*

$$(u_0, v_0, \theta_0) \in H^4[0, L] \times H_0^4[0, L] \times H^4[0, L],$$

problem (8.1.9)–(8.1.15) *admits a unique global solution*

$$(u(t), v(t), \theta(t)) \in C([0, +\infty), H^4[0, L] \times H_0^4[0, L] \times H^4[0, L])$$

satisfying for any $t > 0$,

$$\|u(t) - \bar{u}\|_{H^4}^2 + \|u_t(t)\|_{H^3}^2 + \|u_{tt}(t)\|_{H^1}^2 + \|v(t)\|_{H^4}^2 + \|v_t(t)\|_{H^2}^2 + \|v_{tt}(t)\|^2$$
$$+ \|\theta(t) - \bar{\theta}\|_{H^4}^2 + \|\theta_t(t)\|_{H^2}^2 + \|\theta_{tt}(t)\|^2 \le C_4, \tag{8.5.1}$$

$$\int_0^t \Big\{ \|u - \bar{u}\|_{H^4}^2 + \|u_t\|_{H^4}^2 + \|u_{tt}\|_{H^2}^2 + \|u_{ttt}\|^2 + \|v\|_{H^5}^2 + \|v_t\|_{H^3}^2 + \|v_{tt}\|_{H^1}^2$$

$$+ \|\theta - \bar{\theta}\|_{H^5}^2 + \|\theta_t\|_{H^3}^2 + \|\theta_{tt}\|_{H^1}^2 \Big\}(\tau)d\tau \le C_4. \tag{8.5.2}$$

Proof. See, e.g., [102] and [111]. $\qquad\square$

Lemma 8.5.2. *Under the assumptions of Theorem 8.1.4, there are positive constants $C_4 > 0$ and $\gamma_4^{(2)} = \gamma_4^{(2)}(C_4) > 0$ such that, for any fixed $\gamma \in (0, \gamma_4^{(2)}]$, it holds that for any $t > 0$,*

$$e^{\gamma t}\|v_{tt}(t)\|^2 + \int_0^t e^{\gamma\tau}\|v_{ttx}(\tau)\|^2 d\tau \le C_4 + C_4 \int_0^t e^{\gamma s}\left(\|v_{txx}\|^2 + \varepsilon\|\theta_{ttx}\|^2\right)(\tau)d\tau, \tag{8.5.3}$$

$$e^{\gamma t}\|\theta_{tt}(t)\|^2 + \int_0^t e^{\gamma\tau}\|\theta_{ttx}(\tau)\| d\tau \le C_4 + C_4 \int_0^t e^{\gamma\tau}\left(\|\theta_{txx}\|^2 + \varepsilon\|v_{ttx}\|^2\right)(\tau)d\tau$$

$$+ C_4 \int_0^t e^{\gamma\tau}\left(\|v_x\|^2 + \|v_{tx}\|^2 + \|\theta\|^2 + \|\theta_t\|^2\right)\|\theta_{tt}\|^2(\tau)d\tau. \tag{8.5.4}$$

Proof. Differentiating (8.1.10)–(8.1.11) with respect to t, we can get

$$\|v_{tt}(t)\| \le C_4(\|\theta_x\| + \|u_x\| + \|v_{txx}\| + \|\theta_t\|_{H^1} + \|f_r\|_{L^2[0,L]} + \|f_t\|_{L^2[0,L]}), \tag{8.5.5}$$

$$\|\theta_{tt}(t)\| \le C_4(\|\theta_t\|_{H^2} + \|v_x\| + \|v_{tx}\| + \|\theta_x\|_{H^1} + \|g_r\|_{L^2[0,L]} + \|g_t\|_{L^2[0,L]}). \tag{8.5.6}$$

Next, differentiating (8.1.10) twice with respect to t, multiplying the result by $e^{\gamma t}v_{tt}$, integrating over $[0, L]$, and using Young's inequality, we obtain

$$\frac{d}{dt}\left(e^{\gamma t}\|v_{tt}(t)\|^2\right) + (C_1^{-1} - C_1\gamma)e^{\gamma t}\|v_{ttx}(t)\|^2 \tag{8.5.7}$$

$$\leq C_4 e^{\gamma t}\left(\|v_{xx}\|^2 + \|\theta_x\|^2 + \|v_{tt}\|^2 + \|v_{txx}\|^2 + \|\theta_{tx}\|^2 + \|\theta_{ttx}\|^2 + \left\|\frac{d^2 f}{dt^2}\right\|^2\right).$$

Now let us integrate (8.5.7) with respect to t, noting that

$$\frac{d^2 f}{dt^2} = f_{rr}v^2 + f_{rt}v + f_r v_t + f_{tt}.$$

Then it follows from (8.1.31) that there exists a positive constant $\gamma_4^{(1)} = \gamma_4^{(1)}(C_4) \leq$ $\min\left(\frac{1}{2C_1^2}, \gamma_2, \alpha_4\right)$ such that, for any fixed $\gamma \in (0, \gamma_4^{(1)}]$,

$$\int_0^t e^{\gamma s}\left(\|f_r\|_{L^2[0,L]}^2 + \|f_{rr}\|_{L^2[0,L]}^2 + \|f_{rt}\|_{L^2[0,L]}^2 + \|f_{tt}\|_{L^2[0,L]}^2\right)(s)ds \leq C_4$$

which, together with (8.5.5) and Theorems 8.1.2–8.1.3, yields (8.5.3).

By the same method, differentiating (8.1.11) twice with respect to t, multiplying the resultant by $e^{\gamma t}\theta_{tt}$, integrating the result over $[0, L]$, and using Young's inequality, we have

$$\frac{d}{dt}\left(e^{\gamma t}\|\theta_{tt}(t)\|^2\right) \leq C_4 e^{\gamma t}\Big\{ -\|\theta_{ttx}\|^2 + \varepsilon\|v_{ttx}\|^2 + \|\theta_{tt}\|^2 + \|v_{tx}\|^2$$

$$+ \|\theta_t\|^2 + \|v_x\|^2 + \|\theta\|^2 + \left\|\frac{d^2 g}{dt^2}\right\|^2\Big\}$$

$$+ C_4 e^{\gamma t}\left(\|v_x\|^2 + \|v_{tx}\|^2 + \|\theta\|^2 + \|\theta_t\|^2\right)\|\theta_{tt}\|^2. \qquad (8.5.8)$$

Finally, integrating (8.5.8) with respect to t and noting that

$$\frac{d^2 g}{dt^2} = g_{rr}v^2 + 2g_{rt}v + g_r v_t + g_{tt},$$

we can derive there exists a positive constant $\gamma_4^{(2)} = \gamma_4^{(2)}(C_4) \leq \gamma_4^{(1)}$ such that, for any fixed $\gamma \in (0, \gamma_4^{(2)}]$,

$$\int_0^t e^{\gamma s}\left(\|g_r\|_{L^2[0,L]}^2 + \|g_{rr}\|_{L^2[0,L]}^2 + \|g_{rt}\|_{L^2[0,L]}^2 + \|g_{tt}\|_{L^2[0,L]}^2\right)(s)ds \leq C_4$$

which, together with (8.5.6) and Theorems 8.1.2–8.1.3, gives (8.5.4). $\qquad \square$

Lemma 8.5.3. *Under the assumptions of Theorem 8.1.4, there are positive constants $C_4 > 0$ and $\gamma_4^{(4)} = \gamma_4^{(4)}(C_4) > 0$ such that, for any fixed $\gamma \in (0, \gamma_4^{(4)}]$, it holds for any $t > 0$ that*

$$e^{\gamma t}\|v_{tx}(t)\|^2 + \int_0^t e^{\gamma \tau}\|v_{txx}(\tau)\|^2 d\tau \leq C_4 + C_4\varepsilon \int_0^t e^{\gamma \tau}\|\theta_{txx}\|^2(\tau)d\tau, \qquad (8.5.9)$$

$$e^{\gamma t}\|\theta_{tx}(t)\|^2 + \int_0^t e^{\gamma\tau}\|\theta_{txx}(\tau)\|^2 d\tau \tag{8.5.10}$$

$$\leq C_4 + C_4 \int_0^t e^{\gamma\tau}\Big[\varepsilon\|v_{txx}\|^2 + (\|v_x\|^2 + \|\theta\|_{H^1}^2 + \|v_{xx}\|^2)\|\theta_{tx}\|^2\Big](\tau)d\tau.$$

Proof. Differentiating (8.1.10) with respect to t and x, multiplying by $e^{\gamma t}v_{tx}$, integrating with respect to x by parts, and using Young's inequality, we have, for any $\varepsilon > 0$,

$$\frac{d}{dt}\Big(e^{\gamma t}\|v_{tx}(t)\|^2\Big) + (C_1^{-1} - C_1\gamma - \varepsilon)e^{\gamma t}\|v_{txx}(t)\|^2 \tag{8.5.11}$$

$$\leq C_4 e^{\gamma t}\left(\|v_{xx}\|^2 + \|\theta_x\|_{H^1}^2 + \varepsilon\|v_{txx}\|^2 + \|v_{tx}\|^2 + \|v_x\|_{H^2}^2 + \varepsilon\|\theta_{txx}\|^2 + \left\|\frac{d^2 f}{dtdx}\right\|^2\right).$$

Integrating (8.5.11) with respect to t and noting that

$$\frac{d^2 f}{dtdx} = f_{rr}ur^{1-n}v + f_{rt}r^{1-n}u + f_r v_x,$$

it follows from (8.1.31) that there exists a positive constant $\gamma_4^{(1)}$ such that, for any fixed $\gamma \in (0, \gamma_4^{(1)}]$,

$$\int_0^t e^{\gamma s}\Big(\|f_r\|_{L^2[0,L]}^2 + \|f_{rr}\|_{L^2[0,L]}^2 + \|f_{rt}\|_{L^2[0,L]}^2\Big)(s)ds \leq C_4$$

which, together with Theorems 8.1.2–8.1.3, yields (8.5.9) if we take $\varepsilon \in (0,1)$ so small that $0 < \varepsilon \leq \min(1, 1/(4C_1))$ and $0 < \gamma < \min(\gamma_4^{(2)}, 1/(4C_1^2)) \equiv \gamma_4^{(3)}$. In the same manner, we can easily show that there exists a positive constant $\gamma_4^{(4)} \leq \gamma_4^{(2)}$ such that for any fixed $\gamma \in (0, \gamma_4^{(4)}]$, (8.5.10) holds. $\square$

Lemma 8.5.4. *Under the assumptions of Theorem 8.1.4, there are positive constants $C_4 > 0$ and $\gamma_4^{(4)} = \gamma_4^{(4)}(C_4) > 0$ such that, for any fixed $\gamma \in (0, \gamma_4^{(4)}]$, it holds for any $t > 0$ that*

$$e^{\gamma t}\Big(\|v_{tt}(t)\|^2 + \|\theta_{tt}(t)\|^2 + \|v_{tx}(t)\|^2 + \|\theta_{tx}(t)\|^2\Big)$$

$$+ \int_0^t e^{\gamma\tau}\Big(\|v_{txx}\|^2 + \|\theta_{txx}\|^2 + \|v_{ttx}\|^2 + \|\theta_{ttx}\|^2\Big)(\tau)d\tau \leq C_4. \tag{8.5.12}$$

Proof. Adding (8.5.9) to (8.5.10), we have

$$e^{\gamma t}\Big(\|v_{tx}(t)\|^2 + \|\theta_{tx}(t)\|^2\Big) + \int_0^t e^{\gamma\tau}\Big(\|v_{txx}\|^2 + \|\theta_{txx}\|^2\Big)(\tau)d\tau$$

$$\leq C_4 + C_4 \int_0^t e^{\gamma\tau}\Big(\|v_x\|^2 + \|\theta_x\|_{H^1}^2 + \|v_{xx}\|^2\Big)\|\theta_{tx}\|^2(\tau)d\tau. \tag{8.5.13}$$

Using Gronwall's inequality and Theorems 8.1.2–8.1.3, we can show that, for $\gamma \in (0, \gamma_4^{(4)}]$,

$$e^{\gamma t}\left(\|v_{tx}(t)\|^2 + \|\theta_{tx}(t)\|^2\right) + \int_0^t e^{\gamma\tau}\left(\|v_{txx}\|^2 + \|\theta_{txx}\|^2\right)(\tau)d\tau \le C_4. \qquad (8.5.14)$$

Multiplying (8.5.3) and (8.5.4) by ε, and then adding the results to (8.5.14), we obtain

$$e^{\gamma t}\left(\|v_{tt}(t)\|^2 + \|\theta_{tt}(t)\|^2 + \|v_{tx}(t)\|^2 + \|\theta_{tx}(t)\|^2\right)$$

$$+ \int_0^t e^{\gamma\tau}\left(\|v_{txx}\|^2 + \|\theta_{txx}\|^2 + \|v_{ttx}\|^2 + \|\theta_{ttx}\|^2\right)(\tau)d\tau$$

$$\le C_4 + C_4(\varepsilon)\int_0^t e^{\gamma\tau}\left(\|v_x\|^2 + \|v_{tx}\|^2 + \|\theta_x\|^2 + \|\theta_t\|^2\right)\|\theta_{tt}\|^2(\tau)d\tau. \qquad (8.5.15)$$

Applying Gronwall's inequality to (8.5.15), and using Theorems 8.1.2–8.1.3, we get (8.5.12). $\qquad\square$

Lemma 8.5.5. *Under the assumptions of Theorem 8.1.4, there are positive constants $C_4 > 0$ and $\gamma_4 < \gamma_4^{(4)}$ such that for any fixed $\gamma \in (0, \gamma_4]$, it holds that, for any $t > 0$,*

$$e^{\gamma t}\|u_{xxx}(t)\|_{H^1}^2 + \int_0^t e^{\gamma\tau}\|u_{xxx}(\tau)\|_{H^1}^2 d\tau \le C_4, \qquad (8.5.16)$$

$$e^{\gamma t}\left(\|v_{xxx}(t)\|_{H^1}^2 + \|\theta_{xxx}(t)\|_{H^1}^2 + \|u_{txxx}(t)\|^2 + \|v_{txx}(t)\|^2 + \|\theta_{txx}(t)\|^2\right)$$

$$+ \int_0^t e^{\gamma\tau}\left(\|v_{tt}\|^2 + \|\theta_{tt}\|^2 + \|\theta_{txx}\|_{H^1}^2 + \|v_{txx}\|_{H^1}^2 + \|u_{txxx}\|_{H^1}^2\right)(\tau)d\tau \le C_4,$$

$$\qquad (8.5.17)$$

$$\int_0^t e^{\gamma\tau}\left(\|v_{xxxx}\|_{H^1}^2 + \|\theta_{xxxx}\|_{H^1}^2\right)(\tau)d\tau \le C_4. \qquad (8.5.18)$$

Proof. Differentiating (8.1.10) with respect to x, and using equation (8.1.9), we get

$$\beta\frac{d}{dt}\left(\frac{u_{xxx}}{u}\right) + R\frac{\theta u_{xxx}}{u^2}$$

$$= (r^{1-n})_{xx}v_t + 2(r^{1-n})_x v_{tx} + r^{1-n}v_{txx} + \beta\left\{3\frac{(r^{n-1}v)_{xxx}u_x}{u^2} + 3\frac{(r^{n-1}v)_{xx}u_{xx}}{u^2}\right.$$

$$+ 6\frac{(r^{n-1}v)_x u_x^3}{u^4} - 6\frac{(r^{n-1}v)_{xx}u_x^2}{u^3} - 6\frac{(r^{n-1}v)_x u_{xx}u_x}{u^3}\right\} + R\left\{\frac{\theta_{xxx}}{u} - 3\frac{\theta_{xx}u_x}{u^2}\right.$$

$$\left. - 3\frac{\theta_x u_{xx}}{u^2} + 4\frac{\theta_x u_x^2}{u^3} + 2\frac{\theta u_{xx}u_x}{u^3} + 2\frac{\theta_x u_x^2}{u^3} + 4\frac{\theta u_x u_{xx}}{u^3} - 6\frac{u_x^3\theta}{u^4}\right\} - (f_{rr}u^2 + f_r u_x)$$

$$:= E(x, t) \tag{8.5.19}$$

with

$$\|E(t)\| \le C_4 \left(\|v_t\|_{H^2} + \|u_x\|_{H^1} + \|v_x\|_{H^2} + \|\theta_x\|_{H^2} + \|f_{rr}\|_{L^2[0,L]} + \|f_r\|_{L^2[0,L]} \right).$$

It follows from Theorem 8.1.3, Lemma 8.5.4 and conditions (8.1.20) and (8.1.29) that for any fixed $\gamma \in (0, \gamma_4^{(4)}]$,

$$\int_0^t e^{\gamma \tau} \|E(\tau)\|^2 d\tau \le C_4, \quad \forall t > 0. \tag{8.5.20}$$

Multiplying (8.5.19) by $e^{\gamma t} \dfrac{u_{xxx}}{u}$ and integrating the result over $[0, L] \times [0, t]$, we obtain

$$e^{\gamma t} \|u_{xxx}(t)\|^2 + \int_0^t (C_2^{-1} - C_2 \gamma) e^{\gamma \tau} \|u_{xxx}(\tau)\|^2 d\tau \le C_4 + C_4 \int_0^t e^{\gamma \tau} \|E(\tau)\|^2 d\tau,$$

so if we take $\gamma_4^{(5)} = \min(\frac{1}{4C_2^2}, \gamma_4^{(4)}) > 0$, then for any fixed $\gamma \in (0, \gamma_4^{(5)}]$, we get

$$e^{\gamma t} \|u_{xxx}(t)\|^2 + \int_0^t e^{\gamma \tau} \|u_{xxx}(\tau)\|^2 d\tau \le C_4. \tag{8.5.21}$$

Differentiating (8.1.10) with respect to x, we can deduce that

$$\|v_{xxx}(t)\| \le C_4 \left(\|v_{tx}\| + \|\theta\|_{H^2} + \|v\|_{H^2} + \|u\|_{H^2} + \|f_r\|_{L^2[0,L]} \right) \le C_4. \tag{8.5.22}$$

Next, differentiating (8.1.10) with respect to x twice, we have

$$\|v_{xxxx}(t)\| \le C_4 \left(\|v_{txx}\| + \|\theta\|_{H^3} + \|v\|_{H^3} + \|u\|_{H^3} + \|f_{rr}\|_{L^2[0,L]} + \|f_r\|_{L^2[0,L]} \right). \tag{8.5.23}$$

By the same method, differentiating (8.1.11) with respect to x once and twice, respectively, we get

$$\|\theta_{xxx}(t)\| \le C_4 (\|\theta_{tx}\| + \|v\|_{H^2} + \|\theta\|_{H^2} + \|u\|_{H^2} + \|g_r\|_{L^2[0,L]}) \le C_4, \tag{8.5.24}$$

$$\begin{aligned} \|\theta_{xxxx}(t)\| \le C_4 (\|\theta_{txx}\| &+ \|v\|_{H^3} + \|\theta\|_{H^3} + \|u\|_{H^3} \\ &+ \|g_{rr}\|_{L^2[0,L]} + \|g_r\|_{L^2[0,L]}). \end{aligned} \tag{8.5.25}$$

By Theorem 8.1.3 and (8.5.13), we have for $\gamma \in (0, \gamma_4^{(4)}]$,

$$e^{\gamma t} \left(\|v_{xxx}(t)\|^2 + \|\theta_{xxx}(t)\|^2 \right) + \int_0^t e^{\gamma \tau} \left(\|v_{xxx}\|_{H^1}^2 + \|\theta_{xxx}\|_{H^1}^2 \right)(\tau) d\tau \le C_4. \tag{8.5.26}$$

Differentiating (8.1.10) and (8.1.11) with respect to t and using Theorem 8.1.2 and Lemma 8.4.1, we can show that

$$\|v_{txx}(t)\| \le C_4 \left(\|v_{tt}\| + \|\theta_{tx}\| + \|u_{tx}\| + \|f_r\|_{L^2[0,L]} + \|f_t\|_{L^2[0,L]} \right)$$
$$\le C_4, \tag{8.5.27}$$

$$\|\theta_{txx}(t)\| \le C_4 \left(\|\theta_{tt}\| + \|v_{tx}\| + \|u_{tx}\| + \|g_r\|_{L^2[0,L]} + \|g_t\|_{L^2[0,L]} \right)$$
$$\le C_4. \tag{8.5.28}$$

By (8.5.23), (8.5.25)–(8.5.28), Theorems 8.1.2–8.1.3 and conditions (8.1.21)–(8.1.22), we have, for $\gamma \in (0, \gamma_4^{(5)}]$,

$$e^{\gamma t} \left(\|v_{xxxx}(t)\|^2 + \|\theta_{xxxx}(t)\|^2 \right) + \int_0^t e^{\gamma \tau} (\|v_{xxxx}\|^2 + \|\theta_{xxxx}\|^2)(\tau)d\tau \le C_4. \tag{8.5.29}$$

Further, differentiating (8.5.19) with respect to x, we get

$$\beta \frac{d}{dt} \left(\frac{u_{xxxx}}{u} \right) + R \frac{\theta u_{xxxx}}{u^2}$$
$$= \frac{(r^{n-1}v)_{xxxx}u_x}{u^2} + \frac{u_{xxx}(r^{n-1}v)_{xx}}{u^2} - 2\frac{u_{xxx}u_x^2}{u^3} + 2\frac{\theta u_{xxx}u_x}{u^3} - \frac{\theta_x u_{xxx}}{u^2} + E_x$$
$$= E_1(x,t), \tag{8.5.30}$$

with

$$\|E_1(t)\| \le C_4 \left(\|u\|_{H^3} + \|v\|_{H^4} + \|\theta\|_{H^4} + \|v_t\|_{H^3} + \left\| \frac{d^3 f}{dx^3} \right\| \right). \tag{8.5.31}$$

It follows from Theorem 8.1.3, Lemma 8.5.4 and conditions (8.1.21) and (8.1.30) that there exists a positive constant $\gamma_4^{(6)} = \gamma_4^{(6)}(C_4) \le \gamma_4^{(5)}$ such that, for any fixed $\gamma \in (0, \gamma_4^{(6)}]$,

$$\int_0^t e^{\gamma \tau} \|E_1(\tau)\|^2 d\tau \le C_4, \quad \forall t > 0. \tag{8.5.32}$$

Differentiating (8.1.10) with respect to t and x, we can derive

$$\|v_{txxx}(t)\| \le C_4 \left(\|v_{ttx}\| + \|\theta_{txx}\| + \|f_r\|_{L^2[0,L]} + \|f_{rt}\|_{L^2[0,L]} + \|f_{rr}\|_{L^2[0,L]} \right.$$
$$\left. + \|v_t\|_{H^2} + \|v\|_{H^3} + \|\theta\|_{H^2} \right). \tag{8.5.33}$$

Multiplying (8.5.30) by $e^{\gamma t} \dfrac{u_{xxxx}}{u}$, integrating the resultant over $[0, L] \times [0, t]$, and using Theorem 8.1.3, Lemma 8.5.4, (8.5.21), (8.5.29), (8.5.32) and (8.1.21)–(8.1.22), we obtain, for $\gamma \in (0, \gamma_4^{(6)}]$,

$$e^{\gamma t} \|u_{xxxx}(t)\|^2 + \int_0^t e^{\gamma \tau} \|u_{xxxx}(\tau)\|^2 d\tau \le C_4, \quad \forall t > 0. \tag{8.5.34}$$

Differentiating (8.1.10) and (8.1.11) three times with respect to x respectively, we get

$$\|v_{xxxxx}(t)\| \le C_4 \left(\|v_t\|_{H^3} + \|v\|_{H^4} + \|\theta\|_{H^4} + \|u\|_{H^4} + \left\| \frac{d^3 f}{dx^3} \right\| \right), \qquad (8.5.35)$$

$$\|\theta_{xxxxx}(t)\| \le C_4 \left(\|\theta_t\|_{H^3} + \|v\|_{H^4} + \|\theta\|_{H^4} + \|u\|_{H^3} + \left\| \frac{d^3 g}{dx^3} \right\| \right). \qquad (8.5.36)$$

Differentiating (8.1.11) with respect to t and x, we get

$$\|\theta_{txxx}(t)\| \le C_4 \left(\|\theta_{ttx}\| + \|\theta\|_{H^3} + \|\theta_t\|_{H^2} + \|v_t\|_{H^2} + \|v\|_{H^2} + \|u_{tx}\| + \left\| \frac{d^2 g}{dtdx} \right\| \right).$$
$$(8.5.37)$$

By (8.1.9), we obtain

$$\|u_{txxx}(t)\| \le C_4 \|v(t)\|_{H^4}, \quad \|u_{txxx}(t)\|_{H^1} \le C_4 \|v(t)\|_{H^5}. \qquad (8.5.38)$$

Similarly to (8.4.26), we conclude that there exists a positive constant $\gamma_4 \le \gamma_4^{(6)}$ such that, for any fixed $\gamma \in (0, \gamma_4]$, (8.5.18) holds.

Finally, combining (8.5.5)–(8.5.6), (8.5.29), (8.5.32), (8.5.35)–(8.5.38) and the embedding theorem, for any fixed $\gamma \in (0, \gamma_4]$, we obtain (8.5.17).   □

This completes the proof of Theorem 8.1.4.   □

8.6 Bibliographic Comments

In the case when $f = g \equiv 0$, Fujita-Yashima and Benabidallah [138, 139] established the global existence of solutions to problem (8.1.9)–(8.1.15), Jiang [54] proved the large-time behavior of global solutions in H^1, Zheng and Qin [150] obtained the global existence of universal attractors in H^1 and H^2, and Qin et al. [108] established the exponential stability of global solutions in H^4.

In the case when $f \neq 0$, $g \neq 0$, Qin and Huang [102], and Qin and Wen [111] proved the global existence of solutions in H^i ($i = 1, 2, 4$) to problem (8.1.9)–(8.1.15), but the asymptotic behavior remained open. In this chapter, we established the asymptotic behavior and the exponential stability of solutions to problem (8.1.9)–(8.1.15).

We also refer the reader to the related results in Cho, Choe and Kim [12], Qin and Muñoz Rivera [109], Xu and Yang [136], Yanagi [137], Zheng [148], Zheng and Qin [149], and Zimmer [151].

8.6 Bibliographic Comments

Bibliography

[1] S.N. Antontsev, A.V. Kazhikhov and V.N. Monakhov, Boundary Value Problems in Mechanics of Nonhomogeneous Fluids, Studies in Mathematics and its Applications, **22**, North-Holland, Amsterdam, 1990.

[2] E. Becker, Gasdynamik, Teubner, Stuttgart, 1966.

[3] H. Bellout, F. Bloom and J. Nečas, Young measure-valued solutions for non-Newtonian incompressible viscous fluids, Comm. Partial Differential Equations, **19** (1994), 1763–1803.

[4] H. Bellout, F. Bloom and J. Nečas, Existence, uniqueness and stability of solutions to the initial-boundary value problem for bipolar viscous fluids, Differential Integral Equations, **8** (1995), 453–464.

[5] F. Bloom and W. Hao, Regularization of a non-Newtonian system in an unbound channel: existence of a maximal compact attractor, Nonlinear Anal., Ser. A: Theory Methods, **43** (2001), 743-766.

[6] J.W. Bond, K.M. Watson and J.A. Welch, Atomic Theory of Gas Dynamics, Addison-Wesley, 1965.

[7] H. Cabannes, Theoretical Magnetofluiddynamics, Academic Press, New York, 1970.

[8] G.-Q. Chen, Global solutions to the compressible Navier-Stokes equations for a reacting mixture, SIAM J. Math. Anal., **23** (1992), 609–634.

[9] G.-Q. Chen, D. Hoff and K. Trivisa, Global solutions of the compressible Navier-Stokes equations with large discontinuous initial data, Comm. Partial Differential Equations, **25** (2000), 2233–2257.

[10] G.-Q. Chen and D. Wang, Global solutions of nonlinear magnetohydrodynamics with large initial data, J. Differential Equations, **182** (2002), 344–376.

[11] G.-Q. Chen and D. Wang, Existence and continuous dependence of large solutions for the magnetohydrodynamics equations, Z. Angew Math. Phys., **54** (2003), 608–632.

[12] Y. Cho, H.J. Choe, and H. Kim, Unique solvability of the initial boundary value problems for compressible viscous fluids, J. Math. Pures Appl., **83** (2004), 243–275.

[13] P.C. Clemmow and J.P. Dougherty, Electrodynamics of Particles and Plasmas, Addison-Wesley, New York, 1962.

[14] J.P. Cox and R.T. Giuli, Principles of Stellar Structure, I, II, Gordon and Breach, New York, 1968.

[15] C.M. Dafermos, Global smooth solutions to the initial-boundary value problem for the equations of one-dimensional nonlinear thermoviscoelasticity, SIAM. J. Math. Anal., **13** (1982), 397–408.

[16] C.M. Dafermos and L. Hsiao, Global smooth thermomechanical processes in one-dimensional nonlinear thermoviscoelasticity, Nonlinear Anal., **6** (1982), 435–454.

[17] P.A. Davidson, Turbulence: An Introduction for Scientists and Engineers, Oxford Univ. Press, Oxford, 2004.

[18] R.J. DiPerna and P.-L. Lions, Ordinary differential equations, transport theory and Sobolev spaces, Invent. Math., **98** (1989), 511–547.

[19] D. Donatelli and K. Trivisa, On the motion of a viscous compressible radiative-reacting gas, Comm. Math. Phys., **265** (2006), 463–491.

[20] B. Dong and Y. Li, Large time behavior to the system of incompressible non-Newtonian fluids in $\mathbb{R}^2$, J. Math. Anal. Appl., **298** (2004), 667–676.

[21] B. Ducomet, Hydrodynamical models of gaseous stars, Rev. Math. Phys., **8** (1996), 957–1000.

[22] B. Ducomet, A model of thermal dissipation for a one-dimensional viscous reactive and radiative gas, Math. Methods Appl. Sci., **22** (1999), 1323–1349.

[23] B. Ducomet, E. Feireisl, H. Petzeltová and I. Straškraba, Existence globale pour un fluide barotrope autogravitant, C. R. Acad. Sci. Paris, Sér. I, Math. **332** (2001), 627–632.

[24] B. Ducomet and E. Feireisl, On the dynamics of gaseous stars, Arch. Ration. Mech. Anal., **174** (2004), 221–266.

[25] B. Ducomet and E. Feireisl, The equations of magnetohydrodynamics: on the interaction between matter and radiation in the evolution of gaseous stars, Comm. Math. Phys., **266** (2006), 595–629.

[26] B. Ducomet and A. Zlotnik, Stabilization for viscous compressible heat-conducting media equations with nonmonotone state functions, C. R. Acad. Sci. Paris, Ser. I, **334** (2002), 119–124.

[27] B. Ducomet and A. Zlotnik, Stabilization for equations of one-dimensional viscous compressible heat-conducting media with nonmonotone equation of state, J. Differential Equations, **194** (2003), 51–81.

[28] B. Ducomet and A. Zlotnik, Stabilization for 1D radiative and reactive viscous gas flows, C. R. Acad. Sci. Paris, Sér. I, **338** (2004), 127–132.

[29] B. Ducomet and A. Zlotnik, On the large-time behavior of 1D radiative and reactive viscous flows for higher-order kinetics, Nonlinear Anal., **63** (2005), 1011–1033.

[30] B. Ducomet and A. Zlotnik, Lyapunov functional method for 1D radiative and reactive viscous gas dynamics, Arch. Ration. Mech. Anal., **177** (2005), 185–229.

[31] S. Eliezer, A. Ghatak and H. Hora, An Introduction to Equations of State: Theory and Applications, Cambridge Univ. Press, Cambridge, 1986.

[32] C.A. Eringen, Theory of micropolar fluids, J. Math. Mech., **16** (1966), 1–18.

[33] J. Fan, S. Jiang and G. Nakamura, Stability of weak solutions to equations of magnetohydrodynamics with Lebesgue initial data, J. Differential Equations, **251** (2011), 2025–2036.

[34] J. Fan, S. Jiang and G. Nakamura, Vanishing shear viscosity limit in the magnetohydrodynamic equations, Comm. Math. Phys., **270** (2007), 691–708.

[35] E. Feireisl, Dynamics of Viscous Compressible Fluids, Oxford Lecture Series in Mathematics and its Applications, **26**, Oxford University Press, Oxford, 2004.

[36] E. Feireisl and A. Novotný, Singular Limits in Thermodynamics of Viscous Fluids, Advances in Mathematical Fluid Mechanics, Birkhäuser, Basel, 2009.

[37] E. Feireisl and H. Pelzeltová, On the long-time behaviour of solutions to the Navier-Stokes-Fourier system with a time-dependent driving force, J. Dynam. Differential Equations, **19** (2007), 687–707.

[38] E. Feireisl and D. Prazak, A stabilizing effect of a high-frequency driving force on the motion of a viscous, compressible, and heat conducting fluid, Discrete Contin. Dyn. Syst. Ser. S **2** (2009), 95–111.

[39] M. Forestini, Principes Fondamentaux de Structure Stellaire, Gordon and Breach, Le Bris, 1999.

[40] H. Freistühler and P. Szmolyan, Existence and bifurcation of viscous profiles for all intermediate magnetohydrodynamic shock waves, SIAM J. Math. Anal., **26** (1995), 112–128.

[41] J.F. Gerbeau, C. Le Bris and T. Lelièvre, Mathematical Methods for the Magnetohydrodynamics of Liquid Metals, Oxford University Press, Oxford, 2006.

[42] J.P.H. Goedbloed and S. Poedts, Principles of Magnetohydrodynamics: With Applications to Laboratory and Astrophysical Plasmas, Cambridge University Press, 2004.

[43] B. Guo and P. Zhu, Asymptotic behavior of the solution to the system for a viscous reactive gas, J. Differential Equations, **155** (1999), 177–202.

[44] B. Guo and P. Zhu, Partial regularity of suitable weak solutions to the system of the incompressible non-Newtonian fluids, J. Differential Equations, **178** (2002), 281–297.

[45] D. Hoff, Global solutions of the Navier-Stokes for multidimensional compressible flow with discontinuous initial data, J. Differential Equations, **120** (1995), 215–254.

[46] D. Hoff, Discontinuous solutions of the Navier-Stokes equations for multidi-
     mensional heat-conducting fluids, Arch. Rational Mech. Anal., **139** (1997),
     303–354.

[47] D. Hoff and E. Tsyganov, Uniqueness and continuous dependence of weak
     solutions in compressible magnetohydrodynamics, Z. Angew. Math. Phys.,
     56 (2005), 791–804.

[48] L. Hsiao and T. Luo, Large-time behaviour of solutions for the outer pressure
     problem of a viscous heat-conductive one-dimensional real gas, Proc. Roy.
     Soc. Edinburgh Sect. A, **126** (1996), 1277–1296.

[49] X. Hu and D. Wang, Compactness of weak solutions to the three-dimensional
     compressible magnetohydrodynamic equations, J. Differential Equations, **245**
     (2008), 2176–2198.

[50] S. Jiang, On initial-boundary value problems for a viscous, heat-conducting
     one-dimensional real gas, J. Differential Equations, **110** (1994), 157–181.

[51] S. Jiang, On the asymptotic behavior of the motion of a viscous, heat-
     conducting, one-dimensional real gas, Math. Z., **216** (1994), 317–336.

[52] S. Jiang, Global spherically symmetric solutions to the equations of a viscous
     polytropic ideal gas in an exterior domain, Comm. Math. Phys., **178** (1996),
     339–374.

[53] S. Jiang, Global solutions of the Cauchy problem for a viscous polytropic
     ideal gas, Ann. Scuola. Norm Sup. Pisa Cl. Sci. (4), **26** (1998), 47–74.

[54] S. Jiang, Large-time behavior of solutions to the equations of a viscous poly-
     tropic ideal gas, Ann. Mat. Pura Appl. (4), **175** (1998), 253–275.

[55] S. Jiang, Large-time behavior of solutions to the equations of a one-
     dimensional viscous polytropic ideal gas in unbounded domains, Comm.
     Math. Phys., **200** (1999), 181–193.

[56] Ya.I. Kanel', Cauchy problem for the equations of gas dynamics with viscos-
     ity, Siberian Math. J., **20** (1979), 208–218.

[57] S. Kawashima and T. Nishida, Global solutions to the initial value problem
     for the equations of one-dimensional motion of viscous polytropic gases, J.
     Math. Kyoto. Univ., **21** (1981), 825–837.

[58] S. Kawashima and M. Okada, Smooth global solutions for the one-dim-
     ensional equations in magnetohydrodynamics, Proc. Japan Acad. Ser. A
     Math. Sci., **58** (1982), 384–387.

[59] B. Kawohl, Global existence of large solutions to initial-boundary value prob-
     lems for a viscous, heat-conducting, one-dimensional real gas, J. Differential
     Equations, **58** (1985), 76–103.

[60] A.V. Kazhykhov, Sur la solubilité globale des problémes monodimension-
     nels aux valeurs initiales-limitées pour les équations d'un gaz visqueux et
     calorifère, C. R. Acad. Sci. Paris, **284 A** (1977), 317–320.

[61] A.V. Kazhikhov, On the theory of boundary value problems for equations of the one-dimensional time-dependent motion of viscous heat-conducting gases, Boundary Value Problems for Hydrodynamical Equations (in Russian), No. **50**, Inst. Hydrodynamics, Siberian Branch Akad., USSR., (1981), 37–62.

[62] A.V. Kazhikhov, Cauchy problem for viscous gas equations, Siberian Math. J., **23** (1982), 44–49.

[63] A.V. Kazhikhov and V.V. Shelukhin, Unique global solution with respect to time of initial-boundary value problems for one-dimensional equations of a viscous gas, J. Appl. Math. Mech., **41** (1977), 273–282.

[64] A.V. Kazhikhov and Sh.S. Smagulov, Well-posedness and approximation methods for a model of magnetogasdynamics, Izv. Akad. Nauk Kazakh. SSR Ser. Fiz.-Mat., **5** (1986), 17–19.

[65] R. Kippenhahn and A. Weigert, Stellar Structure and Evolution, 2nd edition, Springer Verlag, Berlin, 2012.

[66] O.A. Ladyzhenskaya, The Mathematical Theory of Viscous Incompressible Flow, Gordon and Breach, New York, 1969.

[67] O.A. Ladyzhenskaya, New equations for the description of viscous incompressible fluids and solvability in the large of the boundary value problems for them, Proceedings of the Steklov Institute of Mathematics, Boundary Value Problems of Mathematical Physics, Vol. V, American Mathematical Society, Providence, RI, 1970.

[68] O.A. Ladyženskaja and V.A. Solonnikov, The unique solvability of an initial-boundary value problem for viscous incompressible inhomogeneous fluids, Zap. Nauchn. Sem. LOMI, **52** (1975), 52–109.

[69] O.A. Ladyženskaja, V.A. Solonnikov and N.N. Ural'ceva, Linear and Quasilinear Equations of Parabolic Type, Transl. Math. Monogr., Vol. **23**, Amer. Math. Soc., Providence, RI, 1968.

[70] L.D. Landau and E.M. Lifshitz, Electrodynamics of Continuous Media, Pergamon Press, New York, 1960.

[71] P.-L. Lions, Mathematical Topics in Fluid Mechanics, Vols. 1, 2, Clarendon Press, Oxford, 1996, 1998.

[72] T.-P. Liu and Y. Zeng, Large-time behavior of solutions for general quasilinear hyperbolic-parabolic systems of conservation laws, Memoirs Amer. Math. Soc., **599** (1997).

[73] G. Łukaszewicz, Micropolar Fluids. Theory and Applications, Birkhäuser, Boston, 1999.

[74] A. Matsumura and T. Nishida, The initial value problems for the equations of motion of compressible viscous and heat-conductive fluids, Proc. Japan Acad. Ser. A Math. Sci., **55** (1979), 337–342.

[75] A. Matsumura and T. Nishida, The initial value problems for the equations of motion of viscous and heat-conductive gases, J. Math. Kyoto. Univ., **20** (1980), 67–104.

[76] A. Matsumura and T. Nishida, Initial-boundary value problems for the equations of motion of general fluids, In: Computing Methods in Applied Sciences and Engineering. V. (R. Glowinski and J.L. Lions, eds.), North-Holland, Amsterdam, 1982, 389–406.

[77] A. Matsumura and T. Nishida, Initial-boundary value problems for the equations of motion of compressible viscous and heat-conductive fluids, Comm. Math. Phys., **89** (1983), 445–464.

[78] D. Mihalas and B. Weibel Mihalas, Foundations of Radiation Hydrodynamics, Oxford Univ. Press, New York, 1984.

[79] N. Mujaković, One-dimensional flow of a compressible viscous micropolar fluid: stabilization of the solution, in: Proceedings of the Conference on Applied Mathematics and Scientific Computing, 253–262, Springer, Dordrecht, 2005.

[80] N. Mujaković, One-dimensional flow of a compressible viscous micropolar fluid: the Cauchy problem, Math. Comm., **10** (2005), 1–14.

[81] N. Mujaković, Uniqueness of a solution of the Cauchy problem for one-dimensional compressible viscous micropolar fluid model, Appl. Math. E-Notes, **6** (2006), 113–118.

[82] N. Mujaković, One-dimensional compressible viscous micropolar fluid model: stabilization of the solution for the Cauchy problem, Bound. Value Prob., **2010**, Article ID 796065, 21 pages, 2010.

[83] N. Mujaković and I. Dražić, The Cauchy problem for one-dimensional flow of a compressible viscous fluid: stabilization of the solution, Glasnik Math., **46** (2011), 215–231.

[84] N. Mujaković, Global in time estimates for one-dimensional compressible viscous micropolar fluid model, Glasnik Math., **40** (2005), 103–120. a local existence theorem, Ann. Univ Ferrara, **53**(2007), 361–379. regularity of the solution, Boundary Value Problems, Volume 2008, Article ID 189748, 15pages.

[85] N. Mujaković, Non-homogeneous boundary value problem for one-dimensional compressible viscous micropolar fluid model: a global existence theorem. Math. Inequal. Appl., **12** (2009), 651–662.

[86] T. Nagasawa, On the one-dimensional motion of the polytropic ideal gas nonfixed on the boundary, J. Differential Equations, **65** (1986), 49–67.

[87] T. Nagasawa, On the outer pressure problem of the one-dimensional polytropic ideal gas, Japan J. Appl. Math., **5** (1988), 53–85.

[88] T. Nagasawa, Global asymptotics of the outer pressure problem with free boundary, Japan J. Appl. Math., **5** (1988), 205–224.

[89] T. Nagasawa, On the asymptotic behavior of the one-dimensional motion of the polytropic ideal gas with stress-free condition, Quart. Appl. Math., **46** (1988), 665–679.

[90] Š. Nečasová and P. Penel, L^2 decay for weak solution to equations of non-Newtonian incompressible fluids in the whole space, Nonlinear Anal., **47** (2001), 4181–4192.

[91] A. Novotný and I. Straškraba, Introduction to the Mathematical Theory of Compressible Flow, Oxford University Press, Oxford, 2004.

[92] S.N. Ojha and A. Singh, Growth and decay of sonic waves in thermally radiative mangnetogasdynamics, Astrophys. Space Sci., **179** (1991), 45–54.

[93] M. Okada and S. Kawashima, On the equations of one-dimensional motion of compressible viscous fluids, J. Math. Kyoto Univ., **23** (1983), 55–71.

[94] Paulo R. de Souza Mendes, Dimensionless non-Newtonian fluid mechanics, J. Non-Newton. Fluid Mech., **147** (2007), 109–116.

[95] G.C, Pomraning, The Equations of Radiation Hydrodynamics, Pergamon Press, 1973.

[96] Y. Qin, Global existence and asymptotic behavior for the solution to nonlinear viscous, heat-conductive, one-dimensional real gas, Adv. Math. Sci. Appl., **59** (2001), 119–148.

[97] Y. Qin, Global existence and asymptotic behavior for a viscous, heat-conductive, one-dimensional real gas with fixed and thermally insulated endpoints, Nonlinear Anal., Ser. A. Theory Methods, **44** (2001), 413–441.

[98] Y. Qin, Global existence and asymptotic behavior for a viscous heat-conducting one-dimensional real gas with fixed and constant temperature boundary conditions, Adv. Differential Equations, **7** (2002), 129–154.

[99] Y. Qin, Exponential stability for a nonlinear one-dimensional heat-conductive viscous real gas, J. Math. Anal. Appl., **272** (2002), 507–535.

[100] Y. Qin, Universal attractor in H^4 for the nonlinear one-dimensional compressible Navier-Stokes equations, J. Differential Equations, **207** (2004), 21–72.

[101] Y. Qin, Nonlinear Parabolic-Hyperbolic Coupled Systems and their Attractors, Operator Theory: Advances and Applications, **184**, Advances in PDEs, Birkhäuser, Basel, 2008.

[102] Y. Qin and Lan Huang, Global Well-Posedness of Nonlinear Parabolic-Hyperbolic Coupled Systems, Frontiers in Mathematics, Birkhäuser, Springer, Basel-Boston-Berlin, 2012.

[103] Y. Qin, G. Hu, T. Wang, L. Huang and Z. Ma, Remarks on global smooth solutions to a 1D self-gravitating viscous radiative and reactive gas, J. Math. Anal. Appl., **408** (2013), 19–26.

[104] Y. Qin, G. Hu and T. Wang, Global smooth solutions for the compressible viscous and heat-conductive gas, Quart. Appl. Math., **62** (2011), 509–528.

[105] Y. Qin, L. Huang, S. Deng, Z. Ma, X. Su and X. Yang, On the 1D viscous reactive and radiative gas with the one-order Arrhenius kinetics, preprint.

[106] Y. Qin, X. Liu and X. Yang, Global existence and exponential stability of solutions to the 1D full non-Newtonian fluids, Nonlinear Anal., Real World Appl., **13** (2012), 607–633.

[107] Y. Qin, X. Liu and X. Yang, Global existence and exponential stability for a 1D compressible and radiative MHD flow, J. Differential Equations, **253** (2012), 1439–1488.

[108] Y. Qin, T.F. Ma, M.M. Cavalcanti and D. Andrade, Exponential stability in H^4 for the Navier-Stokes equations of viscous and heat-conductive fluid, Commun. Pure Appl. Anal., **4** (2005), 635–664.

[109] Y. Qin and J. E. Muñoz Rivera, Global existence and exponential stability of solutions to thermoelastic equations of hyperbolic type, J. Elasticity, **75** (2005), 125–145.

[110] Y. Qin, T. Wang and G. Hu, The Cauchy problem for a 1D compressible viscous micropolar fluid model: analysis of the stabilization and the regularity, Nonlinear Anal., Real World Appl., **13** (2012), 1010–1029.

[111] Y. Qin and S. Wen, Global existence of spherically symmetric solutions for nonlinear compressible Navier-Stokes equations, J. Math. Phys., **49** (2008) 023101, 25pages.

[112] Y. Qin, Y. Wu and F. Liu, On the Cauchy problem for a one-dimensional compressible viscous polytropic ideal gas, Port. Math., **64** (2007), 87–126.

[113] Y. Qin and X. Yu, Global existence and asymptotic behavior for the compressible Navier-Stokes equations with a non-autonomous external force and a heat source, Math. Meth. Appl. Sci., **32** (2009), 1011–1040.

[114] O. Rozanova, Nonexistence results for a compressible non-Newtonian fluid with magnetic effects in the whole space, J. Math. Anal. Appl., **371** (2010), 190–194.

[115] P. Secchi, On the motion of gaseous stars in the presence of radiation, Comm. Partial Differential Equations, **15** (1990), 185–204.

[116] M. Sermange and R. Temam, Some mathematical questions related to the MHD equations, Comm. Pure Appl. Math., **36** (1983), 635–664.

[117] J. Simon, Compact sets in the space $L^p(0, T; B)$, Ann. Mat. Pura Appl. (4), **146** (1987), 65–96.

[118] S.F. Shandarin and Ya.B. Zel'dovich, The large-scale structure of the universe: turbulence, intermittency, structures in a self-gravitating medium, Rev. Modern Phys., **61** (1989), 185–220.

[119] V.V. Shelukhin, A shear flow problem for the compressible Navier-Stokes equations, Internat. J. Non-Linear Mech., **33** (1998), 247–257.

[120] W. Shen and S. Zheng, On the coupled Cahn-Hilliard equations, Comm. Partial Differential Equations, **18** (1993), 701–727.

[121] W. Shen, S. Zheng and P. Zhu, Global existence and asymptotic behavior of weak solutions to nonlinear thermoviscoelastic systems with clamped boundary conditions, Quart. Appl. Math., **57**, (1999), 93–116.

[122] Y.D. Shi, Some results of boundary problem of non-Newtonian fluids, Systems Sci. Math. Sci., **9** (1996), 107–119.

[123] S.N. Shore, An Introduction to Astrophysical Hydrodynamics, Academic Press, New York, 1992.

[124] V.A. Solonnikov and A.V. Kazhikhov, Existence theorems for the equations of motion of a compressible viscous fluid, Annual Rev. Fluid Mech., **13** (1981), 79–95.

[125] G. Ströhmer, About compressible viscous fluid flow in a bounded region, Pacific J. Math., **143** (1990), 359–375.

[126] A. Tani, On the first initial-boundary value problem for compressibloe viscous fluid motion, Publ. Res. Inst. Math. Sci. Kyoto Univ., **13** (1977), 193–253.

[127] A. Tani, On the free boundary value problem for compressible viscous fluid motion, J. Math. Kyoto Univ., **21** (1981), 839–859.

[128] E. Tsyganov and D. Hoff, Systems of partial differential equations of mixed hyperbolic-parabolic type, J. Differential Equations, **204** (2004), 163–201.

[129] M. Umehara and A. Tani, Global solution to the one-dimensional equations for a self-gravitating viscous radiative and reactive gas, J. Differential Equations, **234** (2007), 439–463.

[130] M. Umehara and A. Tani, Global solvability of the free-boundary problem for one-dimensional motion of a self-gravitating viscous radiative and reactive gas, Proc. Japan Acad., Ser. A Math. Sci. **84**(2008), 123–128.

[131] A.I. Vol'pert and S.I. Hudjaev, On the Cauchy problem for composite systems of nonlinear differential equations, Math. USSR-Sb., **16** (1972), 517–544.

[132] D. Wang, Large solutions to the initial-boundary value problem for planar magnetohydrodynamics, SIAM J. Appl. Math., **63** (2003), 1424–1441.

[133] C. Wang and H. Yuan, Global strong solutions for a class of heat-conducting non-Newtonian fluids with vacuum, Nonlinear Anal., Real World Appl., **11** (2010), 3680–3703.

[134] F.A. Williams, Combustion Theory, Benjamin/Cummings, Menlo Park, 1985.

[135] L.C. Woods, Principles of Magnetoplasma Dynamics, The Clarendon Press, Oxford University Press, New York, 1987.

[136] C. Xu and T. Yang, Local existence with physical vacuum boundary condition to Euler equations with damping, J. Differential Equations, **210** (2005), 217–231.

[137] S. Yanagi, Existence of periodic solutions for a one-dimensional isentropic model system of compressible viscous gas, Nonlinear Anal., Ser. A: Theory Methods, **46** (2001), 279–298.

[138] H. Fujita-Yashima and R. Benabidallah, Unicité de la solution de l'équation menodimensionnelle ou à symétrie sphérique d'un gaz visqueux calorifère, Rend. Circ. Mat. Palermo, Ser II, **42** (1993), 195–218.

[139] H. Fujita-Yashima and R. Benabidallah, Équation à symétrie sphérique d'un gaz visqueux et calorifère avec la surface libre, Ann. Mat. Pura Appl., **168** (1995), 75–117.

[140] H. Fujita-Yashima, M. Padula and A. Novotny, Équation monodimension-nelle d'un gas visqueux et calorifère avec des conditions initial moins restrictives, *Ricerche Mat.*, **42** (1993), 199–248.

[141] L. Yin, Y. Yu and H. Yuan, Global existence and uniqueness of solution of initial boundary value problem for a class of non-Newtonian fluids with vacuum, Z. Angew. Math. Phys., **59** (2008), 457–474.

[142] H. Yuan and C. Wang, Unique solvability for a class of full non-Newtonian fluids of one dimension with vacuum, Z. Angew. Math. Phys., **60** (2009), 868–898.

[143] J.P. Zahn and J. Zinn-Justin, editors, Astrophysical Fluid Dynamics, Les Houches, vol. XLVII, Elsevier, Amsterdam, 1993.

[144] Ya.B. Zel'dovich and Yu.P. Raizer, Physics of Shock Waves and High-Temperature Hydrodynamic Phenomena, Vol. 2, Academic Press, New York, 1967.

[145] J. Zhang and F. Xie, Global solution for a one-dimensional model problem in thermally radiative magnetohydrodynamics, J. Differential Equations, **245** (2008), 1853–1882.

[146] C. Zhao and Y. Li, H^2-compact attractor for a non-Newtonian system in two-dimensional unbounded domains, Nonlinear Anal., **56** (2004), 1091–1103.

[147] C. Zhao, S. Zhou and Y. Li, Trajectory attractor and global attractor for a two-dimensional incompressible non-Newtonian fluid, J. Math. Anal. Appl., **325** (2007), 1350–1362.

[148] S. Zheng, Nonlinear Evolution Equations, Chapman & Hall/CRC Monographs and Surveys in Pure and Applied Mathematics, 133, CRC Press, Boca Raton, FL, 2004.

[149] S. Zheng and Y. Qin, Maximal attractor for the system of one-dimensional polytropic viscous ideal gas, Quart. Appl. Math., **59** (2001), 579–599.

[150] S. Zheng and Y. Qin, Universal attractors for the Navier-Stokes equations of compressible and heat-conductive fluid in bounded annular domains in $\mathbb{R}^n$, Arch. Ration. Mech. Anal., **160** (2001), 153–179.

[151] J. Zimmer, Global existence for a nonlinear system in thermoviscoelasticity with nonconvex energy, J. Math. Anal. Appl., **292** (2004), 589–604.

[152] A. Zlotnik, Global Lyapunov functionals of the equations for one-dimensional motion of viscous heat-conducting gas, Doklady Math., **66** (2002), 121–126.

Index